THEOLOGY AND BIOETHICS

# PHILOSOPHY AND MEDICINE

*Editors:*

H. TRISTRAM ENGELHARDT, JR.

*The Center for Ethics, Medicine, and Public Issues, Baylor College of Medicine, Houston, Texas, U.S.A.*

STUART F. SPICKER

*School of Medicine, University of Connecticut Health Center, Farmington, Connecticut, U.S.A.*

VOLUME 20

# THEOLOGY AND BIOETHICS

## *Exploring the Foundations and Frontiers*

Edited by

EARL E. SHELP

*Institute of Religion, and*
*Center for Ethics, Medicine, and Public Issues,*
*Baylor College of Medicine,*
*Houston, Texas, U.S.A.*

D. REIDEL PUBLISHING COMPANY

A MEMBER OF THE KLUWER  ACADEMIC PUBLISHERS GROUP

DORDRECHT / BOSTON / LANCASTER / TOKYO

**Library of Congress Cataloging in Pulication Data**
Main entry under title:

Theology and bioethics

(Philosophy and medicine ; v. 20)
Includes bibliographies and indexes
1. Medical ethics. 2. Medicine – Religious aspects.
3. Bioethics. 4. Science – Religious aspects.
I. Shelp, Earl E., 1947– . II. Series.
[DNLM: 1. Bioethics. 2. Ehics, Medical. 3. Religion and Medicine. W3 PH609 v.20 / W 50 T391]
R725.5.T57 1985 174'.2 85–11723
ISBN 90–277–1857–1

---

Published by D. Reidel Publishing Company,
P.O. Box 17, 3300 AA Dordrecht, Holland.

Sold and distributed in the U.S.A. and Canada
by Kluwer Academic Publishers,
190 Old Derby Street, Hingham, MA 02043, U.S.A.

In all other countries, sold and distributed
by Kluwer Academic Publishers Group,
P.O. Box 322, 3300 AH Dordrecht, Holland.

Printed in The Netherlands.

## TABLE OF CONTENTS

## SECTION III: RELIGIOUS REASONING ABOUT BIOETHICS AND MEDICAL PRACTICE

# FOREWORD

We who live in this post-modern late twentieth century culture are still children of dualism. For a variety of rather complex reasons we continue to split apart and treat as radical opposites body and spirit, medicine and religion, sacred and secular, private and public, love and justice, men and women. Though this is still our strong tendency, we are beginning to discover both the futility and the harm of such dualistic splitting.

Peoples of many ancient cultures might smile at the belatedness of our discovery concerning the commonalities of medicine and religion. A cursory glance back at ancient Egypt, Samaria, Babylonia, Persia, Greece, and Rome would disclose a common thread – the close union of religion and medicine. Both were centrally concerned with healing, health, and wholeness. The person was understood as a unity of body, mind, and spirit. The priest and the physician frequently were combined in the same individual.

One of the important contributions of this significant volume of essays is the sustained attack upon dualism. From a variety of vantage points, virtually all of the authors unmask the varied manifestations of dualism in religion and medicine, urging a more holistic approach. Since the editor has provided an excellent summary of each article, I shall not attempt to comment on specific contributions. Rather, I wish to highlight three broad themes which I find notable for theological ethics.[1]

The first is *faith and ethics*. While these pages note that, as a matter of fact, religious ethicists have contributed mightily to the emergence of bioethics as a discipline in the last two decades, the question persists: just what, if anything, does *theological* ethics have to offer to bioethics and to the practices of health care? Surely, one basic contribution is a critical analysis of the faiths which inevitably shape and guide our scientific and medical activities. By what faith do we perceive the meanings of health and disease? By what faith do we understand the purposes of medical practice? By what faith do we interpret the possibilities of new reproductive technologies? What faith shapes the decisions about the distribution of medical care?

Indeed, there are no faithless ethics, no faithless moral actions. H. Richard Niebuhr gave this truth classic expression:



*E. E. Shelp (ed.), Theology and Bioethics,* vii–xi.

> The faith we speak of . . . is not intellectual assent to the truth of certain propositions, but a personal, practical trusting in, reliance on, counting upon something. . . . Faith, in other words, always refers primarily to character and power rather than to existence. . . . Now it is evident . . . that without such active faith or such reliance and confidence on power we do not and cannot live. Not only the just but also the unjust, insofar as they live, live by faith ([3], pp. 116 f.).

We live by faith because we cannot do otherwise. Every moral decision as well as every act of knowing depends upon some center of value, some power or worth, some object of devotion whose goodness and truth we cannot prove. Beyond all of the logical intricacies of the Is-Ought problem, most theological ethicists have recognized all of this. A common thread which runs through these pages is the commitment to assess critically those operating faiths which shape the biomedical enterprise. That is not only a major contribution of theological ethics to bioethics, it is at the same time a major step beyond one common dualism.

A second common anti-dualistic theme in this volume is the urgent need to overcome the split between *micro-ethics and macro-ethics*. Thus far, ethical reflection has leaned most heavily toward the former, accenting specific treatment decisions faced by patients, families, and medical professionals. If in this individualistic society such focus ought not surprise us, neither is it adequate. It buys into the split between love and justice. It trivializes the pervasively social nature of all personal existence. And it perpetuates the reigning model of our current system: disease cure rather than health care.

If the language of 'crisis' currently is badly overworked, at the very least it is safe to say that America currently faces major systemic problems in health care. They can be seen in three large and visible factors: costs, quality, and accessibility. Costs of medical care have skyrocketed. While the quality of laboratory and crisis medicine is impressive, health and longevity statistics are distinctly unimpressive. Moreover, issues of accessibility and distribution approach a national scandal: in poverty areas chronic illness is 30% more frequent, infant mortality 50% higher, tuberculosis three times more prevalent, and diabetes, hypertension, and vision impairment 50% higher than among the economically secure.

Thus, the issues of macro-ethics are not simply those of the just distribution of currently available medical resources, important though such issues be. They are also questions of appropriate priorities within the health-care budget – artificial hearts or well-baby clinics? They are issues of priority between crisis care and preventive medicine, between investing more in sophisticated technologies for acute disease and promoting a healthy natural and social environment. Such issues obviously shade into the larger context of macro-ethics: what resources should we put into health care itself in comparison with other social expenditures, e.g. education, social welfare, or defense?

Coping with these matters calls for careful understanding of factual data. We need to know, for example, that while our government recently stated that 'only by preventing disease from occurring, rather than treating it later, can we hope to achieve any major improvement in the nation's health', the vast preponderance of national expenditure was going to disease cure and only a miniscule percentage to preventive measures. We need to know that the major advances in health during the last hundred years have been due much more to improvements in general living conditions than to developments in medical treatment itself. Beyond such factual data lie important questions for theological ethics: what is health? what is the value of health relative to other values? is there a right to a decent minimum of health care? what is that decent minimum?

The bridging of micro- and macro-ethics in this field is a matter of both theoretical integrity and practical urgency. That this anti-dualistic, holistic ethical concern is present in so many of the pages which follow will be evident. And it is a cause whose time has come. At this moment an accelerated change in institutionalized medicine is underway. Under the combined impact of burgeoning costs, of the government's application of the diagnosis-related group formulae (DRGs), and of the growth of health maintenance organizations (HMOs), hospitals are struggling for survival and alternatives to traditional hospitalization are growing rapidly. The writers in this book know well that a micro-ethics for biomedicine will no longer suffice.

If all of the basic expressions of dualism are interconnected – as they appear to be – perhaps the most basic expression of all is that dualistic view of the human which pits soul against body, and hence man against woman. This split – philosophically articulated by late Hellenistic Greece, frequently embraced by the Christian church, reinforced in modern science by Descartes, and fundamental to patriarchal practice – has deeply affected us all. That we are late in our challenge of this dualism is all too evident. But that the challenge is now vigorously being made is hopeful, and that is evident in the pages of this volume. Further, there is good reason for hope in the variety of ways in which the body-spirit split is now being challenged in the day-to-day experience of health care:

(1) Increasingly, sick people want not only a medical diagnosis of their specific problems, but they also want to know how their health relates to

the whole of their lives: how they sleep, work, play; what their sexual and family lives are like; what they hate, love, fear, and dream.

(2) Challenges to the spirit-body separation mount from psychosomatic medicine with its growing understanding of how chronic illness is related to chronic stress.

(3) Challenges take the form of a consumer critique as more people realize that the right to significant decisions about their living and dying ought not be taken from them by professionals who see themselves as spirit and mind while viewing their patients as bodies.

(4) Challenges are coming from more and more articulate women who question the ways in which a male-dominated medical world has treated them.

(5) Challenges are coming from post-Einsteinian physics which calls into question the subject-object dualism on which biomedicine is still largely based.

(6) Challenges are coming from a cross-cultural critique wherein it is realized that the ways of conceptualizing health and disease ingrained in a white, male-dominant Western society may not be the final word.

(7) And challenges are coming from theology, with its growing recognition that moving beyond the body-spirit split is basic to the incarnationalist foundations of both Judaism and Christianity as well as imperative for greater human wholeness. Such challenges are accentuated by feminist theology with its grasp of the embodied nature of our existence, the inherently relational nature of life, and the immersion of the human in the natural world – all of which lead to a thoroughgoing critique of dualistic assumptions.

I have identified just three expressions of dualism which are addressed in this book. Other expressions could be noted. Suffice to say, I genuinely celebrate this important volume which presses the growing edges and the center of the companionship of theology and bioethics. I celebrate particularly the vision I see running through these richly varied essays of a more holistic existence fundamental to both personal/social health and to faithful theology and ethics.

*United Theological Seminary*
*of the Twin Cities,*
*New Brighton, MN, U.S.A.*

JAMES B. NELSON

## NOTE

[1] I have elaborated some of these themes more fully in [1], especially Chapter 11 and in [2], especially Chapters 1 and 8.

## BIBLIOGRAPHY

[1] Nelson, J. B.: 1983, *Between Two Gardens,* Pilgrim Press, New York.
[2] Nelson, J. B. and Rohricht, J. S.: 1984, *Human Medicine: Revised and Expanded Edition,* Augsburg Publishing House, Minneapolis.
[3] Niebuhr, H. R.: 1943, *Radical Monotheism and Western Culture,* Harper & Brothers, New York.

# INTRODUCTION

A volume devoted to theology in a series of books titled Philosophy and Medicine might appear misplaced. For a series committed to the philosophical clarification of concepts, the exploration of moral theories, and the analysis of clinical and research medicine to allow Deity and traditions of faith and religious thought within its range is not a special cause for remorse or celebration. Rather, it is as it should be. Theology and philosophy are ancient disciplines, each in its own way concerned to provide ways to construe the world, discover the good, and discern the right. Much of contemporary interest in bioethics has been stimulated by individuals who are theologically educated. These scholars have joined with moral philosophers to generate a worthy and expanding body of literature. Yet moral theologians have been criticized from within their own ranks and by philosophers for failing to adequately ground their normative thought in theology, and to demonstrate the unique contribution a theological morality can make to bioethics in general.

James Gustafson, University Professor of Systematic Theology at the University of Chicago Divinity School, analyzed a failure to connect theology to bioethics in a robust way as a strategic maneuver. Gustafson wrote in 1975, when the bioethics movement was gathering momentum, "Frequently the failure to develop the theological grounds for one's work in medical ethics stems from lack of interest in those grounds on the part of the participants in the discussions of clinical moral issues. Frequently it stems from the effort to be persuasive on such grounds as diverse persons can agree upon; often to introduce theology becomes an unjustifiable reason for one's secular colleagues to discount what one might say about medical ethics" ([1], pp. 1–2). Similarly, philosopher Alasdair MacIntyre of Vanderbilt University, whose discontent includes contemporary moral philosophy (cf. [2]), in a commentary on an issue of *The Journal of Medicine and Philosophy* addressing 'Theology and Medical Ethics', identified three tasks for contemporary theologians writing on medical ethics. "First – and without this everything else is uninteresting – we ought to expect a clear statement of what difference it makes to be a Jew or a Christian or a Moslem, rather than a secular thinker, in morality generally. Second, and correlatively, we need to hear a



*E. E. Shelp (ed.), Theology and Bioethics,* xiii–xxiv.

theological critique of secular morality and culture. Third, we want to be told what bearing what has been said under the first two headings has on the specific problems which arise from modern medicine" ([2], p. 435). The implicit complaint of Gustafson and the agenda forwarded by MacIntyre is only now being answered.

Project X, under the general direction of church historian Martin E. Marty and theological ethicist Kenneth L. Vaux, is studying "ten life 'themes' [well-being, sexuality, passages, morality, dignity, madness, healing, caring, suffering, dying] common to both medicine and religion and exploring them in the light of ten world faith traditions" ([3], p. ix). This is an ambitious and potentially valuable partial response to MacIntyre's recommendation. However, the essays in this volume differ in a significant way from the emphasis of Project X. Here the contributors were challenged to look to the *future* of theology and bioethics, to *transcend* faith tradition and denominational loyalties. These essays are intended to prompt theologians to identify, claim, and communicate the distinctive contributions theology can make to the discussion of moral questions in medicine. It also is intended to challenge the philosophical community to become aware of the integrity of theological scholarship, and to engage in conversation with theological traditions in order that each may learn from the other.

This volume is an initial effort to redress the rather major omissions of theological scholarship in relation to medicine and bioethics. The authors were urged to be provocative, creative, and prospective in their treatment of selected topics central to the continuing bioethical and theological inquiry. More work needs to be done, and perhaps these probings will provide an incentive for others to undertake a task worth pursuing. The essays are presented in three sections. The first section contains six essays that address the relation of theology to science and bioethics. The second section of five essays is concerned with foundational and methodological issues. And the third section contains five examples of religious reasoning about bioethics and medical practice.

The first division of essays on 'Theology, Science, and Bioethics' begins with an historical narrative on 'Religion and the Renaissance of Medical Ethics in the United States: 1965–1975' by LeRoy Walters. Walters observes that prior to 1966 the principal religious texts on medical ethics were produced by Roman Catholics with the exception of early work by Protestants Joseph Fletcher and Paul Ramsey. Though these debates were lively and prefigured more extensive discussions, Walters sees Pope Paul VI's affirmation of the Church's traditional opposition to artificial contraception as the spark that ignited the interest of influential scholars in and extensive institutional support for bioethical research. It was bioethics, the relevance of scientific achievement to moral living, that diverted the attention of religious ethicists from meta-ethics to applied ethics, to an ethics that touched the lives of people in important ways. In Walters' view, it was largely the response of religious ethicists to doctrines seen as out of touch with the contemporary world that led to the renewed interest in the relationship between science and religion.

George P. Schner describes the difficulty in attempts to merge science and theology, and shows that this difficulty does not create an impasse in ethics; rather, it is a condition of its possibility. Schner explores the tasks of theology and science, concluding that the two disciplines are connected but not reconcilable. The connection is in their imaginative constructions of the world and in the language of ethics with which the conflicting rationalities engage one another. Ethics, according to Schner, exists to deal with problems of human freedom. The function of science and theology for ethics is the following: theology describes the human drive toward transcendence, so its kind of truth cannot be bound by criteria of clarity and certainty. Science describes the human drive toward the transparency of reality, so its truths are bound by clarity and certainty. Science and theology bring their distinct domains to bear upon human situations through ethics. Accordingly, science keeps theology honest by challenging it to meet the demands of new representations of the world, and theology keeps science honest by seeing that clarity and certainty do not collapse into ideology. In Schner's opinion, we do not need to accommodate theology and science with one another. In fact, it would not even be desirable, because it is their distinctiveness which allows science and theology to speak meaningfully to human freedom.

Charles Hartshorne extends the examination of science and religion in an essay on 'Scientific and Religious Aspects of Bioethics'. He explores the connection of theological conceptions to bioethical judgments, with a primary focus on abortion. Hartshorne objects to the concept of God as unmoved mover, holding that conceptions of this sort fail to explain the nature of life. God, for Hartshorne, is an enhanceable actuality whose divine life is enriched by the activity of humans. The future, even for God, is not wholly determined, so every creature has some capacity for self-determination and love which are characteristic of the Creator. The possible particular goods are left to the creatures to actualize. Hartshorne is concerned about the impact of population control on human well-being. Abortion is one means by which the human good can be served insofar as it limits population growth. In his examination of abortion, Hartshorne criticizes biblical literalists for failing to appreciate the difference between actuality and potentiality in the Divine, and therefore they fail to appreciate its importance for issues in bioethics. Hartshorne argues that the issue of abortion turns upon considerations of personhood, which he believes a fetus does not possess. The function of religion in this and other discussions in bioethics is to provide ultimate ideals or principles which are contingent upon the circumstances of any age. Science's role is to tell us about particular circumstances. He concludes that extreme pro-life advocates are not to be taken seriously as scientists, philosophers, or theologians until they take seriously the relevant difference between potential and actual.

H. Tristram Engelhardt, Jr., in a commentary on Hartshorne's essay, describes natural theology, its world view, and its consequent implications for bioethical decisions. The natural theologian tells us that God is nameless, i.e., not bound by any particular cultural or religious persuasion, but rather belongs to all creation as all creation belongs to God. Engelhardt sees Hartshorne as a natural theologian whose task is to view God apart from cultural representation, much as science is designed to see the world. The paradigmatic scientific relation is that of observer to object with no concern for the ethnicity of the observer. Natural theology operates upon a similar paradigm. Hartshorne's concern is with the relation of the human to Deity, a relation which turns upon human agency in the face of a non-determinate future. Hartshorne's bioethical considerations focus precisely on these relations. The real issue is meta-ethical. It is one of the relation of the God-concept to rational ethical decision making. For Engelhardt God is known through reason, not through any particular revelation. Where no common grace occurs to settle disputes, their resolution must be in rational argument. This critical function of natural theology allows it to diagnose unrecognized cultural prejudices which have been effective in the construction of some of our moral guidelines, and gives us critical, rational perspectives on health-care morality.

William Frankena's essay directs the discussion away from the relation of theology and science to the relation of theology to ethics in general. Whereas Hartshorne considered the relation of theology to ethics primarily with respect to the issue of abortion, Frankena, like Engelhardt, speaks to the broader issue of 'The Potential of Theology for Ethics'. His focus is the ways in which ethical theories deal with the is-ought gap. It is generally held that we cannot reach logically sound conclusions about what ought to be from what is, and Frankena shows how this problem is at work in theologically grounded ethics. Theology, he says, is based on two sorts of beliefs: (1) *is* beliefs, which are the theological world-view, and (2) *ought* beliefs, which are theological ethical precepts. Two questions are posed: (1) What can the theological world-view do for ethics? (2) What can a theological ethic do for ethics? These questions are considered and Frankena concludes that the is-conditions of a theological world-view cannot provide objective ethical principles. He notes that there is a sense in which a theological world-view can have potential for ethics, but it would be limited to those who embrace that world-view. Having reached this pessimistic conclusion, Frankena admits that ethical insights and ideas can come from anywhere, including religious sources, even theology.

Basil Mitchell is more optimistic than Frankena about the contribution of theology to ethics and bioethics. Rather than relegating theology to a possible source of moral insight, Mitchell holds that "moral reasoning finds its proper autonomy only in a full theological context and, . . . while theology has its implications for ethics, these must be assessed and understood from a distinctively ethical point of view." Mitchell finds traditional medical ethics compatible with traditional Judeo-Christian moral values and norms, particularly the notion of the sanctity of life. He worries that the emerging emphasis on patient autonomy threatens to reduce medical morality to an expression of personal preference or to become a matter of utilitarian calculus. Not being happy with either eventuality, Mitchell defends the view that ethical insights of the Judeo-Christian tradition, even though there is no consensus regarding them, are fundamentally important to the rational development of moral norms that guide the conduct of medicine.

Engelhardt advances the claim that the importance of religion to life, individual and social, has been negated by the unleashed forces of the Renaissance and Enlightenment except for special private communities which peacefully co-exist with other secular communities in a pluralist society. A residual of respect for religion and the contribution it might make to social values and norms is evidenced, for example, by religious representation on governmental committees charged to study moral issues in medical research and care. Yet, for religious moral commitments to be persuasive in the society Engelhardt describes, religious warrants will need support by secular arguments in order to be adopted. The potential value of theology for bioethics is the provision of meaning to experiences of pain, suffering, and death, a task that secular or philosophical ethics cannot do, and one that theological bioethics has not done because it cannot provide "a particular account of the meaning of life and the world that can be justified in terms of general rational considerations". The future of theology in bioethics, according to Engelhardt, is twofold: (1) it can conserve the theologies of particular faith communities, and (2) it can undertake a new search for meanings that are not confined to a single tradition or deity. The present failure of theology in bioethics can be of life-living importance to bioethics if theologians will do that which theology is best equipped to do – provide "indications of ultimate [transcendence] meaning and purpose", not bound by tradition or denomination, for life, suffering, and death.

The second section of essays examines the 'Foundations and Frontiers in Religious Bioethics'. The lead essay by Richard McCormick addresses the foundational element of Christian ethics in relation to bioethics. McCormick notes that much of the literature on bioethics has focused on considerations of right and wrong. In this essay he is concerned with the categories of goodness and badness, "on the personal transformative influences of the faith experience and what this means for bioethics." McCormick offers an interpretation of the Christian moral life that has charity or love as the substance of revelation in Christ and the central virtue of the Christian moral life, the virtue upon which all other virtues depend. The person transformed by faith, according to the author, expresses a love "shaped by the absoluteness and ultimacy of the God-relationship", of which Christ is the standard. Other basic human goods are subordinate to the God-relationship and can be sought in love within their context as subordinate. McCormick discovers resources for the moral life in the narratives of public worship in which the practice of following Christ is told and Christ is made present in liturgy. In a concluding section, relevant Christian themes and perspectives are related to health care practitioners and contexts.

Whereas McCormick found the constant of Christian morality in the transformative event of faith, Langdon Gilkey relates the changing frontiers of theology to issues in bioethics. Gilkey is concerned with twentieth century crises of faith, as it were, not only in religious belief, but in the belief structures of science and Western culture. The frontier is described as a 'time of trouble', following Arnold Toynbee, and is brought to fruition, in part, by (1) the shrinkage of the world-dominant power of the West, (2) a decline in the pull of Western ideas, values, and religion, (3) a science and technology, once the pride of the West, that generates problems in the forms of atomic weapons, dehumanizing social hierarchies, and other similar consequences of material advancement, and (4) a decline in the ideologies of progress and the Marxist social vision. In short, there has been a radical change in Western 'futures'. Gilkey sees the frontier as an historically extended phenomenon, whose early conceptual framework was such that science was believed to need no external moral guidelines. Now, he believes, the secular culture is itself threatened by external (viz., theological) criticism. The guidelines of science have turned out not to be enough to secure deeper human values, and it is this turning of the tables that represents the new frontier of theological ethics, i.e., to provide new foundations for courage, hope, and normative obligation. The development of this frontier stands upon a religious base which vests positive value in both humanity and nature. The course of the future interrelation of science and faith with culture lies in the recognition that principles of value are as important as those of knowledge. Theology, Gilkey believes, with both Eastern and Western world-views in mind, will mediate the development of a new awareness which concentrates, on the one hand, upon the unity of our spiritual and organic lives, and, on the other hand, upon issues of value relevant to the life sciences.

Douglas Sturm's essay reminds us that concerns for methodology ought not be overlooked as scholars go about the business of addressing urgent issues. He poses three methodological questions for bioethics: questions of scope, focus, and grounding. With regard to scope, Sturm sees its basis in what he calls a 'perception of social reality', and is defined as a 'range of issues appropriate to the discipline'. The scope of bioethics is microcosmic, particularized by personalist views of bioethics, and macrocosmic, particularized by contextualist views of bioethics. The personalist view is constituted by attention to individual cases of patient-physician relations, and the contextualist view finds its interest in social and historical patterns. Focus, his second methodological question, determines importance and centrality. He characterizes the theater of focus as that of structural social science. Sturm contrasts Parsons' structural functionalist approach with the more radical view of Freidson's 'social construction of illness'. Grounding, according to Sturm, is split into 'popular' and 'philosophical-theological' understandings of the foundation of bioethics. The popular ground for these moral considerations stems from sociopolitical ideology and problems of health care availability. The theological understanding seeks to deal not only with physical needs, but spiritual needs as well. Sturm then describes the most favorable conditions for bioethics as the following. The personalist-contextualist dilemma is solved by attention to the interacting conjunction of the two. That is, the consequence which follows from this conjunction is that bioethics be construed in scope as social ethics. The social focus gives bioethics a particular task: to show how cases of practice conjoin with social structure and historical process. And finally, the ground for bioethics is in the more profound covenantal relationship which encompasses the philosophical principles of liberty, equality, and common good.

The feminist movement is on the frontier in theological as well as in political circles. Margaret Farley in the next essay considers the implications of feminist theology for a range of moral issues related to the biological sciences, technology, and medicine. Her focus is the development and use of reproductive technologies. She addresses three major feminist themes that bear directly on bioethical problems: (1) patterns of human relation, (2) the phenomenal consequences of human embodiment, and (3) human assessment of the meaning and value of the world of nature. Oppression is a central issue in Farley's discussion. She analyzes the oppressive dimensions of the three realms identified above and provides a feminist theological response. According to Farley, feminist ethics in general focus upon female experience, social and moral responsibilities to women disadvantaged by social structure, and the efficacy of women as moral agents whose unique interpretive perspectives allow them important moral insights. Farley brings this perspective to bear on *in vitro* reproductive technologies. Potential moral problems and advantages that this technology poses, from a feminist viewpoint, are discussed. The essay concludes that feminist bioethics undoubtedly will expand in scope, but will remain interpretively linked to issues of relation, embodiment, and environment.

The second section of essays closes with a contribution by Mark Juergensmeyer. Juergensmeyer is concerned about 'Doing Ethics In a Plural World'. As an example of the kind of ethical issue with which this essay is concerned, the author points to abortion. Though it has been practiced in nearly every known human society, and though it has been the subject of moral debate in every society which has practiced it, there is yet little intercultural agreement concerning the reasons why the abortion issue poses moral problems. It is therefore not surprising that moral reaction to abortion and similar issues varies widely from culture to culture. The author urges that the correct approach to cross-cultural ethical studies of such issues would involve both a descriptive and a normative component. These can be characterized, if somewhat oversimply, in terms of a division of labor among investigators. The anthropologist/sociologist/specialist in comparative studies would describe how various moral traditions develop and function; and the policy-maker/concerned ethical actor/moral authority would decide, on the basis of the descriptions provided, how best to deal, in practical situations, with problems posed by intercultural ethical variation. Some examples of recent descriptive work are discussed; and as an example of normative application of such description Juergensmeyer discusses the life of Ghandi, whose rich knowledge of ethical and religious traditions other than his native Hinduism enabled him to evaluate the moral impact of his actions from many different perspectives.

The third and final group of essays provides examples of 'religious reasoning about bioethics and medical practice'. Stanley Hauerwas is concerned with the link between the Christian church and the practice of medicine. Hauerwas rejects two ways of looking at the connection between medicine and religion. The first is that the physical should be left to physicians and that the spiritual should be left to the clergy. The objection is that this strict dichotomy would relegate God to the gaps in scientific theory. The second option is a resacralization of medicine, that is, making a religion of medicine. The objection here is that it perverts medical activity by expecting a promise not simply of health but of happiness as well. Questions then remain. How ought religion to speak to medical ethics? He thinks that consequentialist and deontological approaches are inappropriate since theology is not necessary to either. The question is then reformulated, 'how does the church speak to the practice of medicine?' Hauerwas's answer is that the church is to be a people who are faithful to one another by a willingness to be present to one another despite human vulnerabilities. The church is not needed to bolster moral claims, but to share in the vigil with those in pain for the long haul. It is this sharing vigil that suggests another, transcedent presence. It is how to do this, being present to weakened people, that the church can teach medicine. This is why, according to Hauerwas, medicine needs the church.

James Childress, similar to McCormick in the second section, examines love or agape in Christian biomedical ethics. However, Childress analyzes the principle of justice in relation to the principle of love as a way to demonstrate the 'impossibility' of resolving some bioethical disputes "without attention to their broader theological, metaphysical, and anthropological contexts." Two areas of controversy – distribution of burdens in non-therapeutic research and distribution of the benefits of medical care – are used to illustrate this thesis. His main concerns about the efficacy of agapeic principles in general are shown in a discussion of the Parable of the Good Samaritan, which he finds incomplete as a paradigmatic case of agapeic responsibility. He outlines several basic conceptions of the structure of agape, exemplified in the work of Ramsey, McCormick, Outka, and others, in relation to considerations of justice. The problem, for Childress, is whether agape and justice are distinct, and, if they are, how they may be in principle accountable to one another for the production of morally significant rights and responsibilities in health care. Childress concludes that since most of the work regarding these issues has been normative in character, it fails to speak to the more profoundly relevant theological, metaphysical, and anthropological problems, which will have to be considered systematically in the future.

The next essay by Ronald Green provides a critical assessment of contemporary Jewish bioethics. Green observes that the tradition is complex but seems subject to more flexible interpretation than tends to be presently given, and that though conservative authors claim to be reporting the views of the text, they seem not to admit that all reporting is interpretive. Accordingly, Green thinks that contemporary Jewish bioethical scholarship should be treated with caution since the humanistic intent of the Talmud appears to be neglected by present-day writers. He examines a series of issues to illustrate his difficulties with what he considers an unresponsiveness by conservative authors to the changed conditions that confront the tradition. Green doubts that current conservative writing is representative of the Jewish tradition as a whole. The conservative trend is thought by Green to be a result of a felt need on the part of Talmudic scholars to distinguish themselves from those who have less interest in Halakhic study. The maintenance of a tradition is difficult in the face of rapid secular change, yet Green is aware, as conservative scholars often are not, that viable religious traditions do not exist apart from social forces. If the tradition is to remain respected, in Green's judgment, it must deal with the human side of the issues it faces.

David Smith proposes that loyalty or fidelity is a core moral concept with important implications for medical practice. Josiah Royce's understanding of loyalty is summarized and extended to a consideration of loyalty with regard to the basis of medical ethics and to the problem of suffering. Next, Smith reviews loyalty as understood by H. Richard Niebuhr who, in concert with Royce, thought loyalty to be the essence of morality with implications for several bioethical issues. The third and last interpreter of loyalty that Smith reviews is Paul Ramsey, who has placed loyalty or 'covenant fidelity' at the center of his medical ethics. In a concluding section, Smith finds the interpretations of Royce, Niebuhr, and

Ramsey to be incomplete: Royce makes a God of one's cause, Niebuhr extracts God from the world, and Ramsey makes a God of the patient. Smith presents his own thought regarding loyalty in order to demonstrate the relevance of a fidelity ethic and to suggest a form such an ethic might take in general and in medical practice.

The final essay in the third section, and in the volume, is by Paul Lehmann. Lehmann provides a theological perspective on bioethics that centers around the issue of taking responsibility for life. He is not satisfied with Enlightenment principles and doctrines of freedom as a way out of our bioethical dilemmas. More adequate, according to Lehmann, is an approach that views bioethical questions in the light of three theological concepts and their corresponding ethical norms. The concepts are providence, eschatology, and destiny. The norms are freedom, responsibility, and justice. Lehmann explains how these concepts and criteria illuminate the ethical quandaries that can attend biomedical endeavors. Two specific issues – abortion and genetic engineering – are discussed at length in order to demonstrate how responsibility for life would be taken when approached in the manner that he proposes.

It should be clear that this collection only has begun to do the work in theology and theological ethics that properly ought to be done if these disciplines are to make a distinctive contribution to the expanding literature in bioethics. Why this work has not been more forthcoming is anyone's guess. Perhaps the explanation of Gustafson quoted above is adequate. It may be, however, that it has not been done because it is difficult to do. A more kind explanation would be that everyone has assumed that the theological roots are or should be sufficiently known as not to warrant a careful, precise explication. Whatever the reason or reasons, the task can be neglected no longer. No one can be certain what the result of the labor will be. Nevertheless, philosophers, theological scholars, religious lay people, physicians, and the public reasonably ought to expect theological presuppositions to be declared and defended where they are operative.

At the conclusion of MacIntyre's response cited above, he raised a question of the credibility of the theological enterprise. Until MacIntyre's question is answered, his view that what theologians say will be greeted by an unbelieving world as fatally arbitrary is probably correct. His challenge in 1979 was proper then and reasonably could be issued again. In MacIntyre's words, "The theologians still owe it to the rest of us to explain why we should not treat their discipline as we do astrology or phrenology. The distinctiveness and importance of what they have to say, if it is true, make this an urgent responsibility" ([2], p. 443). This collection as a whole is a first, partial, and exploratory effort to begin the process of establishing the credibility and contribution of theology to bioethics both now and in the future.

Edited volumes are the work of many individuals. The editor's name appears on the cover but it is the contributors who enable an idea to become a reality. For their commitment, labor, charity, cooperation, and kindness, as editor I express my gratitude to the contributors, individually and collectively. H. Tristram Engelhardt, Jr., and Stuart F. Spicker, general editors of the Philosophy and Medicine series, characteristically provided their good offices, counsel, and support throughout the production of the volume. Jay Jones, my research assistant, and Audrey Laymance, my manuscript secretary, carefully attended to the physical preparation of the text. Susan M. Engelhardt, once again, proofread every typed and printed word with an eye for errors that would put an eagle to shame. Other staff members in the Center for Ethics, Medicine, and Public Issues contributed to the project in various important ways. All of these people, so important to a volume but too little recognized, should know of my appreciation for them and their work. Finally, I thank the staff at D. Reidel Publishing Company who have competently assisted me in bringing this volume to completion.

*Institute of Religion, and*
*Center for Ethics, Medicine, and Public Issues*
*Baylor College of Medicine,*
*Houston, Texas, U.S.A.*

EARL E. SHELP

## BIBLIOGRAPHY

[1] Gustafson, J. M.: 1975, *The Contributions of Theology to Medical Ethics*, Marquette University Press, Milwaukee.
[2] MacInyre, A.: 1979, 'Theology, Ethics, and the Ethics of Medicine and Health Care', *Journal of Medicine and Philosophy* **4** (December), 435–443.
[3] Marty, M. E. and Vaux, K. L. (eds.): 1982, *Health/Medicine and the Faith Traditions*, Fortress Press, Philadepia.

# SECTION I

# THEOLOGY, SCIENCE, AND BIOETHICS

LEROY WALTERS

# RELIGION AND THE RENAISSANCE OF MEDICAL ETHICS IN THE UNITED STATES: 1965–1975

In this essay I will seek to justify three claims: (1) that the field of medical ethics underwent a qualitative shift in the United States in the decade between 1965 and 1975; (2) that several thinkers interested in the relationship between religion and science and between religion and ethics contributed significantly to this shift; and (3) that two Catholic laymen, Daniel Callahan and André Hellegers, and one Protestant theologian, Paul Ramsey, were among the principal institutional and intellectual architects of the recent renaissance in medical ethics.

These claims are, for the most part, modest. It is true the term 'renaissance' conveys the notion that the new era was an improvement over the pre-1965 era – a view that can be, and in fact is, debated in some quarters. However, by 'renaissance' I do not mean to convey more than a new sense of excitement in the field, a willingness to examine new issues and to employ new methods of inquiry, and a significant increase in the number of academics and citizens interested in problems of medical ethics. By concentrating on the religious contribution to this renaissance, I do not wish to deny that nonreligious thinkers and impulses played a major role in the renaissance. Several journalists, physicians, lawyers, and politicians who made no explicit religious appeals contributed significantly to the increased interest in medical ethics.[1] What I hope to establish is that religious thinkers and themes also played an important role – perhaps even the principal role – in this renaissance during the 1965–1975 decade.

## PRECURSORS OF THE RENAISSANCE: 1946–1966

Through, 1966 textbooks of medical ethics by Catholic moral theologians constituted the principal religious literature on medical ethics in the United States.[2] Prototypical for this literature was Charles J. McFadden's widely used textbook. Originally published in 1946 as *Medical Ethics for Nurses* [22], McFadden's text went through five further editions between 1949 and 1967 under the title *Medical Ethics* [23]. Other Catholic moral theologians who published major medical ethics texts during this period



*E. E. Shelp (ed.), Theology and Bioethics, 3–16.*

were Gerald Kelly [20], Thomas O'Donnell [26], and Edwin Healy [17].

While there was some diversity in the issues treated by the various Catholic authors, the general approach to medical ethics was based on the standard textbooks of Catholic moral theology written by such authors as Alphonse Liguori, Gury, and Lehmkuhl. Indeed, Catholic textbooks of medical ethics were, from the earliest syntheses in the late nineteenth century on (cf. [10], [1]), discussions of selected themes from the larger general moral theology textbooks. A structural analysis of the medical ethics texts indicates that most of the questions discussed could be related either to the Decalogue or the Sacraments, as the following outline of topics illustrates ([40], p. 291):

I. The Precepts of the Decalogue

  A. The fifth precept

    1. Mutilation
    2. Extraordinary means
    3. The principle of totality
    4. Sterilization
    5. Transplant surgery
    6. Psychosurgery
    7. Experimentation on human beings
    8. Abortion
    9. Ectopic operations
    10. Euthanasia

  B. The sixth precept

    1. Masturbation
    2. Contraception

  C. The eighth precept

    Professional secrecy

II. The Sacraments

  A. Marriage

    1. The role of sex in marriage
    2. Impotence
    3. Sterility tests
    4. Artificial insemination

  B. Extreme Unction

  C. Care for the dying

Into the relatively settled world of Catholic medical ethics there burst in 1954 a disruptive Protestant challenge, Joseph Fletcher's book entitled *Morals and Medicine*. Originally prepared in 1949 as the Lowell Lectures at Harvard University, Fletcher's work questioned many of the presuppositions of the Catholic consensus. Fletcher acknowledged his indebtedness to the previous work of Catholic scholars on medical ethics and viewed his work as an initial attempt to compose a Protestant counterpart:

> To my knowledge, nothing of this kind has been undertaken by non-Catholics as yet, and certainly this book is at the most only a modest contribution to the *ethics* of medicine, not to its theology. I hope, of course, that the ethical judgments I have reached are within the range and provision of Christian theology, but that would be all that could be claimed for them ([14], p. xix).

Rather than organizing his work around precepts of the Decalogue or the Sacraments, Fletcher chose to discuss a series of human rights to be enjoyed and asserted by every individual, as the following chapter titles from *Morals and Medicine* indicate:

> Medical Diagnosis: Our Right to Know the Truth
> Contraception: Our Right to Control Parenthood
> Artificial Insemination: Our Right to Overcome Childlessness
> Sterilization: Our Right to Foreclose Parenthood
> Euthanasia: Our Right to Die ([14], p. ix).

Fletcher's central thesis was that human beings should freely make choices based on their knowledge of the options available to them. Human freedom is, according to Fletcher, a liberation from many aspects of nature. In his words, "there is no possible ground for supposing that a scrutiny of nature's ordinary and average phenomena can reveal either the will of God or a norm for men" ([14], p. 215).

Predictably, Fletcher's book was noticed by Catholic moral theologians. Indeed, it was the object of several extended reviews in Catholic journals. [3] Among the more sympathetic reviewers was James Royce, who wrote:

> The book is a real and somewhat subtle challenge to the Catholic position on [several] points, and a healthy prick to the smug armchair complacency with which some seminary professors are liable to cling to principles more trite than tried by contact with human problems. It is a "must" for any teacher of ethics . . . but . . . contains vicious errors which forbid its being placed on open library shelves ([33], p. 538).

Fletcher's *Morals and Medicine* also provoked what might be called a mediating Protestant response by a Princeton theologian who had not previously written on medical ethics. In a little-known article published in 1956 and entitled 'Freedom and Responsibility in Medical and Sex

Ethics: A Protestant View' [27], Paul Ramsey discussed Fletcher's book chapter by chapter. Ramsey's approach was mediating in the sense that he tried to interpret Protestant views to Catholic readers and in fact agreed with Catholic medical ethicists on such topics as artificial insemination by donor and active euthanasia. Many of the themes which appear in Ramsey's later works on medical ethics were first announced in this 1956 essay.

During the period from 1956 to 1966, one would expect a lively debate on problems of medical ethics to have ensued between Catholic and Protestant theologians. However, for unknown reasons, the discussion did not continue, and the renaissance was delayed by ten years. Catholic theologians went back to revising their medical ethics textbooks, and Protestant theologians like Fletcher and Ramsey turned their primary attention to other topics like the civil rights movement and war. From 1962 through 1965 the Catholic Church, including its theologians, was preoccupied with the Second Vatican Council, called into being by Pope John XXIII. There the foundations were laid for greater diversity of approach in Catholic theology, including Catholic moral theology, and for increased openness of Catholic thinking to the insights of non-Catholic religious traditions.

Within the shadow of Vatican II, a small group was established by Pope John XXIII to study a particularly controversial problem in Catholic medical ethics – contraception. The seven-member group, officially called the Pontificial Study Commission on Family, Population, and Birth Problems, met for the first time in the fall of 1963, shortly after the untimely death of John XXIII. A year later, the group was significantly enlarged by John's successor, Pope Paul VI; in its new incarnation the Papal Commission on Birth Control was comprised of 58 members, including experts from moral theology, reproductive biology, medicine, demography, economics, sociology, and pastoral care ([34], p. 78). The augmented commission, in which lay members now outnumbered the clergy, was in some respects a precursor to the national and presidential commissions established in the United States in the 1970s.

One of the new members appointed to the Papal Commission on Birth Control in 1964 was Dr. André E. Hellegers, a young Dutch-born obstetrician-gynecologist who was at the time involved in research at Johns Hopkins University. Hellegers played a major role in the work of the commission, serving as a member of the executive committee and as secretary to the pastoral section ([34], p. 84–103). To those who later learned to know him, it was clear that Hellegers had relished his time on the commission, especially the opportunity to interact with moral theologians, social scientists, and other physicians and scientists.

On June 28, 1966, the Papal Commission on Birth Control presented its twelve-volume report to Pope Paul VI. Two short documents, 'An Outline for a Document on Responsible Parenthood' and a 'Pastoral Introduction' to the outline, summarized the commission's majority view that contraception was not intrinsically evil ([34], pp. 94–98). Then the commission, the Catholic Church, and – to some extent – the world as a whole watched and waited for a response to the report by Pope Paul VI.

One of the U.S. Catholic laymen most interested in the Pope's response was a Georgetown- and Harvard-trained philosopher, Daniel Callahan. From 1961 to 1968 Callahan was associated with the leading American Catholic intellectual journal, *Commonweal,* and was in fact its executive editor during the latter stages of this period. From this vantage point he was in an excellent position to observe the Second Vatican Council and to provide informed commentary on such topics as reform in the Catholic Church, Protestant-Catholic relations, and the relationship between religion and culture.

Although Callahan was a remarkably productive editor and author in the early and middle 1960s, his writings from that time provide only hints of his later interest in medical ethics. His major books through 1966 bore such titles as *The Mind of the Catholic Layman* (1963) [3], *Honesty in the Church* (1965) [5], and *The New Church* (1966) [6]. However, there is in Callahan's *Commonweal* essays and editorials of this period a thin thread of interest in abortion and birth control. Perhaps the most significant single essay by Callahan on these topics was a 1964 contribution to a *Commonweal* symposium on 'Responsible Parenthood'. There he critiqued the recent textbook on *Marriage Questions* by the Jesuit moral theologians John C. Ford and Gerald Kelly [15]. While expressing admiration for Ford and Kelly's conscientious attempt to steer a middle course between archreactionaries and pioneers in Catholic moral theology, Callahan attacked their work as a totally inadequate response to the major changes occurring in the Church and in the world in which the Church existed. The traditional teaching on contraception, Callahan argued, was a symbol of a much more fundamental problem:

> The birth control question is, above all, a test case for the Church's understanding of itself and especially of its understanding of the development of doctrine. That means it is a test case for the contemporary renewal of the Church. Its importance lies in the direct confrontation of the theological methods and inclinations of another generation with those now emerging . . . .

> If there is a tragedy latent in this drama, then it lies in the desperation of good people trying to use old tools to cope with new material. It lies also in the desperation of married couples trying to relate old certainties to new uncertainties, holding on by their fingertips to a sandstone ledge they had been asssured was made of granite. In each case, the natural inclination is to panic. That is a sensible response, but not the only one possible. A better one would be for the Church, in its teaching authority and in its members, to immerse itself in the present. No theologian today can be expected to be understood if he continues to argue that the primacy of the species takes precedence over the personal good of individuals. He will not be understood if he argues that biological values take precedence over personalist values. He will not be understood if he says that one must accept a doctrine or a law on the basis of authority alone. The problem is not that these things are necessarily wrong. They are incomprehensible, flying in the face of everything contemporary man has learned about himself, about his conscience, about nature, and about value. They have been taught him by the Church as much as by the world. If the concept of a "living magisterium" has any meaning at all, then it must at least mean this: one way to remain faithful to the past is to affirm the present. That is the demand which our life here today has directed to the Church's exercise of its authority. That is the cutting edge of renewal ([4], pp. 322–323).

Ironically, John Ford was to emerge as the principal conservative spokesman on the Papal Commission on Birth Control. Indeed, he authored the minority report of the Commission which defended the Church's traditional ban on artificial modes of contraception ([34], p. 90). Thus, the period from 1966 to 1968 can be viewed as one in which Catholic conservatives like John Ford vied with liberals like Hellegers (on the Commission) and Callahan (outside the Commission) in an effort to influence Pope Paul VI on the birth control question.

## EARLY DEVELOPMENTS IN THE RENAISSANCE: 1967–1969

In the three years that followed 1966, the institutional foundations for the renaissance of medical ethics were laid. During this period André Hellegers moved from Johns Hopkins University to Georgetown University. In turn, Hellegers convinced Paul Ramsey to spend the spring semesters of 1968 and 1969 at Georgetown, with financial support from the Joseph P. Kennedy, Jr. Foundation, to do research and writing on medical ethics while in residence at a medical school and hospital. And in 1969 Daniel Callahan co-founded a new research center in Hastings-on-Hudson, New York – the Institute of Society, Ethics, and the Life Sciences (now the Hastings Center). These institutional endpoints were intimately related to a series of other developments, both intellectual and institutional, in the years 1967 to 1969.[4]

The first meeting to involve large numbers of future contributors to the medical-ethics renaissance was the International Conference on Abortion held in Washington, D.C., in September 1967. Co-sponsored by the Joseph P. Kennedy, Jr. Foundation and Harvard Divinity School, the conference included among its participants Charles Curran, Arthur Dyck, James Gustafson, André Hellegers, Richard McCormick, John Noonan, Gene Outka, Ralph Potter, and Paul Ramsey. The first documentary product of the conference was a paperback book entitled *The Terrible Choice: The Abortion Dilemma* [12], published in 1968. Two years later a scholarly work based on the conference, but incorporating supplementary material as well, was published by Harvard University Press [25].

Several months after the abortion conference Paul Ramsey began the first of his two spring semesters at the Georgetown University Medical School. Since 1960 Ramsey had focused his primary scholarly attention on ethics and war – first nuclear war, later the U.S. intervention in Vietnam. Beginning in 1966 one sees indications of a rekindled interest by Ramsey in the kinds of problems he had addressed in his 1956 response to Fletcher's *Morals and Medicine.* In 1966 Ramsey had published an essay on 'Moral and Religious Implications of Genetic Control' [28], and prior to the International Conference on Abortion in 1967 he had authored a discussion of abortion entitled 'The Sanctity of Life – In the First of It' [29]. Now, in early 1968, Ramsey undertook the first of two semesters devoted to full-time research on problems in medical ethics.

On July 29th, 1968, shortly after the completion of Ramsey's first semester at Georgetown, Pope Paul VI released his long-awaited response to the work of the Papal Commission on Birth Control. As is well known, in the encyclical *Humanae Vitae* Paul VI rejected the recommendations of the Commission's majority report and reaffirmed the Catholic Church's traditional opposition to artificial contraceptive measures ([34],p. 105).

André Hellegers, who had been a vigorous lay participant in both the process and products of the Papal Commission, was deeply disappointed by the Pope's decision. In an essay published in 1969 Hellegers wrote:

> For the scientist the encyclical presents a number of puzzling aspects: in the first place comes the absence of scientific evidence for, or indeed of scientific thought in reaching, the conclusions which the encyclical draws. Secondly, the scientist is struck by the absence of biological considerations in the entire encyclical. It is striking that the first section which deals with 'New Aspects', and which alludes to demographic, sociological, and educational problems, nowhere acknowledges that there might have been new biological facts of importance discovered since the encyclical *Casti Connubii.* Thus paragraphs 2 and 3 of the encyclical are written as if no biologist had ever been appointed to the Papal Commission. Equally interesting, but more ominous in this context, is paragraph 6. Here it is made clear that nothing that a present or future scientist could possibly contribute in terms of scientific data could have any pertinence to the subject, if certain criteria of solutions would emerge which departed from the moral teaching of marriage proposed with constant firmness by

the teaching authority of the Church. To the scientist it is difficult to see why the Papal Commission should have been called at all. The teaching proposed with constant firmness by the Church was well known before the Commission was appointed, and it did not require the energy and financial expenditures involved in bringing several dozen consultants to Rome to gather information if, *a priori,* such information was to be eliminated if it led to different conclusions than in the past.

The implications of this paragraph extend far beyond the subject of contraception. The wording of the paragraph is of cardinal importance for the relationship between science and theology. The paragraph implies that theology need not take into account scientific data, but shall reach its conclusions regardless of present or future facts. Had the encyclical stated that the data, advanced by the commission, were wrong or irrelevant, or were insufficient to warrant a change in teaching, that would have been one thing. It is quite another thing to imply that agreement with past conclusions is the *sine qua non* for acceptance of a study. Such wording pronounces the scientific method of inquiry irrelevant to Roman Catholic theology ([18], pp. 216–217).

In Hellegers' comments on the relationship of science and theology, one can perhaps already trace the determination to find a non-ecclesiastical forum for the ongoing exploration of problems at the interface of biology, medicine, and moral theology.

Daniel Callahan, who had earlier expressed misgivings about the Catholic Church's traditional teaching on birth control, was equally distressed by *Humanae Vitae.* In fact, *The Catholic Case for Contraception* [9] was edited by Callahan for the express purpose of supporting the moral arguments of couples who dissented from *Humanae Vitae.* In his introduction to this 1969 work Callahan wrote:

It is impossible to exaggerate the surprise the encyclical caused. It flew in the face of the Pope's own commission, whose conclusions were specifically rejected by the Pope. It flew in the face of an emergent consensus of theologians. It flew in the face of a number of bishops who had asked the Pope not to issue such an encyclical and who had already told their people they should do as their informed consciences dictated. Finally, and most importantly, it flew in the face of a great mass of married lay people. On the basis of their own marital experience and fortified by their knowledge of a change in the thinking of many bishops, priests and theologians, they had decided they could morally use contraceptives for the sake of responsible parenthood ([9], p. ix).

For Daniel Callahan 1969 was clearly the watershed year in the movement toward medical ethics. During that year he began research, with financial support from the Population Council, on a major work on the medical, ethical, and legal aspects of abortion [8]. During the same year he also published an essay entitled 'The Sanctity of Life' [7] – a creative synthesis of themes from moral philosophy, Protestant ethical thought, and Catholic moral theology which in many ways prefigured the medical-ethics literature of the early 1970s. Late in 1969 Callahan and a psychiatrist-friend, Willard Gaylin, founded the Institute of Society, Ethics, and the Life Sciences, which was to emerge as one of the major institutional contributors to the renaissance of medical ethics.

Meanwhile, at Georgetown University Paul Ramsey was completing his second spring semester of research in medical ethics. In retrospective reflections on the time spent at Georgetown Ramsey wrote:

I was appointed the Joseph P. Kennedy, Jr. Foundation Visiting Professor of Genetic Ethics at the Medical School of Georgetown University. This was a research appointment for two spring semesters in 1968 and 1969. It enabled me, a Protestant Christian ethicist, to be located in the middle of a medical school faculty – not on its periphery – and to begin some serious study of the moral issues in medical research and practice. The word *genetic* in that title was a term of art, invented to avoid calling me "Visiting Professor of Obstetrics and Gynecology", referring to the department where I was administratively located. Dr. Paul Bruns, Chairman of the Department, and Dr. André Hellegers, Professor of Obstetrics and Gynecology, arranged biweekly conferences for my instruction. On these occasions a physician who was a member of the faculty of Georgetown Medical School would present an analysis of some preappointed topic and his point of view on such issues as medical experimentation involving human subjects, researches upon fetal life, the definition of clinical death, the patient's right to be allowed to die, organ transplantation, genetic counseling, etc. Thus, I could put my questions to experts in many fields of medicine, overhear discussions among them, and begin to learn how teachers of medicine, researchers, and practitioners themselves understand the moral aspects of their practice. Participating in these meetings were also theologians from seminaries and departments of religion in the Washington area and scientists from other departments of Georgetown University and from the National Institutes of Health, Bethesda, Maryland.

From each meeting and from subsequent conferences with individual doctors I came away with a year's work to do before I, a layman, could venture to say anything about a single medical ethical question. Still I am grateful for the "cultural shock" as well as for the instruction I received ([30], pp. xix–xx).

When Paul Ramsey did begin to speak on the basis of his medical-school immersion at Georgetown, his remarks took the form of the Lyman Beecher Lectures at the Divinity School and School of Medicine at Yale University. Those who heard Ramsey lecture in April of 1969 were dimly aware that a new genre of medical-ethics literature was being created and that a new branch of applied religious ethics could no longer be ignored.

## THE FLOWERING OF MEDICAL ETHICS: 1970–1975

The documentary crystallization of Paul Ramsey's Lyman Beecher Lectures at Yale was a 1970 book entitled *The Patient as Person: Explorations in Medical Ethics* [30]. Perhaps more than any other single work, *The Patient as Person* helped to bring the field of medical ethics to the attention of a broader academic and professional public. Such claims cannot be demonstrated with ease, but one indicator of the impact of Ramsey's book is that in the 1978 *Encyclopedia of Bioethics* [31], *The Patient as Person* was the work most frequently cited in authors' bibliographies.

Several characteristics distinguished *The Patient as Person*. While the work was explicitly by a Christian ethicist, it employed categories that had both religious and secular significance – for example, covenant – to communicate with its audience. Ramsey's book appreciatively borrowed from Catholic medical-ethics textbooks yet was bound by neither their structure nor their authority. The primary focus of *The Patient as Person* was concentrated on the micro-level, on the relationship between individual clinicians or researchers and their patients or subjects, rather than on the health-care or health-research system. The footnotes of Ramsey's work demonstrated a thorough familiarity with medical and legal materials, as well as with writings from his own field of theology. And finally, *The Patient as Person* avoided the topics of contraception and abortion, which were and are frequently perceived as parochial. Instead, Ramsey chose topics of significant general interest: informed consent, research involving human subjects (especially minors), the definition of death, appropriate care for dying patients, organ donation and transplantation, and the microallocation of scarce medical resources.

Ramsey's book simultaneously symbolized and popularized the renaissance of medical ethics in the United States. Between 1970 and 1975 there was a significant increase in the volume of English-language literature on medical-ethics topics. And people who either had strong religious interests or were theologically trained played a principal role in the flowering of the field. It was perhaps Protestants who contributed the greatest volume of literature in medical ethics. Among the Protestant authors were Roy Branson, Frederick Carney, James Childress, Arthur Dyck, John Fletcher, Joseph Fletcher, James Gustafson, Stanley Hauerwas, Michael Hamilton, James Johnson, Karen Lebacqz, William F. May, James Nelson, J. Robert Nelson, Gene Outka, Ralph Potter, Paul Ramsey, Charles Reynolds, Roger Shinn, David H. Smith, Harmon Smith, Kenneth Vaux, Robert Veatch, LeRoy Walters, Preston Williams, and J. Philip Wogaman. Catholic authors contributing to the medical-ethics renaissance in the United States included Daniel Callahan, John Connery, Charles Curran, John Dedek, Germain Grisez, Bernard Haering, André Hellegers, Albert Jonsen, Daniel Maguire, William E. May, Richard McCormick, Albert Moracewski, John Noonan, Edmund Pellegrino, and Warren Reich. And Jewish scholars contributing to the U.S. renaissance included J. David Bleich, David Feldman, Solomon Freehof, Ronald Green, Hans Jonas, Leon Kass, Fred Rosner, and Seymour Siegel.

Few patterns emerge from these lists of scholars. However, among the Protestants there were clearly two generations, the older of which was comprised of faculty members trained at Boston University, Duke, Harvard, Union Theological Seminary, the University of Chicago, or Yale, and the younger of which was trained at Harvard,[5] Princeton, Union, or Yale.

The institutional developments of 1970–1975 also included a significant religious component. Daniel Callahan continued to develop the Institute of Society, Ethics, and the Life Sciences as a major U.S. research center in medical ethics. Neither the Institute nor Callahan's own writings during the years 1970–1975 were characterized by explicit appeals to religious categories or sources of authority. Thus, while Callahan's concern about religious questions played a significant role in his early work on medical ethics, contraception, and abortion, his interest in medical ethics also represented a desire to move away from religion and back to his academic field of philosophy. For example, his major 1970 book on abortion was primarily philosophical in thrust, although it did take seriously several major religious traditions.[6] It must therefore be conceded that the Institute at Hastings and Callahan's own writings from 1970 on constitute partial counterexamples to the thesis of this essay – or at least evidence of an early secular trend in the renewed study of medical ethics. Yet theologically-trained people played a central role on both the staff (Bruce Hilton, Robert Veatch) and the board (James Gustafson, Paul Ramsey) of the Institute.

At Georgetown University, André Hellegers employed the success of Paul Ramsey's research as a basis for proposing the establishment of a permanent research institute on medical ethics. In the spring of 1971 Hellegers succeeded in convincing the Joseph P. Kennedy, Jr. Foundation, which had earlier supported Paul Ramsey's work, to provide initial funding for the Joseph and Rose Kennedy Institute for the Study of Human Reproduction and Bioethics at Georgetown University. Hellegers immediately sought to recruit an ecumenical group of scholars trained in Christian ethics to the Institute. Between 1971 and 1975 Hellegers was able to attract Catholic theologians Charles Curran, Bernard Haering, Richard McCormick, and Warren Reich, and Protestant theologians Roy Branson, Frederick Carney, Stanley Hauerwas, Gene Outka, Ralph Potter, David H. Smith, and LeRoy Walters to Georgetown. While most of these theologians did their research as visiting scholars during sabbaticals from their home institutions, several remained as long-term researchers at the Institute.

By 1975 the Hastings Center and the Kennedy Institute were providing strong institutional support for the revitalized study of medical ethics (now often called 'bioethics' to distinguish the field from the type of medical ethics that is oriented exclusively toward problems of physicians). The Hastings Center was publishing the leading medical-ethics journal, the *Hastings Center Report,* as well as a periodic bibliography. Hastings Center research groups on genetics, population control, death and dying, and behavior control were attracting an interdisciplinary group of scholars from numerous academic institutions and government agencies. By 1975 the Kennedy Institute had established a bioethics library and information retrieval system, was beginning the work of editing the comprehensive *Encyclopedia of Bioethics*, and had initiated a program of graduate education in bioethics in collaboration with Georgetown University's Philosophy Department.

Perhaps the clearest evidence of the dominant role of religious ethics in the years 1970 to 1975 was the early work of the National Commission for the Protection of Human Subjects of Biomedical and Behavioral Research. The eleven-member commission included two ethicists, Albert Jonson and Karen Lebacqz, both of whom had been theologically trained. In 1974 and 1975, during the commission's work on its first topic, fetal research, it solicited testimony from seven witnesses trained in ethics. Of these, five (Joseph Fletcher, Richard McCormick, Paul Ramsey, Seymour Siegel, and LeRoy Walters) came from backgrounds in religious ethics [39].

## CONCLUSION

In the years since 1975 moral philosophers have become increasingly involved in discussions of medical ethics. Indeed, in a provocative essay Stephen Toulmin has argued that applied medical ethics helped to rescue moral philosophy from the arid metaethical debates that had preoccupied the field for more than half a century [37]. In this essay I have sought to establish a distinct but related point, namely that religious ethicists and thinkers interested in the relationship between science and religion were major contributors to a renaissance of medical ethics in the United States in the decade from 1965 to 1975. If my thesis is correct, and if the renaissance described in this essay predated the renewal of interest in medical ethics by moral philosophers that Toulmin describes, then one might profitably explore whether – in this instance only – religion played at least a minor role in the salvation of moral philosophy.[7]

*Kennedy Institute of Ethics,*
*Georgetown University,*
*Washington, D.C., U.S.A.*

## NOTES

[1] See especially [2], [16], [32], [35], [36], and [38].
[2] For a detailed history of Catholic medical ethics in North America through 1960, see [19].
[3] See the highly critical comments on Fletcher's work by Joseph Farraher [13], Gerald Kelly [21], Francis Connell [11], and Joseph Mangan [24].
[4] Two other contemporary institutional developments that cannot be described in detail here were the establishment of the Institute of Religion at the Texas Medical Center in 1967, with theologian Kenneth Vaux as its faculty member in ethics, and the founding of the Society for Health and Human Values in 1969. The Society's founding members included two chaplains (Samuel Banks and E. A. Vastyan) and three education officers in Protestant denominations (Robert Bluford, Robert A. Davis, and Ronald McNeur).
[5] I am indebted to my colleague Robert Veatch for the observation that Protestant ethicists Arthur Dyck and Ralph Potter began offering a graduate seminar on 'Religious Ethics and Population Control' at the Harvard Center for Population Studies in 1966. That seminar and the interest of Potter and Dyck in questions of abortion, family planning, and population policy, in turn, oriented a generation of Harvard graduate students toward the field of medical ethics.
[6] For the perspective presented in this and the preceding sentence, I am indebted to a personal communication from Daniel Callahan.
[7] I wish to thank Daniel Callahan, Paul Ramsey, Robert Veatch, and research assistant Carrie Ure for their helpful comments on an earlier draft of this essay.

## BIBLIOGRAPHY

[1] Antonelli, G.: 1891, *Medicina Pastoralis,* F. Pustet, Rome.
[2] Augustein, L.: 1969, *Come, Let Us Play God,* Harper & Row, New York.
[3] Callahan, D.: 1963, *The Mind of the Catholic Layman,* Charles Scribner's Sons, New York.
[4] Callahan, D.: 1964, 'Authority and the Theologian', *Commonweal* **80,** 319-323.
[5] Callahan, D.: 1965, *Honesty in the Church,* Charles Scribner's, New York.
[6] Callahan, D.: 1966, *The New Church: Essays in Catholic Reform,* Charles Scribner's, New York.
[7] Callahan, D.: 1969, 'The Sanctity of Life', in Cutler, D. R., ed., *The Religious Situations: 1969,* Beacon Press, Boston, pp. 297–339.
[8] Callahan, D.: 1970, *Abortion: Law, Choice and Morality,* Macmillan, New York.
[9] Callahan, D. (ed.): 1969, *The Catholic Case for Contraception,* Macmillan, New York.
[10] Capellmann, C. F. N.: 1877, *Pastoral-Medicin,* R. Barth, Aachen.
[11] Connell, F. J.: 1955, 'A New Work on Morals and Medicine', *American Ecclesiastical Review* **132,** 38–44.
[12] Cooke, R. E., *et al.*: 1968, *The Terrible Choice: The Abortion Dilemma,* Bantam Books, New York.
[13] Farraher, J. J.: 1966, 'Notes on Moral Theology', *Theological Studies* **16,** 233–269.
[14] Fletcher, J.: 1954, *Morals and Medicine,* Princeton University Press, Princeton, N.J.
[15] Ford, J. C. and Kelly, G.: 1964, *Contemporary Moral Theology,* Vol. 2, *Marriage Questions,* Newman Press, Westminster, Md.
[16] Freund, P. A. (ed.): 1969, 1970, *Experimentation with Human Subjects,* George Braziller, New York.

[17] Healy, E. F.: 1956, *Medical Ethics,* Loyola University Press, Chicago.
[18] Hellegers, A. E.: 1969, 'A Scientist's Analysis', in Curran, C. E., ed., *Contraception: Authority and Dissent,* Herder & Herder, New York, pp. 216–236.
[19] Kelly, D. F.: 1979, *The Emergence of Roman Catholic Medical Ethics in North America: An Historical-Methodological-Bibliographical Study,* Edwin Mellen Press, New York and Toronto.
[20] Kelly, G.: 1949–1954, *Medico-Moral Problems,* 5 parts, Catholic Hospital Association of the United States and Canada, St. Louis.
[21] Kelly, G.: 1955, 'Medico-Moral Notes', *Linacre Quarterly* **22,** 55–61.
[22] McFadden, C. J.: 1946, *Medical Ethics for Nurses,* F. A. Davis, Philadelphia.
[23] McFadden, C. J.: 1967, *Medical Ethics,* F. A. Davis, Philadelphia.
[24] Mangan, J. T.: 1955, 'Morals and Medicine [Review]', *Theological Studies* **16,** 307–310.
[25] Noonan, John T., Jr. (ed.): 1970, *The Morality of Abortion: Legal and Historical Perspectives,* Harvard University Press, Cambridge, Massachusetts.
[26] O'Donnell, T. J.: 1956, *Morals in Medicine,* Newman Press, Westminster, Md.
[27] Ramsey, P.: 1956, 'Freedom and Responsibility in Medical and Sex Ethics: A Protestant View', *New York University Law Review* **31,** 1189–1204.
[28] Ramsey, P.: 1966, 'Moral and Religious Implications of Genetic Control', in Roslansky, J. D., ed., *Genetics and the Future of Man,* North Holland, Amsterdam, pp. 107–169.
[29] Ramsey, P.: 1967, 'The Sanctity of Life – in the First of It', *Dublin Review* **511,** 1–21.
[30] Ramsey, P.: 1970, *The Patient as Person,* Yale University Press, New Haven.
[31] Reich, W. T., ed.: 1978, *Encyclopedia of Bioethics,* 4 vols., Free Press, New York.
[32] Rosenfeld, A.: 1969, *The Second Genesis: The Coming Control of Life,* Prentice-Hall, Englewood Cliffs, N.J.
[33] Royce, J. E.: 1955, 'An Insidious Treatise', *America* **92,** 538.
[34] Shannon, W. H.: 1970, *The Lively Debate: Response to Humanae Vitae,* Sheed & Ward, New York.
[35] Taylor, G. R.: 1968, *The Biological Time Bomb,* New American Library, Cleveland.
[36] Torrey, E. F., ed.: 1968, *Ethical Issues in Medicine: The Role of the Physician in Today's Society,* Little, Brown, and Company, Boston.
[37] Toulmin, S.: 1982, 'How Medicine Saved the Life of Ethics', *Perspectives in Biology and Medicine* **25,** 736–750.
[38] U.S. National Commission for the Protection of Human Subjects: 1975, *Research on the Fetus,* 2 vols., U.S. Government Printing Office, Washington, D.C.
[39] U.S. Congress, Senate, Committee on Government Operations, Subcommittee on Government Research: 1968, *National Commission on Health Science and Society,* Hearings on S.J. Res. 145, 90th Congress, 2nd Session.
[40] Walters, L.: 1974, 'Medical Ethics', in *New Catholic Encyclopedia,* Vol. XVI, Supplement 1967–1974, Publishers Guild/McGraw-Hill, Washington, D.C., pp. 290–291.

GEORGE P. SCHNER

# THEOLOGY AND SCIENCE: THEIR DIFFERENCE AS A SOURCE OF INTERACTION IN ETHICS

The question of a relationship between theology and science in general (medical and life sciences in particular) could be answered in a variety of ways. The parameters of such a discussion are exceedingly large, such as not only to complicate the conceptual clarifications needed but to render conclusions impossible or at best remote or inconsequential. The very existence of a collection of articles as in this book is evidence of the possibility of asking such a question and the need to pass beyond the pragmatics of solving problems in an *ad hoc* fashion. However, both theology and science are nests of concepts. They are not simple logical systems; rather, they consist of knowledge with a complex history that is internally composed of a variety of details in need of clarification, and externally already related to each other such that posing the question of relationship requires a historical perspective as well as logical clarity.

A provocative reply to the question would be to say that there is no relationship at all. Neither theologian nor scientist operates out of any real interaction with each other. Moreover, scholars and experts in both disciplines cannot engage in or profit from a conversation with one another unless the hard truth of this position is considered first and the attempts to avoid it through a kind of concept management are confronted. Insight into what kind of relationship can exist between theology and the sciences must begin with a frank exposition of what constitutes their difference before efforts to relate them can proceed with intellectual honesty.

Let me begin with a classic formulation of the Christian intellectual enterprise. For the Christian theologian, the word of God is uncompromised and heeds no other voice of rationality. It has its own intelligibility as demanded by faith itself. This definition is itself a metaphor, though a highly refined one. Moreover, as it grows in scope the metaphor incorporates a great variety of interpretive devices and contents and in so doing claims ultimately to give them their meaning and proper use by reference to a single symbol. Once addressed by the transcendent, Christians must, when it is a matter of meaning and responsibility, seek



*E. E. Shelp (ed.), Theology and Bioethics,* 17–25.

their identity and the norms for action in the person and work of Christ. On the other hand, the scientist of whatever sort is committed to an uncompromising pure inquiry such as Descartes developed in his *Discourse on Method*. By choosing this reference point, I am consciously speaking of 'modern science', or simply of modern consciousness. This should be kept in mind when the term recurs in the text. In keeping with the demands of rationality which makes scientific inquiry possible, there can be no hesitation before a dogmatic enunciation of the structure of meaning which is not yet again subject to the investigative scrutiny and evaluation of the human inquirer. In fact, the very investigation itself may generate data formerly unknown, and thereby call the previous stages of an investigation into question, both as to method and content.

However, when the preoccupation of any discipline becomes the investigation of the internal relations of its own metaphor and thus deserves the name 'ideology', the person of 'religious' conviction can begin the process of critique. In the case of contemporary technological and scientific culture a theologian can begin with a reassessment of the traditional Christian categories of God as creator, saviour, and enlightener and attempt to show that the present culture is in need of what the Christian believer offers as the vocabulary for understanding the world, namely, the needed corrective for the ills of the day. Even without a revision of Christian vocabulary and a corresponding application to the scientific vocabulary, theoreticians of science are themselves attempting to reconsider the enterprise as it understands itself, and introduce transcendence as a necessary element of the scientific method itself. More broadly, a case can be made to show that every human enterprise as a form of knowledge is inextricably linked with human interests, with values either hiddenly operative or reflectively articulated.

There are those, then, who attempt to redefine either the scientific or religious nest of concepts so as to accomplish a rapprochement which overlooks the difference that I have suggested must be attended to. Let me observe such efforts at work in several ways in the theological scholarship of the present.

A point of accommodation which takes as theme the driving force of scientific work is the effort to articulate a theological *method*. In such an effort the theologian can attempt to prove that there is an underlying rationality, a 'set of recurring operations' which unifies all the sciences, including theology, such that there cannot be in principle any basic opposition between them. Vocabularies may differ, but a reconciliation is possible by a return to the invariant foundations of all. Unity is provided by an appeal to subjectivity itself. In this way theology and the sciences can converse first through the common vocabulary of an anthropology or theory of knowledge, and then in differing but not incompatible nests of concepts. What happens to the radicality of theology given its nature as reflection on the incursion of the transcendent into human life, always instanced in this or that concrete religion, and the graciousness of this transcendent intervention which constitutes salvation, is as problematic as the necessary inhibiting of pure inquiry and the admission of sinfulness which the scientist must adopt.

Another search consists in the somewhat inconsistent scramble for *religious experience* which, once found, is thought to give an empirical basis to the work of theology in much the same way that any of the sciences can partially state its definition from a denotation of its field of inquiry. A naive objectivism is unacceptable within the realm of scholarship, and theology embarrasses itself when its devotees lapse into this search. The religious dimension of reality is present in all experience or none at all, and though it is intentional it is not reducible to either an object or a merely subjective condition. Moreover, this dimension, like all dimensions of human experience, must always be content with experience as mediated, with the acquisition of religious symbols within a community of believers. If the search for method is inadequate, so is the search for experience. The choice of experience as foundational, given classic utterance in Schleiermacher's *On Religion: Speeches to its Cultured Despisers,* whether treated as a critical quest for the transcendental experience which founds a given religion or a more positivist quest for certain specific and thematic states of the self, is another effort to locate an element common to both science and theology.

Third, there is always the possibility of accommodating the religious enterprise of reflection to a scientific model by so emphasizing the constructive work of the *imagination* in theology that an 'as if' interpretation is seen to be the gist of theology. While it is necessary or at least advisable as a cultural achievement, it is not contradictory in any hard sense with the interpretations of any other discipline. The interconnected language games appropriate to a particular form of life, to use the Wittgensteinian terminology, form a whole that is as useful as any other such cluster. Any questions of truth, however, are set aside as incapable of an answer; interpretations abound, pragmatic usefulness is the criterion, and conflicts remain unresolvable. The result is twofold: both the hidden assumptions that act as norms not subject to scrutiny and the bland equality of all conceptual schemes mask the hidden human interests that control the preferences.

Now consider the accommodation from the side of science. It would not be impossible for a theoretician of science to construct an account of scientific method and its rationality which included a teleology of movement towards the transcedent, an inherent dynamism of self-transcendence which intimates a movement towards another dimension. What is to be noted is that the very terminology required to do justice to such a construction would be religious, even theological, language. Such a theory of scientific endeavour must face the criticism belonging to that long tradition of empiricism expressed in modern history by David Hume in the *Dialogues on Natural Religion* and ending in the positivism of A. J. Ayer, throughout which the use of a covertly Christian view of reality, an optimism both metaphysical and epistemological, is deemed at best emotive, in fact nonsensical. A question will remain: what authorizes the metaphor used by the scientist who chooses to ground (already the use of a religious metaphor?) the interpretation of the scientific enterprise? Such usages come easily, as Hume put it, "to people of a metaphysical head" and the preference for one analogy or another is ultimately unfounded:

It were better, therefore, never to look beyond the present material world. By supposing it to contain the principle of its order within itself, we really assert it to be God; and the sooner we arrive at that Divine Being, so much the better. When you go one step beyond the mundane system, you only excite an inquisitive humour which it is impossible ever to satisfy. ([1], p. 34)

In a contemporary text, Anders Jeffner speaks of the 'metaphysical optimist' who chooses to answer in the affirmative three fundamental questions, and so adopt three ultimate norms:

The first says that we ought to see the world as ultimately understandable; the second that we ought to accept certain non-scientific explanations; the third prescribes how we ought to choose between explanations. In ethics, I think, we encounter certain basic norms, which we simply have to accept or reject and which we cannot give any further reasons for or against. If we are right here, an analogous situation exists in the theoretical field. We can regret this situation, but we cannot alter it. ([2], pp. 130–31)

I will return later to Jeffner's reference to ethics. For the moment, let let me interpret the conclusion to his work as suggesting that, taken from the point of view of language, the introduction of religious terminology cannot be an empty usage. If it were, it would undermine its very purpose, and resolve into the 'as if' imaginative construction that finally relinquishes any claims to statements of truth. That is not to say that the scientific or theological construction is any less free of the ultimately fruitful ambiguity that marks all human constructions. Nor am I advocating in a hidden fashion a return to a correspondence theory of truth which presumes that language simply states 'the way things are' if it is properly functioning. Theological language itself must accept the symbolic character of its efforts, equally accepting the concealing as well as the revealing that occurs in such language use.

There can be, then, at least three efforts to accommodate theology to a scientific model. One can search for a method which is ultimately a scientific one because it is the function of the same subjectivity at work in science; one can search for experience so as to have data for investigation as does science; and, one can speak of theology as the work of the imagination, a great 'as if' to be dealt with pragmatically. The inherent difficulty with such efforts can be located in the contrast of scientific rationality, the demand for the transparency of reality to the inquisitive eye of the researcher, with the contemplative rationality of the believer who, in the first instance, must exercise a *potentia obedientialis* in face of the encountering transcendent. A larger question yet to be addressed asks why these kinds of accommodation are attempted. It may well have to do with the inclusiveness of the religious metaphors, or a failure of will in face of the dominance of the scientific metaphor. Probably it is an interplay of both.

The exercise of scientific inquiry itself marks a stage in the development of human consciousness which cannot be set aside, as if one could leap-frog back to a former moment, in our case returning to a romantic, pre-technological consciousness, which in fact never exists except as an idealized past to be longed for. The religious attitudes which are the companion of such a stage of the development of consciousness are marked by its own preoccupations: method, experience, imagination. This trio is obviously the same as the group of terms I noted as the preoccupation of contemporary Christian theology. Let me consider them again, this time as logically related to each other.

The list is not merely an indication of theology's attempt to encounter the contemporary age, but is its product. The preoccupation with methodology is symptomatic of the preoccupation with technique. If only we develop the correct techniques, so the argument goes, the necessary advances can be made and the long-sought-for truths will be disclosed. Whatever the technique, however, that upon which the methods are used will become the focus. For example, the one element not structurally bound by B. J. F. Lonergan's theological methodology is the moment of religious conversion. Whether religious conversion is truly different from moral or intellectual conversion is problematic in his work, but that element, as outside the technique, is properly the next moment of investigation.

What then is religious experience? The term is necessary but problematic, since it cannot designate anything. In the controversy with scientific positivism of all kinds, the reflective believer can make the effort to locate a terrain of 'experience' for the science of theology to investigate. The resulting death by a thousand qualifications of such experience, evidenced by the futile distinction between ordinary versus extraordinary experience, lays bare the inevitable interpretive character of all such empirical description. This last point can be structurally occluded. For example, Peter Berger's appeal for a return to experience in the inductive model for religion is partly dependent upon his pejorative definition of tradition as distortive, and his seemingly naive notion of experience as literally a retrieval without interpretation. The nest of problems involved in an appeal to experience was seen by Hume in the *Dialogues* but was successively obscured by the rationalist philosophy of the 18th century until in the 19th century the romantics gave pride of place to the imagination.

The recognition of the role of the imagination in the récent decade, after the brief encounter with the positivism of revelation which is often attributed to Karl Barth, continues the liberal theological project, albeit chastened by the 'hermeneutics of suspicion' which generally is nodded to in the cipher-litany '. . . as shown by Nietzsche, Marx, and Freud'.

If there is to be an interaction between theology and science, it is in the reshaping of the imaginative constructions which constitute the felt and reflected world of our age. It is not an invariant structure of operations nor a self-validating experience, but the work of the imagination understood as an inventive activity of the self not to be regimented or exhausted by a technique, nor confined in its operation to a single realm of so-called religious experience. The habituation of humankind to the life of technique causes the experience of things to be characterized by a sense of disposal and control, of creativity as opposed to inventiveness, of uniformity instead of difference and hierarchy.

That is not to say that *discipline,* in the sense of rules adopted with responsibility, does not exercise a constitutive function. On the contrary, a connection must be made explicit between the use of both rules of faith such as the creeds, and, in a differing but related manner, scriptural texts, and the exercise of the imagination for the purpose of the religious use of language and the specific theological uses in the construction of sentences about doctrinal matters. If this connection is made, then on the one hand imagination will be saved from being psychologized and at the mercy of a romantic idealization, prevented from being invoked as a free agent of inventive activity without norms, an empty productive mechanism; and on the other hand, the rules themselves will not be taken as empty ideological categories or mistakenly understood as propositions or descriptions of states of affairs.

The resemblance of this compatibility to the Kantian dictum about blind sensibility and empty categories of the understanding is not by chance, given the similar Kantian effort to account for the role of imagination precisely as the unifying factor of these two elements of cognition. However, resemblance does not involve more than a family likeness in this case, quite removed from a transcendental analysis.

The third element of this discussion must be the matter of the *ethics* which becomes concerned with science and the medical sciences in particular, whether the directly patient caring and patient involved or the more laboratory and experimental ones. It is not my intention to introduce specific details pro or con in regard to any question of detail, but to reflect broadly upon ethics as the sphere in which conflicting rationalities engage one another. Of equal importance are the specific doctrinal issues, in the present case, of the Christian articulation of belief concerning the person and work of Christ, the creative and eschatological work of God, and the immanent work of grace.

It would be my contention that ethics is the proper sphere in which the encounter between the rationality of science and that of theology can take place. Moreover, the encounter will be structured by a narrative logic and a dialectic of positions, both characteristics being essential to the health of the dialogue partners. In as much as theology as a discipline articulates the drive of human beings toward the transcendent and therefore toward mystery, it does so not in an abstracted way, but within the human community of a particular religious tradition. It cannot adopt the definition of truth as clarity and certainty. That is not to say that theology cannot be clear in what it says, rigorous in its use of logic, and persuasive in its arguments. It cannot hold, however, that reality is ultimately perspicuous.

In as much as science articulates the human drive for a transparency of reality and for the exercise of control over it, not in an abstract fashion but in the context of the *polis,* it must adopt the definition of truth as clarity and certainty and eschew reverence before reality and deference before any particular interpretation of reality. This is as it should be: science is not religion, nor religion science. Their interaction is not accomplished by a reduction of one to the other, but rather is dependent upon the way in which they are brought to bear upon human situations in and through what we call ethical thought and language.

Both theology and science are subjected to the narrative logic of the human situation. Perhaps Christian theology seems more adaptable to such a logic, given its foundation in the life and work of Christ, and this person himself as founded in an already dynamic Godhead, a trinity of persons forming a community. Yet science is not unaware of such a logic when it reflects upon the dynamism of experimentation, the story character of scientific discovery, and the retrieval of myth as the past and future of its own enterprise. However, it is not a matter of an internal accommodation of either to an abstractly imposed norm which will accomplish the reconciliation of the two. Both must maintain strictly their essential identity and therefore their essential difference.

It is in the sphere of the ethical, the moment of dilemma and freedom, that both are engaged. Both are challenged by the reality of human freedom, and in turn challenge each other and that freedom itself. In the challenging, honesty is demanded of both, complimentarity is achieved, and critique is exercised. Let me deal with each relation in turn.

Theology will be kept honest by science, not by adopting its criteria of truth, conforming its own methods to that of scientific research, or subjecting conclusions to the scrutiny of science. Rather, as science alters the world itself and the metaphors available for understanding it, theology is engaged. Whether the issues be genetic engineering, nuclear technology, sex-change operations, or discoveries within depth psychology, theology must reassess the metaphors in which it expresses its anthropology and, because of the commitment to a doctrine of Incarnation, its *theo*logy proper. It does not reassess the normative character of the person and work of Christ. It reconsiders its own interpretative schemes, how it is to understand the human person, the constructing of a world of meaning, and the orientation to the future in human community.

Similarly, theology challenges science to honesty. It does not do so by insisting that science be religion. On the contrary, it is when science *becomes* religion, that is, when it becomes ideological, that it abandons its own inner dynamic toward truth, namely truth as certainty and clarity. Science needs the reminder that any given position of truth which it attains must be subjected to scrutiny so that the drive to clarity will continue. It cannot be satisfied with any particular achievement, but must continue its search for the consequences and development of what it has achieved.

It is the very existence of religion and science side by side in difference that constitutes their mutual challenge. The inner logic of each must be preserved, the unique criteria of both maintained. When both together challenge human freedom to action, the rapprochement can occur.

To return to the comment of Anders Jeffner on the encounter of what I will call the moment of 'faith' in reflection on ethical matters, and in reflection upon theoretical matters as well: the contrast of faith and reason is an Enlightenment problem transportable to the twentieth century but increasingly difficult to maintain. The search for presuppositionless knowledge, for pure reason, has passed through a long history of critique. Boldly stated, science, like theology, has its own position of faith to which it appeals. When Jeffner observes that ethics resolves into certain basic norms for which reasons cannot be given, I observe that it is precisely the appeal to scientific thought on humanity and nature and theological thought on humanity and nature that is appealed to. If and when such an appeal is made self-consciously, then several things can happen.

First, human freedom is engaged and therefore the quest for truth is engaged. The possibility of a decision specifying action requires both transcendence and transparency. Second, the complex moment of choice will have as its focus, not the accommodation of scientific and theological reasoning to one another through a reduction or subordination of one to the other, but the use of both for the construction of the human narrative, that is, quite simply, *life*. In this way the logics of theology and science must encounter the logic of life. Third, the subsequent encounter of both in and through the human complex of taking a decision gives the occasion for the engagement of both in dialogue.

My provocative reply to the initial question of this essay has resulted, in its unfolding, in what might be called a classic position. In a certain sense there is no such thing; there is no going back to former things. There is only the bringing forward of such things into the present. The rapprochement between science and theology will be crafted by those who, in reflection upon both, confront the pressing matters that exist for human freedom. These moments of testing demand internal honesty and insightfulness from both, each in their different ways.

*Regis College, Toronto, Ontario, Canada*

BIBLIOGRAPHY

[1] Hume, D.: 1948, *Dialogues Concerning Natural Religion,* Hafner Publishing Co., New York.
[2] Jeffner, A.: 1972, *The Study of Religious Language,* SCM Press Ltd., London.

CHARLES HARTSHORNE

# SCIENTIFIC AND RELIGIOUS ASPECTS OF BIOETHICS

## GENERAL CONSIDERATIONS

Both 'scientific' and 'religious' mean various things to various people. In this essay the results, as well as the methods, of empirical science, especially biology, will be relevant. As for 'religious', I shall emphasize what Judaism and Christianity have in common rather than what separates them. I feel close also, in some respects, to Buddhism, and to the Bengali branch of Hinduism (Sri Jiva Goswani), also to Iqbal in Islam; but these relationships will not be of central importance on this occasion. The two Great Commandments referred to by Jesus, viz., love God with all thy mind and heart and soul and love thy neighbor as thyself, define what, for me, is the religious attitude.

I wish to make clear at the outset that it is God that I worship, not any book. Scriptural literalism may for many be religious, but for me it is idolatry. Human hands wrote those scriptures, whether in Greek, Hebrew, Sanskrit, Arabic, or English, and any divine inspiration had to pass through human minds and brains. In addition, the writers in many cases were giving advice or commandments to societies whose problems were, in part, very different from ours. There is not a scintilla of evidence that they had any notion, for instance, of the changes modern technologies have brought about in the death rate, with consequent need for changes in birth rates.

As for what is meant by loving God without reservation, I take this formula indeed literally. But something like it is in many ancient books besides the Hebrew and Christian Testaments. And such philosophical rationality as I have tells me it is the most reasonable of all comparably fundamental beliefs. With Paul Tillich, I take the formula as an implicit definition of what should be meant by 'God'. The word stands for the 'One Who is Worshipped', and 'worship' is unqualified devotion, or the love which in principle includes *all* one's concerns or interests. The only way I can see to understand this is to take God as cherishing every creature, finding in it any value we could find and incomparably more be-



*E. E. Shelp (ed.), Theology and Bioethics*, 27–44.

sides. So, of course, we should value our neighbor, since he or she is valuable in the same way as we are, namely, valuable to God. We can love God with all our being, since everything that could possibly interest us is interesting to God, who is no mere 'unmoved mover' but rather the One to whom, as some medieval mystic said, 'all the forms of being are dear'. I take this to mean that, from the creatures, God acquires values not possible without them. This does not mean that this or that particular creature is indispensable to God, but only that it adds something to the divine life. As I have explained in many writings, this does not prevent God from being 'perfect' in whatever sense this term has a coherent meaning. If defined as the actuality of all possible value, 'perfection' is the attempt to praise God by talking nonsense. Possible value is inexhaustible by any actuality. There can be and is a Being (rather, *the* Being) such that no *other* being could be greater; but this Being can itself be enhanced. It is the all-surpassing, self-surpassing reality. It need not and cannot be simply unmoved.

With my rejection of the 'unmoved mover' as synonym for deity goes an equally emphatic denial of 'omnipotence' as usually understood. God has, in ideal degree, power over all things; but it does not follow that whatever happens divine wisdom must have decided that it would happen and divine power have seen to it that it did happen. In that case the all-loving Being would be the real doer of all deeds, however shocking or unfortunate; and the creatures deceive themselves in thinking that they decide their conduct, since it is rather God who eternally decides all actions. I can only regard this doctrine as absurd and impious. It interprets divine love as denying to us precisely that which love of the highest kind would be least likely to deny, power to make our own decisions. The doctrine expresses the worship of power, not of love. How tragic that for nearly twenty centuries such a view passed as the highest human wisdom! On the contrary, the sane view of divine power is that of supreme freedom influencing and inspiring, but not strictly determining, lesser forms of freedom.

It happens that a poet whose piety is manifest expressed this view with perfect clarity a century ago. Of the idea that if he, Sydney Lanier, wrote a poem praising God, it is really God who wrote the poem (that is, decided what it was to be), he declares,

> It is not true, it is not true . . .
> Who made a song or picture, he
> Did it, and not another, God nor man.
> My Lord is large, my Lord is strong:
> Giving, he gave: my me is mine.

How poor, how strange, how wrong,
To dream He wrote the little song
I made to Him with love's unforced design
. . . Each artist – gift of terror – owns his will.
([2], p. 12).

Two things need to be added to the sketch of a theology just given. We must, as Fausto Sozzino (Socinus), Gustav Theodor Fechner, Lewis (Juels) Lequier, A. N. Whitehead, and others have done (in several centuries and countries), revise the idea of omniscience, not to allow for 'ignorance' of the future in God, but to allow for the truth that the future is not a set of fully definite events, but a set of probabilities and possibilities for the future constituted by a given present. As Berdyaev hints, there must, hard as this is to conceive, be a divine time and a divine present, past, and future: God knows each aspect of the creative process as it is, the past as fully definite, the present as becoming definite, and the future as destined to become definite. In all this there is no ignorance, properly so-called, just as there is no defect of power in God's *not* being what many have meant by 'omnipotent', or in *not* being what many have meant, or thought that they meant, by 'perfect'. What these negations reject is only some human confusions or bits of bad thinking. This is the true 'negative theology'. It is what is positive in our idea of God that is objectively true or that transcends anthropomorphic error. God *is* loving, wise, and good, but is *not* the all-determining, that is, monopolistic decision-maker or the verbally unincreasable maximum of value so much talked about by our ancestors. God is, in concrete actuality, anything but unmoved (as Søren Kierkegaard was one of the first to say, though he also insisted that God was unchanging). Only the divine *kind* of goodness, wisdom, or love, essentially an abstraction from the concrete volitions and cognitions of deity, is unmoved.

The other additional theological point is that to fully escape from the benevolent tyrant conception of deity as sheer determining cause we must admit not only that certain of the creatures have power to decide some aspects of exactly what world is to be for God to know and love, and hence to influence the concrete form taken by the divine knowing and loving, but, that *every* creature must have something of this power. What seem inert bits of matter must be seen as aggregates of entities, each of which has some spontaneity, some initiative, of its own. Quantum physics at least seems to allow a possibility of this, as in half-life laws which tell how long it takes for half of a multitude of uranium atoms, for example, to change into atoms of lead, but says nothing about just when a particular atom will do so. Our notion of mere lifeless, unspontaneous matter can historically and psychologically be shown to derive from the inability of our senses to exhibit distinctly and directly the microstructure of the world, in which, as science has shown, is the real detailed dynamism of inanimate nature. Plato distinguished 'self-moving' and besouled substances from those moved only from without and, therefore, soulless. Modern science knows only self-moved (among the genuinely single or dynamically unitary) entities. Materialism and dualism alike arise mainly from a deficiency of our senses, deficient only for abstruse scientific and philosophical purposes, not for the adaptive needs of human animals with primitive technologies and sciences.

By generalizing creaturely freedom from strict determination by causal conditions, we show the classical problem of evil to arise from a groundless assumption about causality. If every singular, genuinely unitary creature has some freedom, then what happens can never be something which in detail God has willed or decided; rather, countless multitudes of agents have, in however primitive a fashion, willed or decided aspects of the happening; but none of them, and not God, has willed or decided the happening in its concrete wholeness. There is no scapegoat to be blamed or excused for evils, not even a divine scapegoat. "Why did God do this to me?" is never a correct question when misfortune strikes. God has not done it, a virtual infinity of agents have done it. God is to be valued not as the One who deals out goods and evils but as the One who makes it possible for the blows of fortune, the chance intersections of creaturely decisions, to make, on the whole, a harmonious cosmos, infinitely better than no cosmos and such that the risks of discord and suffering are justified by the opportunities for harmony and happiness. With little freedom, as in insects, risks are also slight; with great freedom, as in our species, they are not slight, but great also are the accompanying opportunities. Cosmic order is the probability of good on the whole. But just what goods there are in detail is up to the creatures. This precisely is the meaning of their existence. They are humble cocreators with God and for God.

Let us now turn to the scientific aspect of our bioethical problems. Biologists see, what many others may scarcely realize, that no species can go on indefinitely multiplying its numbers as the human species has been doing in its recent history. The whole stellar universe known to us is too small to hold the numbers of people that would result in a few thousand years. Already shortages, not only of food but of clean air and water, easily available minerals and energy sources, are beginning to cause, or must soon cause, great hardships to millions or billions. It is irrelevant that if we were all angelically wise and good, present supplies would perhaps suffice present populations. Human beings have never been thus wise or good. When more food, etc., adequately distributed, is available, still, if present practices of reproduction are continued and there is avoidance of all-out war, doubling or quadrupling of the population will create fearful hardships. An ever more crowded world is not likely to be a peaceful one, or one with the respect for life that opponents of birth control favor.

It is mathematically certain that, in the not very long run, one or both of two things must happen: death rates must go up or birth rates must go down. Do those who put an absolute priority on not interfering with the birth rate want death rates to increase? Or do they fancy that there is a third possibility?

The idea is abroad that if we use artificial pregnancy prevention methods or abortions we are assuming the role of God or violating the laws of nature. Just what laws? The natural balance of births and deaths was upset many decades ago by discoveries and uses of scientific hygiene. These were human actions and decisions. Already we have assumed powers over life and death that were previously unknown on this planet. And when, under conditions obtaining before modern technology, persons would soon die, we now keep some of them alive (perhaps in a coma, perhaps with dead brains) for months; is that taking the role of deity? If not, why is artificial birth control or abortion viewed as taking it? Is the religious view defensible that technology can be used to worsen the problems it creates but not to alleviate them? In the past some Catholic theologians held that a fetus has a soul only after the third month, thus agreeing fairly well in advance with the Supreme Court decision. Some Catholic theologians have also held that it is not sinful to avoid extraordinary artificial means to keep elderly and terminally ill persons barely alive. So there are glimmerings of good sense here and there among official representatives of religion.

## ABORTION

I wonder at the ease with which people assume knowledge of divine thoughts about problems created since the sacred scriptures were written. Also, 'the sacredness of life' is not a phrase I have noticed in the Bible. Mosquitoes are alive, and they are at least multicellular creatures, whereas the fertilized egg cell, which some say is 'the beginning of (a human) life', is a single cell. For some weeks after pregnancy the fetus is not demonstrably more than a cell colony. True, it has some organization, but so do termite colonies; yet they are not individual animals. With the formation of the nervous system there begins to be an individual animal, which presumably begins to have its own individual feelings. But in what sense are they human feelings, valuable as such?

The Greeks knew long ago that it is speech, and the accompanying capacity to use complex symbols and symbols for symbols (e.g., the word 'word'), that makes us superior to the other animals. Clearly, there is no such capacity in a fetus. The nervous system is still too primitive for this to be possible. Ask any physiologist. Until well after a year following the beginning of the mother's pregnancy the baby's nervous system is still too primitive for the baby even to *begin* to learn the distinctively human mode of thinking. It is learning, but not how to think in the human linguistic way.

The most substantive argument of the 'pro-life' enthusiasts is that if we do not respect human life in the fetus we cannot consistently respect it in the infant, since in neither case is the individual capable of learning to speak or think in the human fashion. The difference between fetus and infant is at most a matter of degree. And so is that between infant and child who is able to speak a little. And so on. However, differences of degree are important. Indeed, importance itself is a matter of degree. It is not demonstrable that the ability of apes to use language, or something like it, and think in complex ways differs from that of human beings except in degree. Should our ethics, legal system, and religion stand or fall by the indemonstrable claim that the difference is absolute? We do not allow children to marry, to vote, or to assume political offices, and these are matters of some importance. Yet adults differ from children only in degree. The history of infanticide shows that in nearly every society it has been more or less sharply distinguished from ordinary murder. This was true in ancient Greece, Rome, and China. It has also been true of some modern European countries.

We should love our neighbors as ourselves. Was Jesus thinking of fetuses – or even of infants – as neighbors? I think it is clearly false that an infant has all the value of an adult or even of a normal year-old child. The 'beginning of a human life' is not, by any evidence, the actuality of a being on the human level of value. It is the possibility of such a being, and therefore an important and marvelous creature, but the mother is also marvelous and far more important. So are still other people who may have to help care for or deal with the offspring if it is born and survives into childhood and adulthood.

The difference between possible and actual is either important or unimportant. If it is important, then respect for a human adult is different from respect for a human fetus. If the difference is not important, then it must be taken as a terrible tragedy that millions of fertilizable human egg cells never are fertilized. Is the sheer number of human lives the most important question? Or is the quality of these lives at least as important? If 'religion' is to stand for the short-circuiting of these clearly relevant considerations, how do we answer those who say that religion is the enemy of human enlightenment and survival?

To the question, Is abortion murder?, I hold the affirmative answer to be a gross misuse of a very important word. We are now doing poorly in our efforts to protect adults or children from slaughter; will we be better off if our police must also try to protect fetuses, or even to make sure that no mother kills her own infant? Our prisons have plenty of occupants, but many criminals go free. The police power is overextended, as anyone can see.

I offer a few remarks concerning cloning and gene manipulation. What cloning would do is to change the ratio of genetically unique human individuals to genetically non-unique ones, as in identical twins and the like. That the ratio has always been extremely high is, there is reason to think, a good thing. Of all species in nature the human is the most highly individuated. Uniqueness is a special human value. As one goes down the scale toward atoms, individuals grow more and more alike. Variety is an aesthetic good; the variety of human beings is one measure of the superiority of our species. The notion that many genetic duplicates of Einstein would have been a grand thing is not necessarily a very intelligent idea. One Einstein was perhaps enough.

It is clear that gene manipulation to determine sex would present the species with a fearful problem. Would there be a great deficiency of females? Would couples quarrel bitterly over the matter? Clearly, there is a reason to hesitate before rushing into such a situation. But at least one could hardly be tempted to describe the projected action as murder. The issue is indeed not the same as that of abortion. And I do not at present feel called upon to arrive at a conclusion concerning it.

The cloning and gene manipulation issues seem particularly clear examples of problems not anticipated by biblical writers. As I have said, biblical literalism taken as settling such questions is for me idolatry, i.e., worship of a humanly written book, not of God. Also our Constitution forbids trying to settle political issues by appeal to biblical texts. That many do not support this aspect of the Constitution and wish to force inferences from their idolatry upon the rest of us is a national misfortune which we have to live with as best we can.

What religion does supply is ultimate ideals, such as respect for God and our neighbors, among whom fetuses are not reasonably included. Moreover, even an adult human being is not God, and one is asked to love the neighbor as one loves oneself, not as one loves God. God, not an individual animal, is sacred.

The ultimate ideals are necessary, eternal, and independent of contingent circumstances. But they are very general and their application to specific cases depends on contingent circumstances. Here is where science comes in; for it is science that can best ascertain what the circumstances are. Contingent truths are precisely the subject matter of science. What the human is, in comparison with others, what aspects of nature it depends upon – these are proper topics for science. Pro-life enthusiasts seem to recognize this when they appeal to science to assure us that the fetus is a living human being. But they show an inability to understand what science is when they try to get scientists to answer the legal or moral questions that are really at issue. No one defending a legal right to end a pregnancy denies that the fetus is alive and in some sense human. It is not a canine or feline fetus; it comes from a human mother, and *if properly cared for* it may eventually become an adult human being. This is what we all know, and little science is needed to know it. What is at stake, however, is the *value*, the *importance* of being human in the minimal, largely potential senses specified. This value question transcends natural science, as I think most scientists would admit. To short-circuit consideration of the value question by equating 'human' with 'human in the full value sense' is not a scientific procedure but a political maneuver or semantic trick that can only deceive those not trained to analyze arguments. By some criteria the fetus is human, by some it is only potentially so, and only these latter criteria set the human species above other animal kinds. In this statement I would appeal with some confidence to the great majority of scientists.

Religious fanaticism, whose history is one of the dark aspects of the human story, is older than Christianity, and the newspapers daily remind us that it is still a present danger. Who could count the human adults, and children beyond infancy, who have been slaughtered in the name of some idea of deity? Our country is armed with means to slaughter large fractions of the human species, conceivably even to end human and much non-human life on this planet. Our police and law courts fail to prevent properly so-called murders from happening at a rate far from equalled in other countries – incomparably higher than in Japan, for instance. Yet abortion is common in Japan. Are we in a position to lecture that country about abortion? Do we understand the population problems Japan, Latin America, and much of the rest of the world face? How conceited can we get? Is that what our Christianity does for us?

In sober truth, how can one love a fetus, by all evidence with less actual intelligence than a cat, *as one loves oneself?* I say, it cannot be done. Self-righteousness is not the same as love, whether for God or the creatures. The attitude, 'I know, as you do not, that all killing of human beings is murder in the same utterly wicked sense', does not impress me as proof of respect for other citizens. If the one so viewing those who think the abortion issue to be complex and many-sided is a complete pacifist, advocating complete unilateral national disarmament and disarming of the police as well as abolition of capital punishment, I can see some semblance of consistency in the position. Otherwise I cannot see even that.

The word 'murder', properly used, does not include killing in self-defense, nor military killing of enemy soldiers, nor infanticide by a mother or doctor. These are all significantly different. No science can prove otherwise. Homicide has a number of forms and degrees. Each deserves treatment in its own terms. Nor is it just, or decently respectful of other persons, to lump together all reasons for abortion other than the probable or nearly certain death of the mother as frivolous or selfish. There is always at least some possibility that birth would kill the mother. There is always a more or less serious possibility that it will contribute to the destruction of family life and hasten the death of various persons. The unborn may be a potential genius, says a famous writer (I regret to say, a woman). True, and it may be a potential murderer or some other undesirable kind of person. Potential goods must be balanced against potential ills. What else is reason for?

It is important, too, that responsibility for undesirable pregnancy is not always or even usually solely that of the mother, yet it is only she who will bear the child. And what it is like to do that in various conditions and circumstances, men are in danger of flattering themselves if they think they know. (Not that there is anything very unusual in men so flattering themselves.) Nor does being a woman, even one who has borne children, guarantee that one has a realistic grasp of what pregnancy can be like for some women in some situations. If there is such a thing as original sin, surely it can show itself in our manner of attributing sin to others.

I hold with Nicolas Berdyaev, one of the most creative and eloquent philosophical theologians in this century, that what individuals do with their sexuality is an extremely personal matter, which the crude forces of law and public morality should be very cautious about entering into. I have myself lived in accordance with rather old-fashioned ideals in this sphere and consider myself lucky to have been able to do so. But I realize that many have been less lucky and I hesitate to condemn them.

That Anita Bryant has come to realize that she went too far in attacking the homosexuals is to her credit. But the harm she did them is not easily remedied. I have to regard such over-zealous judges of other people's behavior as public nuisances. I cannot do otherwise. In my view homosexuals are not to be envied, but at least they do not contribute to the population explosion. Nor is there good evidence that (on the average) they do as much to mislead or mistreat the young as heterosexuals do.

It is clearly unreasonable to expect that the discovery of powerful technological means of preventing pregnancy could leave the ethical thinking of our population unaffected. To some extent, at least, we must rethink our morality. I do not like promiscuity and have never practiced it. It is surely questionable as the ideal. But not everyone can live in a manner as close to the ideal as the best or most fortunate of us may do. If religions cannot honestly face and intelligently rethink partly new problems, the future is dark indeed. One must hope that they can find the courage and wisdom to relate ideals that are neither old nor new to the circumstances that are always partly new, and, at present, decidedly so.

## THE DIFFERENCE BETWEEN A FETUS AND A CHILD

To illustrate my conviction that from fetus or mere infant to 'child' already beginning to display human rationality is a very great transformation, I will give a few facts about our daughter, Emily.

Age one year. E. says 'doggie' at the sight of a dog visible from our window. This we think was her first word.

Age two and a half years. To my explorative suggestion that she carry a letter to the mail box (which was on the other side of a busy highway), E. replies, "Cross the street? Too many cars." The question mark was clearly supplied by a rising inflection in the first sentence, and the closing period by a falling inflection. In six words, verb, definite article, singular noun, adverb, adjective, and plural noun, all grammatically employed, this small child had lucidly defined a question and given an entirely reasonable ground for a negative answer. No adult could have done this more understandably or more pertinently to the situation.

Age as above. E. replies to my query about a bird song we were hearing (I had identified the song for her a day or more before), "That's a meadowlark" (which it was). Here are four parts of speech, a pronoun, a verb, an indefinite article, and a singular noun, again with perfect grammar and relevance to the situation.

Age about four years. E. makes a short but rapid speech the exact wording of which I do not recall but again it was in good clear English and concerned God, how busy God must be doing this, that, and the other necessary things for the world. It was a lively display of thought, which, if it came from a somewhat unsophisticated adult, would call for no apology.

Age five or six years. E. remarks about God that she wonders "if there is such a being." (Partly then, and partly a little later, I said something to her about my attitude toward this question, using the quotation, "The pure in heart will see God.")

It is not at all certain that even the trained sign-using chimpanzees, gorillas, and porpoises that we are learning to admire have reached the stage of thought our daughter attained before the age of three. No infant, much less any first or second trimester fetus, is anywhere near that stage. No physiology of a human life in the womb can account for Emily's reaching the stage she so early did. It required immense expenditures of love and effort and all sorts of experiences with a considerable number of adults and other children for it to be possible. An embryo is a required means, *but far from the only required means,* for the making of a person. No mere embryology or physiology can explain personality-production; for that we must look to anthropology, psychology, pedagogy, history, philosophy of religion – all the humanistic studies. The reader will now perhaps see why I am negatively impressed by the idea that the proper conclusion concerning the 'rights' of the 'innocent' in the womb is to be arrived at by looking with horror at moving pictures of a living, dying, or newly dead embryo. There are probably many people rather than few who would have trouble eating meat if they had spent an hour or two in some giant slaughter-house. The cattle also are 'innocent', and they may well suffer on their way to our tables, as much as, or more than, aborted fetuses. Cattle, too, are intricate, wonderfully organized creatures.

It is entirely relevant to take into consideration the possibility of bearing a child that one feels unable or unwilling to bring up oneself for adoption by foster parents. But at least two qualifications are in order. One is that we need careful inquiry into the likelihood that the job thus put off on another or several others will be done with at least minimal care and ability. (Statistics on successes of this kind would be interesting.) I repeat: The embryo is means to, not the actuality of, mature human personhood. The second qualification is that for a pregnant woman to take this course is a major not a minor matter for her. And she is not just another human being, alongside the fetus. She is, in an incomparably fuller sense, a person, even though perhaps somewhat immature in a different comparison.

I am open to conviction that perhaps in a substantial proportion of cases the heroic course just referred to is the most moral one a woman could take. But I would be less easily won over to the idea of some men, who *could not* do what she will be doing, trying to make the decision for her.

A wise colleague of mine tells me that he refuses to debate emotional issues such as abortion unless the rule is adopted that the only vote taken at the end is to determine, not which side do you favor, but which side, if either, you have been *caused by the debate* to favor, although beforehand you were undecided or decided in favor of the other side. The show of hands is to find out how many changed their minds because of what they heard during the debate, and in which direction. I will certainly not again engage in debate on abortion unless some such rule as my colleague's is given. There are too many skillful politicians around for me to care to try to rival them in the remaining portion of my active life.

## DEBATING ABOUT ABORTION

I once, unwisely, engaged in the political type of debate with a leading partisan of the anti-abortion organization. The object in such a debate is to secure a majority vote. The best way to do this may be to cleverly beg the question, exploit ambiguities, or take advantage of tactical mistakes by the opponent. Thus, on my side of the debate we objected to calling abortion 'murder'. The chief opposing speaker cleverly avoided the word 'murder' and made a point of this in rebuttal, thus distracting attention from the fact that, by failing to distinguish between 'human life', or 'individual', and 'human person', the charge of 'murder' had, by implication, been made.

The pro-life people are entirely right and admirable in taking seriously the questions concerning abortion, and particularly right in having strong feelings about the 'rights of persons'. Alas, the entire argument is precisely over that something which makes an animal a person, *when* it does so, and why this something is *important.* And here I judge my opponents curiously materialistic in their thinking. They have seen pictures showing embryos as intricate, organized little animal forms; they know that these creatures already have their own hearts and nervous systems (after a very early stage) and that many or most of them would, in the normal course of events, ultimately function as human persons. But they do not realize nearly as distinctly and vividly – and seem to close their minds resolutely to any such realization – what non-obvious and not merely physical factors are required for this transformation to take place.

Human children, youths, adults – using all these words to mean something more than infants or fetuses – will have to contribute not just a little, but enormous help if the fetus is ever to become a person in the sense which makes our species, in principle, superior to the others. Our prolonged infancy and childhood is not there for nothing, or as a meaningless eccentricity of our species. It is there so that *cultural inheritance,* learning from elders and peers, rather than mere maturing of innate tendencies, may make us people rather than subpersonal animals – this and our own psychological decisions and efforts. Personality is partly made, partly self-made; it is not inborn. Individuality is inborn, but the other animals have that. It is even said sometimes that a dog or horse has 'personality'. What are referred to are singular traits, eccentricities, special attitudes not closely duplicated in other individuals. But if this is full 'personality', then only vegetarians avoid conniving with murder when they eat. I shall perhaps be more impressed by my opponents on this issue when I learn more about their sensitivity to the vegetarian issue. To me they seem knee-jerk responders to a vague idea of 'human life', with no real grasp of what, concretely, human life as a whole is and what makes it valuable.

## PRO-CHOICE: REPLIES TO SOME OBJECTIONS

Some responses to an essay of mine on abortion [1][1] are typically various and show how unsuitable a subject this is for the constitutional amendment advocated by some legislators. Some of the letters do make interesting points, interesting whether or not convincing. Here are some of these points.

(1) St. Paul's 'we' who are 'God's children now' are taken to include fetuses. Also, God is said to address a fetus as 'Thou'. With such liberties in interpreting sacred writings much can be 'proved'. But with our American separation of church and state, the justification for a constitutional prohibition on abortion does not follow.

(2) The arguments for a limited legalizing of abortion are held in prin-

ciple to imply the rightness of killing children. Another writer says that the arguments in principle imply approval of killing 'all those who are not fully human'. The arguments as I see them do justify disapproval of laws absolutely prohibiting, or perhaps even severely restricting, killing extremely immature, *radically far* from fully human fetuses, and I do, by way of honesty, admit that there is no absolute difference between fetuses (in later stages) and some infants, other than the formers' part-of-the-mother's-body status, a not trivial point. But from 'there is no absolute difference' to 'there is no important difference' is an invalid deduction. As A. N. Whitehead well says, the change from a new-born to a two-year-old human offspring is one of the greatest changes in the animal world. I have been told by a Rabbi that a traditional Jewish view is that an infant becomes a person only in the second month after birth.

The Supreme Court decision makes a distinction between the first trimester of pregnancy and the second, and between the second and the third, and does not allow infanticide. I regard the Court's abortion decision as a reasonable compromise. But I also hold that no infanticide (by parents) deserves the horror we all feel about the killing of children (in the distinctive sense which contrasts with infants). However, whereas the fetus is in a very real sense part of the woman's body, the infant is not. To suppose that this makes no important difference is to be blind to the natural rights of women. Abortion done early and well is for the mother much safer than childbirth. There are other relevant differences.

(3) The fetus is said to have a 'soul', and therefore . . . . Contemporary psychology and philosophy are cautious about 'soul', and one religion at least (Buddhism) rejects the term in the meaning assigned to it in the Western tradition. Aristotle (and, I believe, Plato) held that every animal has a soul. It is not mere soul that shows our superiority to other animals, but *rational* soul, soul capable of thinking and thinking about thinking, as only animals with elaborate language abilities can do. A fetus needs much help if it is ever to become an actually rational individual, a person.

(4) It is held that not to decide about abortion is to decide. Yes, if we have no laws forbidding abortion, then indeed certain things will happen. I disapprove of some of these things. But making severe laws against abortion will not prevent some of the worst of these happenings, and in a worse form; for there will then be illegal, and to the mother, dangerous or cruel abortions. Many other not attractive things will happen. We must not confuse decisions made by legislators about what to compel people by police power to do or not to do and decisions about particular

cases made by mothers, doctors, fathers, using their own consciences.

(5) One objector wants evidence for the statement that a fetus does not think and reason in the degree that makes our species superior. Answer: there is good evidence that an infant cannot do this; so we need no additional evidence that a fetus cannot. If we give the infant the benefit of doubt on this head, then we cannot deny it to chimpanzees and porpoises, which give impressive evidence of doing something like human thinking.

(6) We are reminded that the democratic principle is majority rule. However, our Constitution was designed to set limits to the coercion of minorities by majorities. The majority are not, in our democracy, allowed to require any minority to accept a particular religion. The extreme pro-life group exhibits a sort of religion, in part very definitely so for a substantial portion, if not a majority, of its members. Not everything in this country, fortunately, can be settled by majority vote. In addition, a momentary legislative majority vote would not prove that the majority of citizens favored the decision. And above all, we should beware of riding roughshod over even a minority in so emotional and subjective an issue. Individuals, especially women, should decide it in their own cases (not legislators, judges, and police). They may be preached to, but not coerced. Their consciences should be respected. As for fetuses, they have no consciences. In this sense they are indeed 'innocent'. So are rabbits. I would not make this callous-seeming comparison if others had not made a, to me, callous comparison of adults to fetuses. Adults are incomparably more than fetuses.

(7) An objector compares the killing of human beings in the womb to the enslaving of blacks (who also were sometimes arbitrarily killed). This comparison is a monstrous insult to blacks. The superiority of the human species is shown by its use of language, by evidence of reasoning and consciousness of rights and duties. No such reasoning or consciousness can, by anyone in his or her senses, be attributed to fetuses; but we all know today, I take it, that they must be attributed to blacks, who have language and some of whom are obviously more intelligent than some whites. To compare blacks to fetuses is not to contribute to rational discussion.

The sad thing is that the letters elicited by my article are, on the whole, on a higher level than much of the agitation (as seen in letters to newspapers) to try to coerce women and doctors to accept the extreme doctrine that potential personality or intelligence is as good as actual personality or intelligence and its destruction no less serious. I fear somewhat for our country since so many are on such a low level of that reasoning power by which we are human in *value* as well as merely biologically.

(8) One writer mentioned the immortality of the human soul as reason for taking abortion with deadly seriousness. I wonder. If the fetus is immortal, then we could not really end its existence, and its posthumous destiny is in God's hands. Those who know Whitehead's idea of 'objective immortality' or 'everlastingness' can understand in what sense my faith implies an indestructible value in our personal existence. On this topic Whitehead surpasses all others. Reinhold Niebuhr told me that he would not say that this version of immortality was necessarily unchristian. Even children can be given some idea of their immortality in the Whiteheadian sense, but not fetuses or infants. Relatively mature human beings can be aware of their indestructibility in the sense in which process theology holds that we are indestructible. But the fetus cannot *think* its immortality but only at most feel it; as perhaps all the animals do. It makes more sense to imagine a whale, with its magnificent brain, thinking its immortality than a fetus or even an infant.

Our human superiority is not that we start life on a high level, but that, from a very low level (a single cell or cell colony), we reach a height no other species, so far as we know, can reach. Each human animal starts with nothing like human intelligence, but, if sufficiently cared for (not otherwise), it eventually reaches the level of mature human thinking and decision-making. Whatever value the fetus has, compared to other animals, consists in its potentiality for eventually living as a person, not in its mere actuality as a tiny, live, but primitive individual animal.

A wise colleague, the logician Norman Martin, has reminded me of the important truth that for moral or legal rules to have much importance in a society they must be supported by the traditions of that society. History shows that on the basis of pure reason, or what can be demonstrated by manipulating abstractions (such as 'the beginning of life'), human beings tend to disagree more or less hopelessly. Cooperation, consensus, is not arrived at by pure thought, as it may be in mathematics. (And even mathematicians have had some disagreement about how to deal with infinities.) It is common memories and a common cultural tradition that makes effective cooperation possible. In every society a distinction has been made between homicide in general, killing of a human individual, and utterly and horribly wrong forms of homicide, that is, murder. The ancient Jews were told "Thou shalt not kill," but this was not, so far as I know, taken by anyone at the time to mean that there were to be no soldiers whose duty might sometimes be to kill enemy individuals. So long as we have soldiers this distinction must be acknowledged. And killing in self-defense is also an agreed-upon exception.

The historically most widely agreed-upon meaning of 'murder' is the slaughter of children or adults, not of infants, and much less of fetuses. In our society, however, infants too are included. We have drawn the line at birth. We feel horror at infanticide, which many societies have not felt.

As I claim to have shown above, there is no absolute rational proof that infanticide ought to be legally prohibited or viewed with the same horror as murder in the universally accepted sense. But our society has somehow made the decision to extend the connotations of 'murder' or utterly wrong homicide to infants that have by natural forces been freed from the womb. To tamper with this decision is to weaken the support of tradition for our moral and legal structures.

It is evident on reflection that tampering with the traditional distinction *either way,* whether by returning to the tolerance of infanticide of many other societies, or by ceasing to tolerate abortion, is morally and legally dangerous. I say that this is shown by reflection, and the reasoning is not complicated. If abortion is *as bad* as the killing of children or adults, then the killing of children or adults is *no worse* than abortion. (This is mere logic, pure reason.) But since the majority of our people, according to the polls, do not view abortion with anything like the horror they feel for ordinary murder, what the absolute pro-lifers are doing is to dilute the force opposing the worst of all crimes. You cannot have it both ways: abortion is as bad as murder, murder is worse than abortion. To upset the traditional distinction between the killing of actual people and the killing of what, by all the evidence, are only potential people (if by people is meant creatures able to think in the human fashion which sets us above the other animals) is to endanger the consensus without which we are all at the mercy of the least scrupulous among us. By calling for a constitutional amendment the admission is in effect made that it is our basic tradition that is being challenged.

In removing the bars to complete citizenship for blacks and women we did, it is true, significantly alter our tradition. But the case of abortion (and the rights of fetuses) is not at all a parallel case. This is to be seen in several ways. First, Thomas Jefferson saw in the case of the blacks that their exclusion was quite wrong, and the wife of John Adams saw that the exclusion of women was wrong. Abraham Lincoln was clear about the utter wrongness of slavery, and Ralph Waldo Emerson, who was almost the conscience of the country for a century, was clear about the wrongness of both exclusions. But who thought of fetuses in this connection? Second, and above all, pure reason makes an incomparably better case for the full humanness of blacks and women than for that of fetuses.

Blacks and women have in high degree the gift of language which sets our species apart. Fetuses have nothing of the kind.

A constitutional amendment in behalf of fetuses is the least reasonable, and by far the most violent, proposed alteration in our tradition that has been offered since the Constitution was written. It threatens to destroy the minimal consensus without which we fall into dismal anarchy and confusion.

It is a sobering reminder of our human limitations that the President of the United States in 1982 has identified the issue of abortion as the question, Is the fetus alive or dead? Every woman who has been pregnant for some weeks knows the answer to that question. Of course, in all normal, healthy cases the fetus is alive; the woman can feel it kicking and the doctor can hear its heart beat. We are given a perfect example of the fallacy of failing to talk to the issue. To be alive is one thing, to be a person is another. Only when the extreme pro-life people stop pretending there is no difference can they be taken seriously as thinkers. At present they are a political threat, but they are not scientists, philosophers, or theologians in intellectually respectable senses.

*University of Texas,*
*Austin, Texas, U.S.A.*

## NOTE

[1] For an interesting discussion of the abortion problem by a liberal Catholic see [4]. Steinfels refers with fairness to my article. For other comments on it see [3].

## BIBLIOGRAPHY

[1] Hartshorne, C.: 1981, 'Concerning Abortion: An Attempt at a Rational View', *Christian Century* **98,** 2, 42–45.
[2] Lanier, S.: 1947, *Hymns of the Marshes, II. Individuality,* edited by S. Young, Charles Scribner's Sons, New York and London.
[3] Steinfels, P.: 1981, 'Challenging Hartshorne on Abortion', *Christian Century* **98** (April), 356–360.
[4] Steinfels, P.: 1981, 'The Search for an Alternative', *Commonweal* **108** (November 20), 660–664.

H. TRISTRAM ENGELHARDT, JR.

# HARTSHORNE, THEOLOGY, AND THE NAMELESS GOD

One of the major contributions of natural theologians to our or any culture is the reminder that God is not the God of any one culture or religion, that God is not a Christian, Jewish, or Hindu God, but a nameless God who belongs to all creation and to Whom all creation belongs. One might, in order to take a hint from outside of the West, recall one of the hymns of the *Rig Veda* in which the prayer is sung, "O Agni, you are Indra" (*Rig Veda* II.1.3). The God who is *the* God exists behind the particular names of the thirty-three devas, or for that matter the names for God provided by any particular culture and its religions. Charles Hartshorne has served this function of the natural theologian well. His appeal to God offers the heuristic power of the natural theologian who attempts to step outside the particular understandings of God forwarded by particular cultures and religions, and to speak of God Itself.

In this way, the natural theologian undertakes an endeavor similar to that of the scientist. Western civilization produced the remarkable undertaking of men and women attempting to reason about the world outside of the constraints of their particular cultures and times. They have attempted to describe the world anonymously so that it would not matter whether it be an American or a Japanese, a Russian or a Chinese, white or black, female or male scientist doing work in physics, chemistry, or biology. In principle, it should not even matter if the scientist is human. It should be enough if the scientist has sense organs sensitive at least to our range of sense stimuli and capable of discursive reasoning – a point often made in science fiction stories. Of course, the goal of stepping outside the conditioning of history, culture, and circumstance is never realized, as historians and philosophers of science remind us. Attempts to stand outside one's culture and to know free of the marks of one's culture and time are doomed. However, such attempts are essential to the notion of science, and for that matter philosophy. Science requires at least as a regulative ideal this reach to an intersubjectivity that hopes to be anonymous and free of the idiosyncrasies of particular cultures and particular persons.



*E. E. Shelp (ed.), Theology and Bioethics,* 45–48.

So, too, the natural theologian attempts to aid us in appreciating what it would be like to speak of God, and of God's viewpoint, outside of the embrace of particular cultures and religions. It is an attempt that can help us to recognize the extent to which particular cultures and religions impute to the mind of God their idiosyncratic views of the good life and of proper conduct.

Hartshorne offers us as case studies issues such as artificial contraception, homosexuality, abortion, and infanticide, which have been condemned by many, if not by most, of the Western religious traditions at least at some time in their recent history. His arguments concerning the divine perspective suggest that there will be little basis in terms of reason alone for holding that artificial contraception or homosexuality is in itself evil, or that abortion and infanticide involve evils on a par with the murder of a person. These conclusions should not be unexpected. The natural theologian's task is, after all, to place persons in the context of their relationship to the Deity. As such, our existence as humans withdraws in significance against our meaning as persons. It is persons who share with the Deity the capacity to know, to understand, and to choose. The modern fables of E.T., Star Wars, and the Return of the Jedi have moral instructive power through enforcing this point. Our singular standing before the Deity is to be derived from our status as persons, a status to be shared with E.T. and Yoda, a status we do not share with infants and fetuses.

Taking the Deity's point of view, at least within a natural theological account, is likely not only to be heuristic, but shocking insofar as we succeed in setting aside our idiosyncratic imputations of moral sensibilities to the Deity. The God who is God is unlikely to share fully the moral sensibilities of our particular cultures and religions. If there is any way to know Its viewpoint, it will be through reasoned arguments, a capacity we share with persons generally.

I will not take issue with Charles Hartshorne's reflections concerning artificial contraception, homosexuality, abortion, and infanticide, for in the main I believe he is correct in the ways he develops his arguments [1]. These topics are best interpreted in this context as examples of the distance in moral perspective that is likely to exist between views of the Deity's sensibilities drawn from particular cultural or religious perspectives, versus views drawn from rational arguments and reasoned analyses. As one comes to examine the rationality of belief and its bearing on God, morals, and medicine, one is likely to find that reasoned arguments concerning God may challenge traditional religious judgments regarding a number of activities in medicine. Much of what is traditionally condemned in the areas of contraception and abortion within our Western religious traditions may not find general rational warrant.

Natural theology thus also reminds us of what it is to resolve a controversy. If the condition for the possibility of resolution presumes generally justified grounds for drawing a conclusion, appeals to special revelation will not suffice. Beyond simple agreement, rational argument is the only means to settle conflicts when common grace is not available to resolve moral disputes. Thus, examining the rationality of belief, an element of rational theology's endeavors, is likely to bring into question many of the dogmas of revealed religions, including moral prescriptions regarding abortion and infanticide as one asks how religious appeals could in principle resolve, or contribute to, the resolution of moral controversies.

As Professor Hartshorne suggests, the natural theologians' inquiries have as a result implications for political controversies as well. The constraints that one will be justified in imposing on a society such as ours, which is not simply Christian or Judeo-Christian, but which spans individuals from numerous religious traditions, as well as those embracing no religious belief at all, will need to be drawn from as general and as non-idiosyncratic a viewpoint as is possible [2]. The laws bearing upon proper moral conduct in medicine may not impose the moral viewpoints of particular religions, even if they constitute a predominant majority, unless they can be independently justified. The authority of public laws, if it is not simply to depend on force, must in this end be drawn from either general agreement, or from rational arguments which justify either the content of laws or the process for their establishment. As a consequence, natural theological perspective in indicating what can be rationally justified with regard to the Deity has implications for political considerations as well. The God who is to be invoked in fashioning public policy cannot be the partisan of a particular religion. In fact, Its dictates must so conform to the constraints of rational argument that Its very existence can be denied while the moral conclusions remain intact. This is likely to have some profound implications for bioethics in that a great deal of Western bioethics has, after all, been directly or indirectly inspired by the work of theologians working within particular religious traditions [3, 4, 5, 6, 7, 8, 9, 10]. Judeo-Christian moral assumptions, as a consequence, have often been incorporated almost unnoticed and uncritically into the characterization of the proper moral sense to be employed in making bioethical choices and into the characterization of what we mean by rational and prudent decision makers in the biomedical context.[1] However, if one stands outside of the assumptions of this particular culture, things might indeed look quite different. Hartshorne's analyses can aid us in diagnosing unrecognized cultural prejudices, thus bringing them under the light of critical analysis as we attempt to understand better what can be rationally justified as moral constraints in health care.

*Center for Ethics, Medicine, and Public Issues,*
*Baylor College of Medicine,*
*Houston, Texas 77030, U.S.A.*

## NOTE

[1] I do not mean to suggest either that the Judeo-Christian heritage has not greatly enriched moral thought or that it does not contain morally defensible conclusions. However, a number of its traditional conclusions have properly been brought into question. As an example of the latter genre of conclusions, one might think of the traditional Christian restraints on contraception, sterilization, and the rights of women.

## BIBLIOGRAPHY

[1] Bondeson, W. C. *et al.*: 1983, *Abortion and the Status of the Fetus,* D. Reidel Publishing Company, Dordrecht, Holland.
[2] Engelhardt, H. T., Jr.: 1985, *The Foundations of Bioethics,* Oxford, New York.
[3] Fletcher, J.: 1960, *Morals and Medicine,* Beacon Press, Boston, Massachusetts.
[4] Fletcher, J.: 1966, *Situation Ethics,* Westminster Press, Philadelphia, Pennsylvania.
[5] McCormick, R. A.: 1973, *Ambiguity in Moral Choice,* Marquette University Press, Milwaukee, Wisconsin.
[6] McCormick, R. A. and P. Ramsey (eds.): 1978, *Doing Evil to Achieve Good,* Loyola University Press, Chicago, Illinois.
[7] Ramsey, P.: 1978, *Ethics at the Edges of Life,* Yale University Press, New Haven, Connecticut.
[8] Ramsey, P.: 1970, *Fabricated Man,* Yale University Press, New Haven, Connecticut.
[9] Ramsey, P.: 1970, *The Patient as Person,* Yale University Press, New Haven, Connecticut.
[10] Smith, H. L.: 1970, *Ethics and the New Medicine,* Abingdon Press, Nashville, Tennessee.

WILLIAM K. FRANKENA

# THE POTENTIAL OF THEOLOGY FOR ETHICS

This is to be an essay on 'the potential of theology for ethics' with special reference to bio-ethics; I am asked to provide a philosopher's 'conceptual exposition and analysis of the relationship, significance, and contribution of theology to bio-ethics' or to 'normative moral judgment in medicine and health care'. The general topic of the relation of religion or theology to ethics is not new, but perhaps it takes on a somewhat different cast when it is posed by way of the question of what theology can or might do for ethics in the context of contemporary discussion in medical ethics and related fields. I take it that the motivation for the choice of this topic is the fact that many people appeal to theology in these discussions, while others do not or even oppose making such an appeal. With this fact in mind I shall write some prolegomena that may be helpful to anyone reflecting favorably or unfavorably on the use of theology in bio-ethical debate or decision-making.

The question of what religion or theology can or might do for bio-ethics is essentially the same as that of what they can or might do for ethics, for bio-ethics and medical ethics are simply ethics as applied to a certain extended family of questions. My discussion will therefore be for the most about ethics in general, rather than about bio-ethics in particular, but because it is to appear in the context of thinking about that family of questions, I shall try to keep them in mind, especially for purposes of illustration.

By 'theology' I mean the doctrinal or intellectual part or aspect of religion, as distinct from its ceremonial, emotional, and institutional parts or aspects. Of course, these parts or aspects will be different for different religions. A theology, then, is the belief-content of a religion, and it may well be that the potential of a theology for ethics will vary from theology to theology. To simplify matters somewhat, I shall usually have in mind theologies of the Judeo-Christian sorts that figure in our own Western bio-ethical thinking, theologies that are theistic and depend heavily on an appeal to a 'special revelation' accepted by faith. A theology of this sort is a revealed rather than a natural theology; it is an intellectual elab-



*E. E. Shelp (ed.), Theology and Bioethics*, 49–64.

oration and defense of a revelation claimed to be made in some scripture or in some person's or community's life, work, or history. By 'ethics' I mean the sort of thinking that seeks to find answers to questions, either general or particular, either social or personal, about what it is right, wrong, good, bad, virtuous, or vicious, to be or to do. It will be clear at once that, for our purposes, a theology will have two parts: (1) an 'ethic' or ethics in the sense just defined, or what is sometimes called a moral theology or a religious value system, i.e., an imperative or Ought-part, and (2) an indicative or Is-part consisting of beliefs about the nature and creation of the universe and about the nature and destiny of man, e.g., that God exists, that man has an immortal soul, or that a fetus is a person.

## KINDS OF ISSUES IN BIO-ETHICS

Our contemporary bio-ethical debates in which theology purports to figure, like those in other areas, involve issues of several sorts, and it will be useful first to sort them out, and then to indicate which of them I shall be mainly concerned with. (a) Some are *factual,* such as medical science can give an answer to, e.g., when does a human fetus become viable? (b) Others are *conceptual* or *definitional,* e.g., is there a distinction between killing and letting die? (c) Still others are *normative* or *evaluative,* or in the broad sense ethical. It is, however, not always easy to place a certain question. Take 'When is a person dead?'. Here there is a conceptual question: 'How is death to be defined?' But there is also a factual one: given a certain definition of death, is this person alive or dead? And the conceptual issue has a normative aspect: when should we, or when is it not wrong, to regard a person as dead (or a fetus as alive)?

Normative questions, however, are not all of the same sort. (d) Some of them are *prudential* rather than moral; these are questions about what one should do from the point of view of one's own interests or health. (e) There are also questions about what one (a nurse or physician) should do in terms of *professional ethics.* (f) Similarly, there are questions about what is right or wrong in terms of the *law,* i.e., of what is legal or illegal. Some people seem to think that, if something is not ruled out by law or by professional ethics, then it is right. But there are further questions to be asked that are more properly ethical or moral. (g) There is the question of what it is *ethically* right, oughty, good, or virtuous to do in a certain situation or kind of situation, e.g., whether it is ever morally right to have an abortion. Here the answer may be either a particular moral judgment or a general moral principle. (h) Another kind of question is that of what should be incorporated in law, or of what the law ought to require, prohibit, or permit, e.g., whether the law should forbid or permit abortions in certain circumstances. There is also a question about what the law *is,* as when the Supreme Court is asked to decide a case involving abortion, but this, though related, is a somewhat different sort of question.

(i) I especially want to point out that there is another question, analogous to that in (h) but different. Each society has a more or less prevailing moral code or 'moral value system', not the same as its code of etiquette. Locke called this its 'law of opinion or reputation', more recent writers call it a 'positive social morality'. It is an informal quasi-legal code of rules, precepts, and ideals that is taught by the society and supported by non-legal sanctions like praise and blame, so that its members come to judge themselves and others in terms of it. One's conscience may be more, but it is at least an internalized social morality of this sort, secular or religious. Now, just as one may question the morality of a law or a legal system, one may also ask about the morality of the moral code of one's society; both sorts of questions have been raised in connection with racial and sexual discrimination and the treatment of animals, not to mention abortion again. Some new moralists and social reformers have even asked whether a society should have, or has a right to have, such a positive social morality at all, but at least there is the question what it should incorporate, what it should require, prohibit, or permit, what rights and duties it should recognize, what virtues it should foster. In fact, perhaps our contemporary discussions in bio-ethics and elsewhere should be viewed in part as attempts to reach a better consensus about what should go into our social morality.

(j) It should also be mentioned that there is another such question, namely, what should and should not be incorporated in the professional ethics of a certain group, e.g., what kind of a clause of confidentiality should it contain?

Ethical questions of kinds (g), (h), (i), and (j) all appear in our discussions in bio-ethics, often without being distinguished clearly or at all, but they are different and the differences ought to be kept in mind. It may be that they should be differently answered at least at some point. The law should not enforce all promises, even though breaking promises is morally wrong; perhaps the same thing is true of social morality. Maybe both law and social morality should permit abortions in certain cases even if it is wrong to have or perform an abortion in such cases. Some moral matters should perhaps be left to an individual and his or her conscience or God.

Obviously, there are many interesting things to be discussed in connection with this table of questions, but I must get on with my topic. One could, I think, ask about the potentiality of theology with respect to all these different sorts of questions, factual, conceptual, or normative, e.g:, with respect to the question of what death is, or to the question whether law or social morality should permit voluntary euthanasia. But there is not time or space for me to discuss them all and I shall concentrate on asking what theology can or might do with respect to answering questions of the kind indicated under (g). This is the central issue under our topic in any case, and, also, much of what I shall say about it will apply in the cases of questions under (h), (i), and (j) as well.

## ONE QUESTION AND ITS PARTS

What then is the potentiality of theology for ethics in the sense of answering questions, general or particular, about how one should conduct oneself? Earlier I said that a theology has two parts, an Is-part and an Ought-part. Let us call the former its world-view and the latter its ethic or ethics (theologians often use the singular but philosophers prefer the plural). Then our question seems to break up into two parts: (1) What can or might a theological *world-view* do for ethics, and (2) what can or might a theological *ethic* do for ethics? Let us first consider the second. This part of the question seems to answer itself: since a theological ethics is an ethics, theology can give us everything we need – the answer to our question is that theology can 'do it all' because theology includes ethics or at least an ethic. This is precisely what many religious people believe to be the case. There are, however, a number of things that need to be said at this point. (a) One is that even if theology does hand us an ethics in this way, all wrapped up and ready to go, it still cannot answer all ethical questions by itself. This is especially clear in the area of bio-medical ethics, where it is obvious that scientific medical knowledge is also needed, e.g., in order to tell whether a certain operation can and should be performed. Of course, it could be claimed that even this sort of knowledge is provided by a special revelation, but this is not plausible and is not what theologians would want to claim, excepting perhaps certain theological proponents of 'situation ethics' who seem sometimes to claim that God stands ready to give the believer the answer in each situation. Otherwise, the most a theological ethics could provide us with would be the basic norms or principles needed to answer our questions, given the factual and scientific knowledge that is also necessary. (b) Even then, of course, it can do the whole job only if it can provide us with all of the basic ethical principles we need, or, in the present context, all of those that are needed in dealing with the questions of bio-medical ethics. Some theologians would claim that it can do this, but others would not. Even if the basic norms have all been revealed, however, it seems clear to anyone who reviews the literature of moral theology that there are many problems about their interpretation and application. Theologians have held many rather different views about such matters and must clean up their own act before it can become clear just how their ethics yields answers to bio-ethical questions.

(c) Moreover, many theologians would admit or even insist that, for all they can do, their ethic leaves us with ambiguities and difficult decisions, especially in the area of bio-ethics. (d) In any case, however, a theological ethic can yield answers to such questions only for one who accepts it, unless it can be established in some objective way.

(e) There is another matter to be mentioned here. The normative part of a theology, whether it is couched in ethical terms (duty, virtue, etc.) or simply in terms of commands ('Thou shalts'), is a guide to living and hence is an ethic or value system in a broad sense, and for a believer it may be the supreme one. It is, however, not *ipso facto* or necessarily a moral one, a morality proper. It may even be immoral, as Plato thought the theological ethics of the traditional Greek religion was. If not, it may still be non-moral. Not all action or life guides are moral, e.g., law, prudentialism, etiquette, codes of honor, or an aesthetic 'life style'. Even if or when these overlap with morality, as law does, they are conceptually different orders of business from morality as such, and it may be that this is true of religion, too. Indeed, part at least of every religious ethic is ognized, and it might all be. The Sabbath day commandment, for instance, is not as such moral; morally one day is in itself the same as any
ance, is not as such moral; morally one day is in itself the same as any
other. In fact, it is not clear that the other commands of 'the first table of the law' are moral, since the reasons given for our obeying them in Exodus 20: 3–7 are purely prudential. Some Christians have actually thought of their ethics as a system of self-interest in this sense, telling us what to do or be in view of our own well-being in the long run encompassing other life as well as this one, which is hardly the moral way of thinking. Others have conceived of 'the moral law' laid down in the Bible simply as a set of divine commands that are normative or oughty because God commands them, not as something he commands because it is right or oughty. On such a view, Christian ethics need not be a system of self-

interest, but it is also not a morality; it is a superhuman legal system in which hating, killing, and causing unhappiness are wrong, not intrinsically, but because God forbids them. It may overlap with morality but is a rather different order of business. That religion is such a different style or way of life from morality is suggested by the fact that words like 'moral', 'morally', and 'morality' do not appear in the King James translation of the Bible, as well as by the fact that Christians and theologians sometimes contrast the religious with the ethical or moral way of life, e.g., Kierkegaard and Tillich. In the seventeenth century Sir Thomas Browne did so in these words,

> . . . if we are directed only by our particular Natures, and regulate our inclinations by no higher rule than that of our reason, we are but Moralists; Divinity will still call us Heathens. . . . I give no almes to satisfy the hunger of my Brother, but to fulfill . . . the Will . . . of my God; . . . I relieve no man upon the Rhetorick of his miseries, . . . for this is still but a morall charity . . . ([5], p. 68).

I shall not dwell on this point here but it is important to remember if one asks about the relation of religion and theology to *morality* and not just to 'ethics' in a broader sense. One cannot simply assume that religion and theology are trying to answer moral questions. There is some point in asking the question in James M. Gustafson's title, *Can Ethics be Christian?* [4].

However this may be, our main question, the one usually discussed under our topic, is not the second but the first of those posed earlier in this section: What can or might a theological world-view do for ethics taken in the wider sense?

## MORE OF THE SAME (OR: SOME CLAIMS)

Once more, then, what is the potentiality of a theological world-view (a TWV) for ethics? The answers to this question will in large part be the same whether the ethics involved is or is not a morality proper, for the issue now is about the relation of theological Ises to Oughts in general. Religious people usually think that their ethics rests on or presupposes their TWV, or that their TWV is the basis and justification of their ethics. It should be pointed out, however, that since Kant some religious thinkers have thought things are the other way around; taking morality as sound in itself but as requiring certain postulates, they have given 'moral arguments' for the existence of God and other metaphysical and theological beliefs. Indeed, many people today, seeing that religion does not have the hold on our minds that it once did, seem to urge as the reason for believing in it the claim that otherwise morality has no adequate basis – forgetting that it can hardly be a satisfactory basis for morality if this is the main reason for believing in it ([4], pp. 116, 179).

It should be observed that, although theologians usually conceive of their ethic as resting on their TWV, even a theologian might hold his Oughts to be autonomous and independent of his Ises. Such a separatist view is not so unorthodox as might be thought. Suppose one holds that the basic principles of ethics must be and have been specially revealed by God. Then God may still be thought of either as *revealing that* certain things are right or wrong, much as he might reveal to us some mathematical truth we have failed to discover, or as simply *commanding* us to do or not do those things, thus revealing his *will* in the way you do if you order me to close the door, but not revealing any *truth*. In either case, the basic principles of ethics need not be regarded as following from any Ises, not even from the purported fact that there is a God who tells or has told us to do or not to do those things. If he is only revealing to us what is really right or wrong independently of his will, it still may be that the principles are logically autonomous, like the axioms of geometry. If he is simply uttering commands, then the principles are imperatives like 'Close the door!' and do not follow logically from any Ises, even though I must believe there is a God and that he commands those things if I am to take the principles seriously. Paradoxical as it may seem, this means that, even if ethics depends *epistemically* on a special revelation, it may still not rest *logically* on any TWV. Or, in other words, the contributions of a TWV to ethics and those of revelation need not be the same.

Actually, if we look over the writings of those who hold that ethics depends on theology or religion, we find several different claims that are often not distinguished by those who make them. It is said, for instance, that "Only the religious belief in a more than human, absolute authority which commands morality can offer a sufficient basis for morality" ([9], p. 182). But this is ambiguous. There are several ways in which morality can be and has been claimed to be dependent on religion: (a) that it is *genetically* or *historically* dependent on religion, i.e., its emergence in the world was tied up with that of religion; (b) that morality is *psychologically* or *motivationally* dependent on religion, i.e. that, without a TWV, people are not adequately motivated to be moral; (c) that a TWV is necessary to show that it is *rational* to be moral, i.e., to be virtuous or do what is right; and (d) that a TWV is needed in order to show *that* something is moral, right, or virtuous in the first place [1], [3]. These claims are all different and ethics may be dependent on religion in any one of these ways without being dependent on it in the others. (b) and (c) are claims made in answer to the ambiguous question, 'Why should I be moral?'; (c) and (d) are claims made in answer to the question of *justification,* which is also ambiguous.

Now, if ethics and morality are dependent on a TWV in way X, then, of course, a TWV has an important potential for them. The four claims are claims that TWVs have corresponding potentials for ethics. But even if a TWV is not *necessary* for ethics in way X, it may still have an important potential in that way, for it may still have helpful resources to be tapped. Either way, it follows that TWVs may be held to have a potential for ethics in four different respects: (a) for *generating* an ethics, (b) for *motivating* us to live by an ethics, (c) for convincing us that it is *rational* for us to live by an ethics, and (d) for showing us *that* something *is* the ethical thing to do or be. However, in the context of bio-medical ethics we are not asking about the genesis of ethics, about its motivation, or even about the rationality of being ethical. We are asking, in certain sorts of situations, *what* it is ethical to be or do. In a way we are abstracting from problems about genesis, and about the motivation and rationality of being ethical, and concentrating on determining what is right or wrong or good or bad to do. I shall therefore center on the potential of TWVs for helping us to determine this, commenting only that, even if a TWV has no potential in this respect, it may still be important when it comes to the question, 'Why be moral?'

We have, then, finally, a two-fold question that will occupy us for the rest of this essay: (1) Is a TWV necessary as a basis for showing that something is right, oughty, virtuous, or the opposite, and (2) even if it is not, has it any potential for helping us to do this? Theologians have usually answered yes to both questions, though they can say yes to (2) even if they say no to (1). Moral philosophers of the last century or so, however, have generally assumed or insisted on a negative answer to both parts of this question – intuitionists, naturalists (descriptivists), and non-cognitivists (emotivists and prescriptivists) alike. They have usually held that meta-ethical analysis will tell us what ethical judgments are, mean, or do: respectively, that such judgments (basic ones) embody irreducible and autonomous ethical intuitions, that they are reducible to empirical or factual statements involving no theological concepts and capable of being verified by ordinary experience or science, or that they embody emotional or volitional *partis pris* determined by one's experience together with one's emotional or volitional nature or by choice. For them, then, normative ethics is a matter of achieving certain intuitions and implementing them by the use of experience and science, or of empirical investigation resting on certain definitions or conceptual findings, or of forming certain attitudes or volitional posits and implementing them.

Theologians reject such lines of thought insofar as they are associated with an extrusion of theology. And, indeed, it is open to them to argue that, at the very least, a TWV may or should play a part in the application of intuitions and definitions, or in the formation and execution of attitudes and choices, and thus to deny that ethics is entirely autonomous in relation to TWVs (and hence wholly secular).

## SOME ANSWERS REVIEWED

Let us now review the main views about what a TWV can or might do for ethics. One would run as follows:

(1) A TWV can do a great deal for ethics if it is true and can be shown to be true. For then, (a) it can establish the basic principles of ethics in a rational and objective way, (b) it can also provide additional premises for deriving further ethical conclusions, e.g., (n) in the following argument where (m) is assumed to be basic and proved:

(m) It is wrong to kill a being with an immortal soul.

(n) Human beings have immortal souls.

(o) Therefore it is wrong to kill a human being.[1]

(2) But all TWVs are false and can be shown to be false.

(3) Therefore TWVs have no potential for ethics in the sense of showing that something is right or wrong, good or bad.

This is one kind of atheistic view. It has a softer and a harder form. The softer form limits itself to what has been said. The harder one adds that an ethics presupposes a TWV and so concludes that ethics is impossible. God is dead, it says in effect, therefore anything goes.

The corresponding theistic view goes like this:

(1) As above.

(2) And a certain TWV is true and can be shown to be true.

(3) Therefore this TWV has a great potential for ethics.

Notice that these two views agree about what they say under (1) and differ only about (2) and therefore in the conclusions they draw. Here I am going to assume that they are both false, because I doubt both that any TWV can be proved and that every TWV can be disproved in any very conclusive way, i.e., I have doubts about (2) in each view. There are also questions about (1) but we shall come to those. One can make my assumption and still be either a theist or an atheist, but here I shall also not assume either of these positions. Likewise, one can make my assumption and be a Christian; in fact, Christians regard some or all of their TWVs as resting, not on reason, but on special revelation and faith. If my assumption is correct, the potentiality of TWVs for ethics is limited in an important sense, for then they cannot *demonstrate* any ethical first principle like (m) or any factual second premises like (n). One can, of course, insist that for one who has faith in a certain TWV, it will still have a considerable potential for ethics, and I shall return to this point in a moment. One can also still hold that, even though a certain TWV cannot be proved or disproved, it may yet be rational to believe in it and to use it in ethics.

This brings us to a third kind of view, which denies (1) and may be held both by atheists and by theists, Christian or non-Christian. It holds, roughly, that ethics is autonomous in relation to TWVs at least in the business of determining *that* something is right, etc., either in general or in particular cases. If this is so, then even if a TWV is true and can be shown to be true, it is still in a certain sense irrelevant to or impotent for ethics. Even for one who believes in it, it will actually be impotent in this respect, however much he may think otherwise. It should be observed that what is claimed to be autonomous here is the *discipline* of ethics. Whether or not the moral *agent* is also autonomous, as Kant and many others have held, is another question. Even so, however, the issues here are complex.

We should first notice that some ethical questions contain a theological reference in the very way they are posed, e.g., 'Ought we to love those who sin against God?' or 'Should we propagate the Gospel in foreign parts?' It does seem that answers to such questions involve theological considerations, at least if they are in the affirmative. The interesting question, however, is about the potential of TWVs for answering ethical questions that do not include any such ostensible references to theology, e.g., 'Should we love our fellow-person?' or 'Is euthanasia wrong?' Most bio-ethical questions are of this sort. One cannot tell whether or not TWVs are relevant to answering them simply by looking at them. Our question is: Are answers to such questions autonomous in relation to TWVs in such a way as to render TWVs impotent as resources for reaching them? One of the more interesting discussions of this matter is that of Basil Mitchell in *Morality: Religious and Secular* [7], and I have been stimulated by it, but it is not so clear as it might be, and I shall try to cover the same ground in my own way, though much more briefly. He contends that every ethics or morality reflects and presupposes a view about man and the world, religious or secular, and tries to persuade us that ours is or should be tied to a religious one. But this is vague and ambiguous.

One thing that he and others argue for is that meta-ethical theories of ethical judgments, including those most insistent on the autonomy of such judgments, namely, intuitionism, emotivism, and prescriptivism, all reflect a particular vision of human beings and their world ([7], p. 68). This is certainly true. However, it does not imply that substantive ethical judgments are not autonomous; it only shows that meta-ethical theories are not. Intuitionists, emotivists, and prescriptivists may still be right in regarding normative ethics as autonomous; in fact, they are right if the vision of the world involved in their meta-ethics is true.

Even if we distinguish between meta-ethics and normative ethics in this way, however – and whether we may or should do so is also an issue in the debate, but one to which I shall not address myself – the matter is not simple. In one sense, to say that normative ethics is autonomous is to say that ethical judgments neither *logically* follow from nor *logically* presuppose Ises of the sort that belong to a view of man and the world, theological or secular. This is a meta-ethical statement *about* normative ethics. However, we must distinguish between *basic* and *derivative* ethical judgments. A derivative ethical judgment may be established by an argument using a more basic ethical one plus an Is-premise; for example, in the illustration used above (o) is a derivative ethical judgment based on a more basic ethical one (m) and an Is (n). Here (o) does presuppose an Is as well as an Ought, and so is not wholly autonomous with respect to Ises. The important question, however, is about basic ethical judgments, for, being basic, they cannot be reached by any such mixed ethical argument. It may still be that they are wholly autonomous, and indeed, a careful proponent of autonomy will assert only that *basic* ethical judgments are logically wholly autonomous.

Even if the most basic principles or judgments of ethics are logically autonomous, and neither follow from nor presuppose Ises, there is here an important point about which autonomists and their opponents can agree, namely, that, except for one or more basic principles or judgments, all the *rest* of ethics depends at least partly on Ises taken from experience, science, metaphysics, or theology. Thus, for example, given the utilitarian view that the one and only basic principle of ethics is that of promoting the greatest general happiness, then the rest of ethics rests on this principle plus premises about what promotes the greatest general happiness, premises that may be taken from experience or science – or from a TWV, if one believes in one. This general point is well made by Mitchell, though he somewhat loses sight of the fact that, so far as it goes, basic ethical premises may still be wholly autonomous ([7], p. 151).

One potential of a TWV for ethics, then, is at least this: it may be able to supply Is-premises for use in deriving further ethical conclusions from basic ethical principles, e.g., for deriving the answer to a bio-ethical question. It must be added, however, that it can only supply such Is-premises for one who takes them as proved or accepts them on faith.

There is another important point, which I think is also in Mitchell's mind, but needs more careful scrutiny. Consider any typical allegedly basic ethical proposition, e.g., the principle of utility, the law of love, or the requirement not to kill an innocent human being, actual or potential, intentionally. When it is fully stated, it will contain (a) ethical terms like 'right', 'wrong', 'virtuous', etc., and (b) other terms or phrases standing for actions, dispositions, situations, etc., e.g., 'the greatest general happiness', 'love of neighbor', 'killing a human being intentionally', etc. The latter terms and phrases may also occur in factual statements or Ises. Hence even basic ethical judgments share some concepts with Ises. Without such concepts ethics would be both blind and empty. Which of these concepts appear in an ethics will depend on what we may call the Is-view of those subscribing to it, and in this sense even the basic judgments of an ethics (and *a fortiori* the derivative ones) will necessarily reflect some Is-view, which may or may not be a TWV. However, this only means that some of the same non-ethical *concepts* appear in one's ethics and in one's view of man and the world. It does not mean that basic ethical judgments follow from or presuppose Is-*propositions.* It certainly does not mean that an ethics presupposes a TWV.

Theoretically, even though 'People ought to respect one another' contains the concepts of people and respect, one could argue that it does not even presuppose that there are people or that they can or do or do not respect each other. An ethics as such might be only an abstract statement of principles independent of any views about the actual world. But, of course, an ethics will normally be thought to *apply* in this world, to us humans and our actions, dispositions, and relations, and so it will in a sense be tied to a certain vision of life and the world – in the sense of *pragmatically* presupposing such a vision, e.g., that there are people and that they do not all always respect one another. 'People ought to respect each other' does in *this* way not only contain concepts like people and respect, it also presupposes Ises about the world. However, it may still be *logically* autonomous in the sense of not logically following from or presupposing any such Ises.

There is a related point to be brought into the discussion. Kant and some others who believe in the logical autonomy of ethics have maintained that, nevertheless, morality has and requires certain 'postulates' consisting of Ises about man and his world. For Kant these center around the dictum that 'Ought implies Can', and include beliefs in free will, God, and immortality, the last two being theological. For others they consist of such beliefs as that man is free in some important sense that may or may not be compatible with determinism, that what he does or makes of himself makes a real difference in the world, and that what ought to be is not already real. I think myself that ethics and morality have some postulates of this sort, and hence that there is a sense in which they are not wholly independent of Ises. Again, however, this does not mean that they are logically non-autonomous in the sense that their basic principles logically follow from or presuppose Ises, for it may still be that the relation of these to Ises is pragmatic rather than logical; and it certainly does not imply that they have theological presuppositions even in a pragmatic sense, though it does leave this possibility open, so that a TWV may have a potential for ethics in this respect.

These points should suffice to show that intuitionists and others who regard the basic judgments of ethics as logically autonomous do not or need not think of them as entirely cut off from any important connection with beliefs about what is, as their opponents often allege that they do or must. Indeed, in one respect they are precluded from thinking this. Anything, X, is right, wrong, good, bad, virtuous, or vicious, not just by accident, but because it has certain non-ethical characteristics that make it so, e.g., the fact that it keeps a promise, respects a patient's privacy, or is an intentional killing of an innocent human being. That is, there are reasons why X is right, etc., if it is, and these reasons consist of facts or purported facts about it. What these characteristics are will be different for different ethics, but all moral and ethical theories involve recognizing some list of features as right-making, wrong-making, and so on. This is a meta-ethical truth about ethical judgments that is acknowledged alike by intuitionists like Moore and Ross, prescriptivists like Hare, and even in a way by emotivists like Stevenson. However, what it comes to is only that, if X has a certain non-ethical property, P, then X is so far right, good, etc., or that X is right, etc., because it has P. But to say this is just to say that what has P is right or whatever, i.e., it is just to accept a certain ethical principle. Hence there is nothing in this truth that shows this principle itself to be non-autonomous.

This last point, however, does make it possible for a theologian to claim that certain theological facts (e.g., facts or purported facts about what God wills or is like) make things right, wrong, etc. St. John claims this when he says, "Beloved, if God so loved us, we ought also to love one another" (I John 4:11). In principle at least, he might be right. But, even if he is right, this does not mean that 'we ought to love one another' follows logically from or logically presupposes 'God loves us'. John may only be asserting the ethical principle that those whom God loves should love one another (or perhaps that we ought to imitate God); he need not be interpreted as asserting a logical truth. He certainly need not be claiming that the principle itself follows from or depends on some fact about God; it may still be autonomous.

## A CONCLUDING SUGGESTION

Actually, I myself doubt that there are any very interesting or important strictly *logical* connections between Ises and Oughts, though I shall not here say more about this much discussed issue [2], [3].[2] It still might be, however, that one who firmly and vividly believes that God is love and loves us, as John did, will also believe that he or she should love his or her neighbor. Kai Nielsen says, ". . . a religious person will automatically go from 'God wills it' to 'I should do it' . . .," even though the inference is not logically required ([8], p. 341). Similarly, a religious person may automatically go from 'God loves us' to 'We should love each other'. If John is not just asserting the ethical principle that those whom God loves should love one another, then he may be going automatically from the Is-premise, 'God loves us', to the different ethical principle 'We ought to love each other'.[3] But he may be doing more than just going 'automatically' from his Is-premise to this principle. He may not merely be caused or impelled, on coming to believe that God is love and loves us, to love others himself and to think that we should all love one another. He may be thinking that one cannot *rationally* believe, fully and vividly, that God is love and loves us and not love others oneself or conclude that we ought to love each other. If he were more philosophical, he might write that, even if his inference is not required by strict logic, there nevertheless is a sense in which it is a rational one, and hence a sense in which his Ought follows from his Is, a sense in which his TWV has ethical implications or potential ([4], p. 164; [7], p. 113).

Would such a view make sense? I believe it would, i.e., I believe that it may be rational to go from an Is to an Ought, or irrational not to, even if doing so is not strictly required logically (which is not to say that it is illogical). How can this be? In what sense can it be rational to make John's inference? I suggest that it may be rational in the sense that it would be made by any rational human, i.e., any one who is clear-headed and cognizant of what is relevant, who also knows that God is love and loves us. Mill thought, in a much-criticized passage ([6], Chapter 4), that, if a certain kind of psychological hedonism is true, then it is rational to think that ethical hedonism is true, even if this does not follow logically, and I believe he may have been right in spite of his critics (thought I do not think such a psychological hedonism is true). And I believe that inferences like St. John's, from a TWV to an ethical conclusion, may likewise be rational in the sense indicated, e.g., inferences from a Christian TWV to conclusions about abortion, euthanasia, and suicide.

If this is correct, then, even if basic ethical judgments and principles are logically autonomous, there is still a sense in which it is rational to infer them from or rest them on Ises, perhaps even those in a TWV. That is, a TWV may then have ethical implications of a sort and hence a potential, not just for furnishing premises for establishing derivative ethical propositions, but even for providing basic ethical ones. Two things remain to be said. One is that, even if a TWV can have this potential for ethics, it can have it only for one who fully and vividly believes in that TWV. The other is that ideas and insights in ethics may come from anywhere, that we need all of the ideas and insights we can get, and that history shows they have often come from religious sources. Presumably theology will always have at least this potential for ethics.

*University of Michigan,*
*Ann Arbor, Michigan, U.S.A.*

NOTES

[1] I use this argument here for illustrative purposes only.

[2] Referring to [3] Gustafson says I hold "that there is no intrinsic relation between one's metaphysical, theological, or other justifications for morality itself, and the morality that is thereby justified" ([4], p. 86). These are his words, not mine; I said only that the relation is not strictly *logical,* and he does not himself claim that it is. What he claims ([4], p. 164) is essentially what I am now allowing and was allowing even in ([3], p. 315).

[3] It may be that he was reasoning as follows:

(a) God loves us.
(b) Those whom God loves ought to love each other (or we ought to imitate God).
(c) Therefore we ought to love each other.

Then he was using an implicitly assumed ethical principle to get from (a) to (c). I am supposing, however, that he was not doing this. If he was, then he was not going directly from an Is to an Ought, and it is the possibility of doing this that is under discussion.

## BIBLIOGRAPHY

[1] Frankena, W. K.: 1961, 'Public Education and the Good Life', *Harvard Educational Review* **30,** 413–426.
[2] Frankena, W. K.: 1969, 'Ought and Is Once More', *Man and World* **2,** 515–533. Reprinted in Goodpaster, K. E. (ed.), *Perspectives on Morality: Essays of W. K. Frankena,* University of Notre Dame Press, 1976, pp. 133–147.
[3] Frankena, W. K.: 1973, 'Is Morality Logically Dependent on Religion?', in G. Outka and J. P. Reeder (eds.), *Religion and Morality,* Anchor Books, Garden City, New York, pp. 295–317.
[4] Gustafson, J. M.: 1975, *Can Ethics Be Christian?,* University of Chicago Press, Chicago.
[5] Keynes, Sr. G.: 1968, *Sir Thomas Browne: Selected Writings,* University of Chicago Press, Chicago.
[6] Mill, J. S.: 1863, *Utilitarianism,* in S. Gorovitz (ed.), *Mill: Utilitarianism,* Bobbs-Merrill, Indianapolis, 1971.
[7] Mitchell, B.: 1980, *Morality: Religious and Secular,* Clarendon Press, Oxford.
[8] Nielsen, K.: 1982, 'God and the Basis of Morality', *Journal of Religious Ethics* **10** (Fall), 335–350.
[9] Schweid, E.: 1980, 'The Authority Principle in Biblical Morality', *Journal of Religious Ethics* **38** (Fall), 180–203.

BASIL MITCHELL

# THE ROLE OF THEOLOGY IN BIOETHICS

To claim at all that there is a role for theology in bioethics is to face a dilemma. Either the principles which apply in the practice of medicine, or in biological research, are such as to approve themselves to all reasonable men, in which case there is no need to 'bring theology into it'; or religion provides one perspective among others from which the moral issues may be viewed, and it is not clear why in a 'plural society' this one should be preferred to the others. Theology is either otiose or intrusive.

An objection can also be put forward from a more practical standpoint. The important thing, it may be said, is to get as high a degree of consensus as possible. In practice, in our culture at least, we can largely agree on what should be done in particular cases, on what codes we should subscribe to, even on what general principles we should rely on. Let us rest content with that. Any attempt to dig more deeply and uncover the roots of our moral intuitions in a religious tradition will risk the discovery of profound and disturbing differences.

The differences, however, have already become apparent, and it is no longer possible, if ever it was, to replace the soil and pretend they do not exist. They relate to the role of medicine and the doctor-patient relationship, to the proper methods of research and even the proper aims of research as well as to the familiar and more specific issues of abortion, euthanasia, and homosexuality. In the case of all of these, not far beneath the surface, will be found disagreements about the scope, character, and content of morality which in turn reflect more or less profound differences about the nature of human beings. People differ also about the significance of this state of affairs. To some it means the breakdown of a sound tradition, to be remedied by seeking to re-establish it; to others it represents the collapse of an obsolete tradition, with all the confusion that attends the emergence of a rational alternative; to yet others it is the inevitable – and desirable – expression of the plural society, to be accepted and accommodated with suitable flexibility.

If in this complex situation the Judaeo-Christian tradition is still treated seriously, it is partly due, no doubt, to the conservatism of the



*E. E. Shelp (ed.), Theology and Bioethics*, 65–78.

medical profession, but also, I suggest to the recognizably fundamental character of the issues themselves. Matters of health and sickness, life and death, sex and proceation are of such evident and serious importance in human life that almost everyone is disinclined to play around with them. If our powers of intervention in these matters are increasing to an awesome extent, we cannot afford to neglect our ethical resources wherever they come from. So, when it comes to the point, there are not many people who want to prevent theology making a contribution to bioethics, even though there are those who view it with a certain wary suspicion. Theologians, they fear, if invited to contribute, will insist on taking over, and their contributions will be determined by dogmatic presuppositions accepted on faith and not open to rational discussion. Not many Christian theologians nowadays actually conform to this stereotype, but it remains an influential one nevertheless – even if chiefly manifested in the complaints of laymen that the theologians are not being dogmatic enough. The ordinary man tends to want some sort of compromise, in which theological voices are allowed to be heard but not allowed simply to dictate the verdict. If they are not allowed to be heard at all, he fears an impoverishment of our ethical thinking; if they are allowed to dictate the verdict, he anticipates the abandonment of common sense. My argument in this chapter will be that in this the ordinary man is right and at the same time authentically Christian. It is among our most powerful intuitions that moral judgments are true or false; that, when true, their truth is discovered, not invented; that they constrain our choices and are not created by them. And it is because they are viewed in this light that we are so insistent that moral thinking has its own integrity and is not just a function of whatever world view, on other than moral grounds, we choose to embrace.

These intuitions amount to the claim that morality is, as J. L. Mackie used to put it, 'required by the universe';[1] and once Aristotle's immanent teleology has been abandoned, it is hard to see how this claim can be maintained except in terms of some form of theism. For if – and, arguably, only if – God has a purpose for us and we can flourish and attain ultimate satisfaction only by seeking to realize that purpose, can the demands of morality have the objective and categorical character we intuitively ascribe to them. It is in that case to be expected too, that men, as created in the divine image, will have some capacity, as men, to recognize what they are meant to be, and how they are to become it; in other words, their moral intuitions will tend to be trustworthy. Thus the initial dilemma is a false one. The conception of an objective morality open in principle to all men as such is itself part and parcel of theism, and the integrity which this status imparts to ethics is something that theologians must acknowledge. The case to be argued is that moral reasoning finds its proper autonomy only in a full theological context and that, while theology has implications for ethics, these must be assessed and understood from a distinctively ethical point of view.[2]

This overall account is subject, notoriously, to an important qualification. Moral insight is subject to error and corruption, and so, equally, is theology. A theistic metaphysic is able to explain both how ethics enjoys a certain objectivity and independence and how religion can deepen and extend our moral awareness. But, nevertheless, it also recognizes the dangers of error and distortion. Man is made in the image of God, but the image has been defaced and must be remade. Hence the appeal to moral intuition is a riskier business than in principle it ought to be, and the appeal from it to theology is not guaranteed to provide the needed corrective.

This admission, once honestly made – as it has to be – may seem to nullify what was said earlier. An objector may retort that, if theology provides access to moral insights not otherwise reliably available, then indeed there is room for a contribution from theology to bioethics; but not if theologians themselves are liable to error and not wholly to be trusted. Surely, in that case, it is best to rely after all on the good sense of the plain man together with the medical man's professional conscience, schooled as it is by experience. It remains a tempting suggestion. Those who have been under the constant pressure of urgent and exacting medical practice have as a rule been forced to identify the underlying ethical issues, appreciate their complexity, and feel for the right solutions; and the Judaeo-Christian ethical tradition has exercised an influence all the more effective for being largely implicit. But the temptation has, once again, to be resisted, because there are, increasingly, other influences at work which, if left unrecognized and unopposed, will drastically modify and perhaps destroy the traditional medical ethic. Hence we are driven back to a debate about fundamentals.

Foremost among the points at issue is whether there is any longer room for a 'medical ethic' at all, in the sense of an ethic of the medical profession as such. The traditional view has been that medicine is concerned with the maintenance of health and the prevention and cure of disease. The authority of the doctor derives from his known dedication to these aims and his mastery of the means of attaining them. It was always understood that the theoretical knowledge and the practical techniques which gave him mastery of the means could be used improperly in pursuit of illegitimate or inappropriate ends. These might be actually bad or just morally neutral. In either case they had nothing to do with health. This traditional doctrine of the doctor's role is now increasingly challenged, so that the relation between doctor and patient becomes assimilated to that between expert adviser and client. The adviser has the technical expertise to bring about certain changes, and it is for the client to decide what changes are desirable, and for the client to take any moral decisions which the situation demands, since these are no longer confined to questions of health and sickness, and it was this restriction which defined the practice of medicine. Associated with this development is a shift in emphasis from the duties of doctors to the rights of patients. Thus, in place of an ethical question for the surgeon whether the mother's physical or mental health requires that she should have an abortion, the question is one of the woman's right to have an abortion if, for whatever reason, she does not want to have a baby. Similarly, there are those who argue that the individual has a 'right to die' which relieves the doctor of the ethical decision whether he should strive to prolong life in terminal cases or allow the patient to die in dignity and peace.

These developments go along with an increased emphasis upon 'autonomy'. Traditionally, the claims of autonomy have been safeguarded by the requirement of informed consent on the part of the patient. Now the initiative passes from the doctor to the patient, whose 'autonomy' becomes the chief consideration; and it is autonomy conceived in a strongly individualist fashion. Thus not only is the decision on abortion to be taken by the potential mother rather than by the doctor, but without consultation with the potential father; or, in the case of young girls, without reference to their parents. (There may always have been reasons in particular cases for not consulting them: the point is that there tends no longer to be even a presumption in favour of consultation.) The assumption seems to be that marriage, and even parenthood, are essentially external relationships which can and should be subordinated entirely in medical matters to the preferences of individuals, who are regarded atomistically.

Since, however, it is obvious that the interests of others cannot be entirely overlooked in practice, and since in theory a morality of personal preference alone would render social organization impossible, resort is had also to some form of utilitarian calculus. Thus, in the case of teen-age pregnancies the overriding interest of society, so it is felt, is that there should be no unwanted babies. Hence contraceptive advice should be freely available from an early age and abortion must be able to provide a safety net. Since the young might be deterred from seeking advice if any moral teachings were associated with it, or if the parents were at all involved, the interest and responsibilities of parents are effectively excluded. Between the 'rights' of the individual and the purely utilitarian concerns of society at large, there is little room left for a professional medical ethic to operate in.

'Autonomy', as it is coming to be understood, tends to separate the individual not only from other people but also from his or her self. The notion that a woman has no obligation to the child she has conceived, which can be regarded as an alien intruder upon her private world, only makes sense if she is able to think of the natural consequences within her own body of acts she has freely performed as in no sense her responsibility, as not belonging to her. And the attitudes involved in some types of genetic research treat a human embryo as if its origin and natural future were alike indifferent. All that matters and is at all significant morally is what it is like *now*. But for any Christian view, and for any view that is consonant with the Judeo-Christian tradition, the embryo must be regarded as, at least, a potential human being: it cannot at any stage be treated simply as a piece of organic tissue comparable to, say, a fish. Respect for autonomy, therefore, in that tradition is tempered by recognition of certain natural processes which the individual undergoes as part of what it is to be a human being and which are not to be subordinated, as a matter of course, to his or her present preferences or to the preferences of others. Autonomy is limited in the early years of life and, here if anywhere, paternalism is in place. The responsibilities of parents and of other adults towards children cannot be replaced by 'children's rights'. There are things that children need to be protected from, and may not be presumed to consent to, precisely because they are children. (Sexual relations between the mature and the immature are a case in point.) Similarly, towards the end of life there must be some resistance to the demands of 'autonomy' in this highly individualistic sense when they claim sovereignty over the time of death. The medical profession is also under increasing pressure to do what the individual wants by way, for example, of cosmetic surgery, regardless of whether it has anything to do with promoting health and preventing disease or correcting palpable defects. The 'autonomous' individual is unwilling to acquiesce in the natural processes of aging and timely death and wishes to invoke the technical resources of medicine to prolong youth indefinitely and, when this can no longer be done, to expedite death. He demands to be the master of his body and the captain of his fate, and experiences the inevitable failure to achieve this as an intolerable affront to his dignity.

The changes in attitude to which I have been drawing attention exhibit a certain consistency of pattern. They derive in fact from certain philosophical conceptions of a very general kind which run counter to the traditional medical ethic and to the entire religious tradition from which it largely springs. Their tendency is to separate the individual at any one time from other stages of his or her natural life and from other individuals to whom he or she is bound by natural or customary ties, so that responsibilities which arise out of one's situation are treated as if they can be accepted or rejected at will. But, since the preferences of individuals cannot all be realized, and it would sometimes be disastrous if they could, the balance is determined by a straightforwardly utilitarian calculation. The examples already given illustrate the pattern clearly. The treatment of under-age pregnancies follows from the widespread tendency to regard sexual behaviour in general as a matter entirely for private decision. Individuals must be free, of course, to adhere to a personal code of restraint in these matters if they so desire, but it is, simply, a personal code, which does not have the sanction of society at large. Any attempt, therefore, to discourage the young teenager from engaging in sexual intercourse is felt to be somewhat hypocritical – as attempting to deny to the young a freedom which adults have no intention of renouncing themselves. It is also likely to be ineffective in the circumstances. But there is, undeniably, a risk of undesirable consequences, above all in the form of unwanted pregnancies. Hence it is necessary to take measures to prevent these consequences by making contraception and contraceptive advice freely available and by having abortion in reserve.

Similarly with genetic engineering. This promises enormously to increase people's range of choice by making it routinely possible to choose the sex of children and to avoid the birth of children liable to suffer from identifiable diseases or disabilities. It may prove possible, too, in the long run to exercise choice in favour of certain characteristics which are reckoned to be particularly desirable and against others. The moral theory under discussion would leave the individual free to make whatever choices he or she preferred, subject only to certain overriding considerations of public policy which would be of a purely utilitarian kind. Given that the eventual consequences are so incalculable, it is not possible to specify in advance what these might be, but the chief criterion would be the need to avoid tangible harm to assignable individuals.

So there are – as these examples have illustrated – ethical options in the field which diverge significantly from the traditional consensus and which require fundamental decisions to be made; for to leave the matter to individual choice, subject only to utilitarian constraints, is itself to make a fundamental decision. In the face of these challenges the traditional approach would be to refuse to 'play God' – and it is significant that this expression is still widely current. Leaving aside the familiar moral issues raised by abortion, it belongs to natural piety to accept the sex of a child as something given – the *gift* of a boy or girl – and not as something to be manipulated. There seems also intuitively to be a significant distinction between breeding out severe diseases or disabilities and breeding in qualities that are desired. The problem is how, if at all, these intuitive notions can be justified, even within a religious ethic. For it can be argued that we are called to co-operate with God in the forwarding of his purposes; and that it is part of his purpose that we should have the freedom so to co-operate. And surely God would wish not only that we should strive to eradicate disability and disease, but that we should seize any opportunity to make men better by making better men. If we wanted to resist this conclusion, it might be further argued, we should never have allowed contraception, for that too involves interference with a natural process.

If a defensible line is to be maintained here, it will have to be something like this. What is 'natural' in the relevant sense is not to be identified simply with a certain physiological process whose integrity must at all costs be maintained, as Roman Catholic moral theologians have sometimes seemed to hold, but with the humanly significant purpose which the physiological process normally promotes and expresses. The 'good' to which marriage tends is the procreation and education of children by parents who are wholly committed to one another and to the children. Contraception conduces to this in that it permits families of manageable size. So, incidentally, does AIH or *in vitro* fertilization, in the case of those who are prevented from conceiving by some physical handicap. (AID, by contrast, violates the requirement that the parents be genuinely and permanently committed to one another.) To eliminate severe congenital diseases is to remove a palpable evil and is on a par with other examples of preventive medicine. But to choose certain positive qualities and develop them is to assume powers for good and evil which usurp the prerogatives of the Creator with consequences that are unpredictable and might be irreversible. At this point the ordinary man's reluctance to 'play God' represents a profound intuition of the limitations and also the dignity of the human. The sacredness of human life forbids manipulating it as well as taking it. Lying behind this intuition is some conception of a created order which ought not to be disturbed except in accordance with the Creator's purposes in it, on pain of severe confusion and dislocation. It is presupposed that through revelation (however this is understood) and through careful reflection upon our experience (the one affecting the interpretation of the other), we can attain to some understanding of these purposes. There is, manifestly, no simple and straightforward way of applying this conception, which will relieve us entirely of the need to assess the long term consequences of our acts and omissions, and a critic may be inclined to say that, since we have to do this in any case, we might as well leave out all reference to theology. But without *some* controlling vision of our relationship to the natural world, we shall lack adequate criteria by which to make our assessment, and shall be led to exploit it excessively or to acquiesce unduly in curable evils. In matters of such urgency and importance which affect what in future we shall make of ourselves and the world in which we live, the choice between alternative philosophies of life is a forced choice. Doubtless in a democratic society there must be compromises, where full agreement is not possible, but the state itself cannot remain entirely neutral.

I have argued so far that the Judaeo-Christian tradition sets certain limits to what is permissible in medical and scientific research and practice of a kind that accords with the pervasive humanism of our culture, but which is threatened by certain trends in contemporary thought. It is because they are so threatened that it is not possible to rely entirely on a traditional consensus to see us through. Our intuitions need defending and developing in accordance with the religious tradition with which they cohere and from which they largely derive. But this approach is open to two criticisms, one of which I shall partly accept, the other of which – the more powerful – I shall feel bound to reject.

The first criticism is simply that, even within the Christian tradition itself, there is no straightforward consensus. We still have to debate and finally to choose. Nowhere is this more evident than in the Roman Catholic Church, in which the Natural Law doctrine in its traditional form is under pressure from forms of personalist philosophy which appear to reflect the secular culture. Protestant thinkers, by contrast, have often rejected Natural Law altogether in favour of a reliance upon Scripture which is itself under attack from biblical critics. As a result, there is no prospect of solving the problems of bioethics by deriving agreed prescriptions ready made from the various Christian ethical traditions as they now stand. Catholic thinkers, both now and in the past, have been too ready to take over themes from contemporary philosophy somewhat uncritically, and Protestant thinkers have not taken philosophy seriously enough. All this is true and, in any case, a new situation confronts us and fresh thinking is required. No doubt some differences will remain and some compromises will be necessary, but thinkers within the Judeo-Christian tradition share a conception of ethics as in some sense 'required by the universe' and have, therefore, a more than merely pragmatic incentive to combine in search of the truth of the matter.

The second criticism is more fundamental and concerns the status of this entire tradition. So far in this paper I have assumed that the other alternatives open to us – if we are not prepared to rest in unexamined intuitions – are either (or both) that morality is in the end an expression of personal preference or that it is a matter of utilitarian calculus. These two are, alone or together, morally unsatisfying and inadequate to the social function of morality; hence I have urged the claims of some version or other of a traditional theological ethic. In doing so, it may be said, I have overlooked another possibility, that of a form of pluralism, which recognizes that moral systems may differ, but holds that every moral system has a social character and develops over time. From this standpoint it is readily acknowledged that every moral system has a structure of injunctions and prohibitions, virtues and vices, designed to protect a way of life which is felt to be valuable. As a way of life it cannot be a purely private thing; it must spread out to embrace a community. But it is, nevertheless, a human invention, albeit a corporate human invention. Since men, *qua* men, have a good deal in common, any morality will exhibit certain common features, certain 'universal values', and these can be given a rational justification as being necessary to the well-being, indeed often to the survival, of any society. But there is an innate plasticity in human nature which permits the development of a wide variety of ways of life from time to time and from place to place. While some of these can be seen to be distinctly inferior to others by criteria which are clear and rationally defensible, this is very often not the case. More often than not the most one can say is that people at different times and places have made their own selection from among the ways of life that are in principle possible and in so doing have forfeited some satisfactions in order to achieve others. Of course, given the state of a particular culture at a particular time, the area of choice is much reduced. In our own democratic societies, for example, we have little opportunity to embrace the aristocratic virtues of personal honour and gracious condescension, and somewhat limited chances of treating women with old world courtesy. Thus a variety of factors, including habitual attitudes formed by past choices, widespread reaction against these when they are felt to be no longer 'authentic', fresh opportunities created by technological innovation, makes it difficult for even a homogeneous tradition to maintain itself indefinitely as the ethic of an entire society. Meanwhile, increased awareness of the variety of possible ways of life intensifies the process of disintegration.

The solution to which the foregoing analysis points is that of an acknowledged pluralism. If some agreed pattern is needed (for morality must have a social dimension) and no single pattern has a stronger rational claim than others, then let separate communities develop, each dedicated to its own pattern, none claiming authority over a national state as such. Religious groups, with their own ethical traditions, will be among these.

This sort of pluralism is subject to a number of problems in practice which cast doubt on the theory underlying it. Because it stresses the importance of a cultural tradition in terms of which individuals can understand themselves, it is opposed to mere permissiveness and rootless individualism. Yet by viewing the preferred tradition as one among a number of alternatives between which, *ex hypothesi,* no rational choice can be made, it weakens the authority which the tradition is able to exert over the individual. Either, then, the tradition is no longer able to perform its integrating function, or, since the need for this is so keenly felt, it is affirmed in a quite non-rational way as having authority simply because it *is* the historical tradition of those who stand within it. The modern world shows a number of striking examples of this very process.[3] If this sort of cultural positivism is to be avoided, there must be some possibility of uncovering the rationale of the tradition in such a way that it can be further developed in response to the challenge of continuous change. And if it cannot be so developed, it must be abandoned.

Nevertheless, it is manifestly true that human nature is capable of assuming diverse forms, and that it is very often not possible to make straightforward comparisons to the clear advantage of one or other of them. Moreover, this variety of cultural form is an enormous enrichment which we should be reluctant to lose. The question, however, is whether the ethical insights of the Judeo-Christian tradition are simply to be identified, as this analysis implies, with the culture of a particular time and place in such a way as to preclude any universal claims and any possibility of rational development.

The natural answer to this question is that throughout its history Christianity has to some extent taken on the character of an existing culture and at the same time powerfully affected it. A Christian knight has, in many respects, much more in common with a Hindu warrior than with a Christian industrialist; yet there are other respects in which the religious continuity is evident. Christianity did not, and does not, simply absorb the surrounding atmosphere; it also changes it. Because of this, there is a continuing problem about which features of the Christian tradition as it has developed in the West are to be regarded as mandatory because fundamental, and which are expressive of aspects of contemporary Western Culture which can and should be dispensed with. These questions are far from easy to answer, but it is important to notice that the problem is essentially the same whether one is trying to graft Christianity on to an alien culture or trying to maintain or develop the Christian strains in our own culture. In both cases Christianity itself is not straightforwardly to be identified with the given culture of a particular time and place, although it needs to be related to it. Indeed, more than that, it needs to be given effective expression within it.

So, although it can happen that in certain respects there is nothing to choose between alternative ways of life – each develops some human potentialities at the inevitable cost of others and the balance is equal; and although Christianity has to come to terms with the existing culture, even if it aims eventually to transform it, there are limits to what it will at any time permit. And the questions which arise in bioethics concern the placing of these limits. Thus the sanctity of human life is a value which those who have been formed by a Christian culture are not prepared to surrender – whether or not they are themselves professedly Christian. They cannot view it as capable of being traded off against some other pattern of human satisfactions. Yet it is not a 'universal value', if by this is meant one everywhere recognized by human beings or plainly required for the survival of any society. If people in our society are often reluctant to admit its Judeo-Christian ancestry, this is because it seems to them so deeply and so unchallengeably human. But there is truth in both claims. Christianity does not confer upon human life a sanctity it would not otherwise possess: it enables us to recognize a sanctity we might not otherwise acknowledge and could not otherwise defend, and it does this by putting man in his full theological context. It is not that men will inevitably, just because they are men and no matter what their basic convictions, come to think of one another in this way – as intrinsically valuable, however unattractive or unappreciated the particular individual may be – but rather that once they have come to share this vision and have enjoyed a life even imperfectly inspired by it, they cannot easily give it up. There is a felt rightness about it which impresses itself like a clear view in daylight. The truth seems obvious, and so in a way it is, but how it can be true, and why it is so obviously true, can be explained and defended in terms of some world-views and not of others.

Many of the bioethical problems cluster round this central concept; they are problems not because there is doubt that the concept applies, but because it is not clear what in particular instances it actually requires. There is, as we have seen, no complete consensus even among Christians about these questions. So further thinking has to be done. And those who have been compelled by the exigencies of practice to make decisions in these matters have a unique contribution to make, whatever their religious beliefs, because they have been forced to concentrate upon the moral features of the situation.

The sanctity of life may, perhaps, be given this privileged status; but is this true of the values that relate to the marriage and the family? Here, surely, it may be thought, patterns are variable and no single one is sacrosanct. The sexual revolution of our times, aided by technical advances in contraception and abortion has, it may be argued, come to stay, and has permanently altered our ideas of the relationship between the sexes and of sexual activity and procreation. The traditional monogamous marriage, in intention always and in effect generally permanent, in which the partners owe one another fidelity and accept responsibility for any children of the union, can reasonably be regarded now as a minority option which no longer merits the support of public policy. The 'autonomous' modern man or woman expects to engage freely in sexual intercourse before marriage without the risk of children and takes it for granted that pregnancy, should it occur unwanted, will be terminated. Marriage is regarded less and less as a status and more and more as a contract to be dissolved if it does not realize the expectations of the partners. This developing pattern has its drawbacks, no doubt, but it also has its advantages. It sacrifices the complete and unreserved trust which the older pattern made possible and with it the guarantee of security for the children, but it enables both parties to develop their careers and realize their potentialities as individuals as the traditional marriage did not. In short, it may be said, it is a clear case of the way in which the plasticity of human nature permits a choice of value-systems, none of which has a claim to be regarded as uniquely suited to the human condition. The Christian churches, therefore, are confronted with a choice. They can either come to terms with this plurality or they can maintain their traditional conception at the cost of forfeiting their controlling influence in the culture.

Perhaps. But these problems are not so easily separated from those which cluster round the sanctity of life. This latter concept is plainly involved in the question of the rights of the fetus – whatever in the end these are held to be. Whether the fetus is regarded as human or as potentially human, as already a human being or as what will, in the normal course of events, become a human being, the fact that it has this status and makes these claims affects the seriousness of what is done when it is conceived. There is a natural sequence of intercourse, conception, birth, education, which is recognizably the 'standard case' of sexual behaviour, upon the maintenance of which the survival of the race and the transmission of all human culture depends. No wonder that in all societies it is marked by solemn rituals. If the fetus has claims, the primary responsibility for seeing that they are met rests upon the parents and it is not one which they are entirely free to accept or not as they choose. There may be circumstances in which they are justified in subordinating its claims to some other, overriding, claims – and it is in deciding this that the most difficult questions arise – but they cannot just deny that the claims exist. If the responsibilities of parenthood are recognized, then at least we are forced to consider how they can best be discharged and in what circumstances they can be transferred to others. Like all responsibilities, they devolve upon persons who, in order to discharge them, must be in a position to know what is actually happening. It is one thing for some part of the normal process of procreation to be interrupted and some technical means of impregnation substituted, as in AIH or *in vitro* fertilization, in order to enable the process as a whole to be satisfactorily completed, when otherwise it could not be. It is quite another thing for the creation of new life to be totally disjoined from the fully personal relationship in which it belongs.

I have tried in this paper to state a case, in general terms, for paying serious attention to theology in bioethics. The case is, inevitably, controversial. We do not have access, in this or any other area that matters, to plain and incontrovertible truths, which straightforwardly authenticate themselves. It is just for this reason that the Christian theologian believes that the argument cannot be artificially cut short at the point where people's convictions, whether religious or secular, become decisively relevant. Of course, by the same token he cannot expect his presuppositions to be accepted without question. Yet, as I have tried to indicate, he has reasons of his own for believing that the argument, if freely and fairly conducted, can lead to truth.

*Oriel College, Oxford, England*

## NOTES

[1] Cf. [1], p. 62.

[2] See [2], pp. 145–156.

[3] A medical example would be the practice of 'female circumcision', whose justification seems to be solely that of long custom.

## BIBLIOGRAPHY

[1] Mackie, J. L.: 1977, *Ethics: Inventing Right and Wrong*, Pelican Books, Gretna, LA, U.S.A.

[2] Mitchell, B.: 1980, *Morality: Religious and Secular*, Oxford University Press, New York.

H. TRISTRAM ENGELHARDT, JR.

# LOOKING FOR GOD AND FINDING THE ABYSS: BIOETHICS AND NATURAL THEOLOGY

Gods. Good Fortune. Evandros and his wife inquire of Zeus Naios and Dione by praying and sacrificing, to what of the Gods or heroes or supernatural powers they may fare better or more well, themselves, and their household both now and for all time.

*An inquiry to the Gods at Dodona*

## 1. A WORLD WHERE THE GODS ARE SILENT

In all the world there are no Gods to speak to us face to face. For the most desperate petitions there is only silence. As the first Russian cosmonaut noted in his encounter with space, God was not to be found. Or if He (She) is, He (She) speaks clandestinely to the mystic and to the poet. While Thales saw that "all things are full of Gods" (quoted in *De Anima* A5, 411a7), the modern world reveals only the finite. Divinity hides. The Gods once brought rain and cured diseases. Now there is meteorology and pharmacology. The presence of the divine has been exorcized. There remain, however, numerous competing, once mutually antagonistic groups of interpreters of sacred texts and traditions who claim divine authority – for themselves, for such texts, or for the statements of long-dead founders who asserted that they had heard the very word of God. These modern religious groups possess often only the remnants of past political powers and structures through which they once wielded immense governmental authority and commanded intellectual attention. One might think here of the Sacred Congregation for the Doctrine of the Faith, versus its antecedent, the Holy Roman and Universal Inquisition, which possessed considerable intellectual authority and police powers. However, organized religions still remain as powerful political constituencies, and theology continues as one of the four faculties of the traditional university, though now divided peacefully into Catholic and Protestant.

In many respects, religion is not taken as seriously as it was in the past. It has been nearly three hundred and fifty years since the Pax Westphalica concluded the Thirty Years War. Heretics are no longer burnt outside



*E. E. Shelp (ed.), Theology and Bioethics*, 79–91.

of Newgate. One need only read *Foxe's Book of Martyrs* [13] or the histories of the persecution of the Jews by Louis the Pious to capture a sense of what it was like to live in a society that attempted to protect religious belief as we might consider protecting public safety. The consequences of sincere, unfaltering belief have been terrible. One might think of the injunction by the Fourth Lateran Council (1215) that secular authorities purge their land of heresy. Catholics who devoted themselves to the extermination of heretics were granted the same indulgences and privileges as those who made a pilgrimage to the Holy Land. (See Lateran Council Constitutions 3 in [3].) In this spirit, St. Thomas Aquinas argued that heretics after a first and a second admonition to abandon their beliefs should be exterminated (Summa Theologica II. 9.3).

We live on the other side of a religious age. Behind us is an era of orthodoxy stretching from Constantine's A.D. 319 decree that soothsayers (Haruspices) be burnt if they approached a neighbor's house (Nullus Haruspex COD. THEOD. IX xvi. I., in [2], p. 25) and Theodosius I's A.D. 381 decree that heretics should be expelled from cities (Nullus Haereticus, COD THEOD. XVI, x. 6 in [2], p. 31). It is an era that did not end until the processes of secularization unleashed by the Renaissance and the Enlightenment led to the American and French Revolutions and their rejections of an established church. The only powerful established church remaining is the secular religion Marxist Leninism, with its equivalent of a theology, dialectical materialism, a set of metaphysical assumptions devoid of scientific basis or logical necessity, and its own form of the Inquisition. However, the Western world now in great measure lacks such commitments.

The central strength and weakness of the West is precisely that it believes in nothing. Its official organs are either non-religious, or the religious aspects have taken on the character of a polite ritual often possessing the metaphysical commitment of Christmas cards displaying Santa Claus with his reindeer. We are the greatgrandchildren of the Enlightenment, living in peaceable, secular, pluralist societies, which impose no concrete view of the good life. One can be a good citizen whether one is a Roman, Greek, or Coptic Catholic, whether one is an Episcopalian, Baptist, or Methodist, whether one is a Jew, Moslem, or Hindu. These once bitter religious divisions, which in the past supported persecutions and cruel deaths, now no longer preclude friendship or marriage. In the West religion has become, if not a private matter, then a matter for special private communities which can no longer impose a test of good citizenship on the basis of orthodox belief.

We are returning to a religious toleration unseen since Emperor Julian suspended, if only briefly, the Christian persecutions of heretical Christians, pagans, and Jews. Julian offers the perspective of a Post Christian pagan. In response to the persecutions by Christians, Julian argued

> it is by reason that we ought to persuade and instruct, not by blows, or insults, or bodily violence. Wherefore, again and often I admonish those who are zealous for the true religion [i.e., pagan beliefs] not to injure the communities of the Galileans [i.e., Christians] or attack or insult them. Nea, we ought to pity rather than hate men who in matters of the greatest importance are in such evil case. [7]

Our contemporary societies have much in common with this sentiment which involves an important modification of the Roman sentiment that one may worship any God one wishes as long as one does not hold it to be the *only* God. With Julian, the Roman insistence that one at least include the emperor amongst one's objects of worship had vanished. It was enough that one's commitment to one God, or to one particular understanding of the requirements of that one God, did not lead to imposing by force an orthodoxy on society as a whole.

Secular pluralist societies are polytheist in this fashion. The state is neutral toward the various, often quite divergent, peaceable religious beliefs of its citizens. Yet it is willing to consult various representatives of such religions, much as one in Rome might have taken auguries at various shrines. Priests, ministers, and rabbis are employed as special custodians of particular understandings of values and of morals. Thus the National Commission for the Protection of Human Subjects of Biochemical and Behavioral Research, though it had no philosopher as a member, included a Catholic and a Protestant theologian (philosophers, as handpersons of theology, served on the staff). So too the Ethics Advisory Board of the Department of Health, Education, and Welfare had a Catholic theologian, and the President's Commission for the Study of Ethical Problems in Medicine and Biomedical and Behavioral Research had a Catholic theologian and a Rabbi. However, the recommendations of these bodies were carefully framed in secular terms with secular arguments. It is as if the representatives of various Gods had been convened to fashion ethical norms in a godless language.

This curious situation stems in part from the ethical tensions intrinsic to a peaceable, secular, pluralist society. Religious ethical arguments are compelled to occur on two contrasting levels: (1) the level of general secular discourse, and (2) the context of particular religious commitments. Because of this contrast, even if one draws ethical motivation from a special religious source, one will need to express the intellectual results of that motivation in the *lingua franca* of a secular pluralist society. If one opposes abortion on religious grounds, one will need to find secular arguments in order to forbid it as part of the public policy of a secular state.

The problem is not simply one of the bounds of religious authority, but one of the ability to forward generally accessible arguments based on religious viewpoints. Indeed, the problem can be put more universally still. The issue is whether there are generally accessible arguments for establishing the need for a recognition of the religious, however generally such is to be stated, as an element of fashioning a secular society. In short, one comes to a very traditional intellectual problem of assaying the extent to which, if at all, a natural theology can be maintained or framed. Is there a generally defensible understanding of the existence, nature, and wishes of the God(s) that does not depend upon prior commitments to the presuppositions of a particular revelation or of a particular religious community? And if so, what would its public policy implications be in general, and for bioethics in particular? In short, after a millennium of orthodoxy, and nearly half a milennium of rejecting that orthodoxy, what possibilities are there for fruitful moral discourse in a natural theological mode?

## THE SEARCH FOR GOD AND THE HUNGER FOR MEANING

Given the history of organized religion's suppression of dissidence, one might reasonably ask why men and women would continue to be engaged in theological questions. Given the intellectual difficulties in establishing the existence of God, and given the historical misuses of religion, one might very well conclude that it would be best if the entire project were abandoned. It is difficult, however, to relinquish the notion of God, for the idea of God and of transcendent religious significance has given meaning to suffering, pain, disease, deformity, and death. A religious account places such circumstances within an economy of pains and pleasures so that suffering properly borne merits salvation, delivers the poor souls from purgatory, releases one from the cycle of rebirth, or represents a divinely sanctioned testing of moral courage. The metaphysical assumptions of religious accounts prevent pain, suffering, and death from being experienced as surd. They can be seen through religion as having a purpose or significance within an over-arching cosmic plan.

Secular philosophical bioethics cannot supply such meaning. Bioethics succeeds in reaching across divergent communities of belief precisely by eschewing the metaphysical commitments by which religions can give meaning to suffering and pain. Philosophical ethics can give an account of why one must gain the informed consent of patients, or of why one ought to provide certain levels of access to health care, but it cannot supply a deep meaning to the suffering, debility, and death of patients. Hospital chaplains, not bioethicists, have traditionally aided patients in translating otherwise surd pains and sufferings into a religious-metaphysical system or viewpoint which can endow such circumstances with significance.

Despite the psychologically consoling contributions made by ministers of particular religions, their status is at best suspect against the collapse of general natural theological presumptions, and the failure of any particular religious viewpoint to be established by grace as a general consensus. And if one reacts to the embarrassing cacophony of claims and counterclaims made by particular religions by holding only a general belief in God, then suffering may take on a provocative or mocking character. In the absence of the concreteness of a particular religious story and its account of the meaning of pain and death, it becomes unclear how a deity could allow the innocent to suffer. One might think here of the vast chain of pain forged as animals struggle, kill, and die, culminating in the suffering of primates and humans. The very ecology of the planet presupposes pain and the large scale loss of life as many are born, but only a few survive to reproduce. The suffering of the innocent is immense. This suffering may then come either to call the existence of God further into question or to give grounds for condemning the deity.

One might think here of the poem by Jules Laforgue (1860–1887), 'L'Impossible', where he writes

> The stars, it is certain, will one day meet, Heralding perhaps that universal dawn Now sung by those beggars with caste marks of thought. A fraternal outcry will be raised against God. [9]

LaForgue can be read as protesting against the failure of the deity to have come to the aid of suffering sentient beings anywhere in the cosmos. A similar theme is also articulated, for example, in Kurt Vonnegut's novels. One might think here in particular of the revelation in *The Sirens of Titan* that human history has been engineered in order to send messages to a stranded extraterrestrial on Titan. Thus, Stonehenge in that extraterrestrial's language meant "replacement part being rushed with all possible speed." [12] Human history and suffering had been subverted by an alien purpose. Because of the absence of an acceptable account of suffering, the search for meaning from God produces an anti-meaning as the particular stories weaken through which actual accounts are given for the significance of suffering, pain, and death. The very project of theology becomes dubious.

This is a major cultural catastrophe. One would have hoped that theological bioethics could have completed philosophical bioethics through providing an account of the meaning of pain, suffering, and death. Since philosophical bioethics in reaching across moral communities eschews particular accounts of the meaning and purpose of life, one might have naively hoped that particular religious accounts could have supplemented the general bioethical account. However, the general bioethical account, in appealing to reason, calls those particular accounts into question. They become suspect because their concreteness depends not upon a general rational set of arguments, but upon particular faith commitments, religious traditions, or cultural assumptions. One appears then to be seeking from the religious perspective precisely what it cannot provide: a particular account of the meaning of life and the world that can be justified in terms of general rational considerations.

Yet this failure to break through to meaning does not make that quest seem any the less alluring. Either the entire universe exists from all time and through all space as a surd accident for which no reason can be given, or there must be a meaning hiding behind it all, however obscure. The very existence of anything feeds one of the deep springs of natural theological reflections. The inquiring mind reels at the thought that all may exist, that the entire universe may exist, for no reason. It appears absurd, endlessly, mindlessly absurd to assume that anything could exist, not to mention the entire universe exist, without some reason to account for it. Things in themselves are deaf and dumb. They can give no reason for their existence. Science can explain *how* entities react and relate, but not *why* anything exists. That things exist, that they exist at all, is the worst of enigmatic puzzles.

One might think here of the classic query of Leibniz, '*Why is there something rather than nothing?* For nothing is simpler and easier than something.' [10] This puzzle is echoed by Heidegger in his *Was ist Metaphysik?* [6] The presence of anything causes a hunger for an explanation for everything. That there is *anything* is so overwhelming that philosophers have traditionally employed the metaphysical principle of sufficient reason to conclude to the existence of God: a self–creating rational being that can answer the cosmic question "Why?" within the bounds of an ever-existing intelligence. An ever-existing intelligence, unlike ever-existing matter, could give an account of itself. It could complete reason's quest.

But what sort of God would that be? What image of God ought properly to direct bioethics and give it content? How are we to appeal to an unknown, unrevealed God in order to account for the meaning of suffering? A contemporary exploration of this problem is offered by Ingmar Bergman's study of suffering in *Cries and Whispers'* portrayal of the cancer patient Agnes. Bergman has the chaplain, who has been shaken by Agnes' suffering, to say the following prayer over her body after she had died. The first part of the prayer is in classic Christian terms. It appeals to a merciful God who allows suffering because of its redemptive significance.

God, our Father, in His infinite wisdom and mercy, has decided to call you home in the flower of your youth. Prior to that, He found you worthy to bear a heavy and prolonged suffering. You submitted to it patiently and uncomplainingly, in the certain knowledge that your sins would be forgiven through the death on the Cross of your Lord Jesus Christ. May your Father in Heaven have mercy on your soul when you step into His presence. May he let His angels disrobe you of the memory of your earthly pain. [1]

Within this traditional context, pain and suffering are not surd.

The second part of the prayer reflects the distance between man and God, between man and meaning. It articulates the contemporary predicament. The chaplain addresses Agnes:

If it is so that you have gathered our suffering in your poor body, if it is so that you have borne it with you through death, if it is so that you meet God over there in the other land, if it is so that He turns His face toward you, if it is so that you can then speak the language that this God understands, if it is so that you can then speak to this God. If it it so, pray for us. Agnes, my dear little child, listen to what I am now telling you. Pray for us who are left here on the dark, dirty earth under an empty and cruel Heaven. Lay your burden of suffering at God's feet and ask Him to pardon us. Ask Him to free us at last from our anxiety, our weariness, and our deep doubt. Ask Him for a meaning to our lives. Agnes, you who have suffered so unimaginably and so long, you must be worthy to plead our cause. [1]

The chaplain, though an old and holy man, is bewildered by the circumstances. Though he can repeat the accepted Christian cant regarding the meaning of suffering, it gives him no solace. He wants an account of our suffering.

The difficulty is finding meaning for suffering, pain, and death without having to retreat into the special exclave of a religious tradition that defends its views in its tradition's terms, and thus avoids the contemporary crises of theological meaning. Such a retreat can be satisfying as long as the particularity of the community into which one retreats does not give one pause. Which is to say, the retreat will succeed as long as the concerns for general rational justification, and the distraction of the numerous other communities competing for one's attention, do not block one's ability to find surety within the confines of a particular community of belief. A direction of advance could be established only by reclaiming some power for natural theology. This is, of course, what Charles Hartshorne attempts in this volume. His undertaking underscores the significance of the task of giving a general theological account of the meaning of human suffering and death. The question is, however, what the character of a successful account would be, and what implications it would have for bioethics.

## THE NATURE OF THE HIDDEN GOD

Immanuel Kant addressed issues similar to those joined by this essay. He asked whether it is possible, without an assumption of the existence of God, to comprehend a harmony between happiness and morality. To that point Kant argued that one must assume the existence of God and immortality, though that assumption could not lead to knowledge. The assumption is embraced, instead, as a postulate that serves as an object of the moral will. [8] For those unconvinced that such postulates are necessary for the moral life, Kant offers no solution. For such critics, Kant would appear at best to assume an answer rather than to find an answer, and at worst to trivialize the suffering of the innocent by a gratuitous assumption of meaning. However, Kant's argument is a testimony to the central importance of the search for meaning for suffering and pain. He makes the helpful suggestion that an assumption of meaning is not central to articulating the moral law, but to directing the moral will.

Kant's approach, however, is set within the traditional Western understandings of God. A radical response to the problem is offered by John Findlay who has argued that though we may be justified in hoping for the existence of numerous saints, gods, and angels, the conditions of the world and the existence of the undefended suffering of the innocent argue against the existence of a single all-powerful God. As Findlay states,

> There may, in fact, be a state in which God or *Brahman* really enjoys in concentrated union the whole of what he has dispersedly undergone or been. But, if such a state is possible, which I am far from affirming, it is not our present state; we live in an world age in which there may be gods, but in which there certainly is no God or none worth having. In our world the many and separate are more in evidence than the One, and the advantage goes to the sort of philosophy that more plainly recognized this fact. ([5], p. 100)

Findlay is arguing that a cosmological proof of the existence of God would at most lead to securing the knowledge of the existence of many gods. To put the matter somewhat bluntly, the sad state of the world suggests that it has been assembled and is overseen by a committee. Findlay appends the additional note that if there is a single, all-powerful God, its moral culpability for the suffering of the innocent would preclude it from being a proper object of worship.

Hartshorne addresses this problem by giving arguments for a Deity who is not in fact all powerful and fully realized. In this way the plurality that Findlay affirms in the divine is realized over time in the history of the Godhead. Moreover, the unity between God and the world adds a mark of plurality. This approach may still not be adequate to the tensions between the unity and diversity of human experience. These suggestions for radically rethinking the meaning of the deity can be pressed even further. One might think here, for instance, of the portrayal of the deity in Olaf Stapledon's classic science fiction novel *Star Maker*. In the novel the deity is portrayed as having an aesthetic, rather than an ethical, regard of the cosmos. It enjoys the aesthetic composition of the suffering and triumph of both the virtuous and the vicious. Stapledon's solution is to extend the radical otherness of God to a morally alien understanding of the significance of suffering and pain. [11]

Traditionalists might rejoin that God's earliest revelations in the Judeo-Christian tradition present Him as escaping human categories. One way of understanding God's answer to Moses' question in Exodus about God's name is, "I will be what I will be." [4] Such a God may simply refuse to give meaning to suffering. One can interpret the passage in the Talmud in this vein where Moses asks about the suffering of Akiba. After seeing the learning of Akiba, Moses asks: " 'Lord of the Universe, Thou hast shown me his Torah, show me his reward'. 'Turn thee round', said He; and Moses turned round and saw them weighing out his flesh at the market-stalls. 'Lord of the Universe', cried Moses, 'Such Torah, and such a reward!' He replied, 'Be silent, for such is My decree'." (Menohoth 29 b) God gives no account of the purpose of Akiba's torture and death. The difficulty is that as one makes God more unknown, the less an appeal to the deity can solve the problem of the meaning of suffering, pain, and death. A radically different deity may solve that problem for the particular individuals to whom it chooses to speak. But how is a culture to respond? How is one rationally to consider the possibilities of a radically different deity who in uniting its unity and plurality speaks at its own whim to whom it chooses and in its own ways? If such grace of meaning is given, it will be lived in a secrecy guarded by the failure of words to speak the nameless.

## THEOLOGY IN A POST-CHRISTIAN AGE

These intellectual and moral concerns lead to a radical reassessment of the tasks of theology. On the one hand, theology may continue as a collection of theologies, where the particular theologies exist as the custodians of traditions and cultural perspectives. Such custodianship can properly be discharged in a conservative fashion. Or, it may be pursued in a liberal fashion by attempting to refashion and rethink past understandings, but within a particular tradition. In either case, such theologies are the theologies of particular faith commitments. They contribute first, foremost, and most directly, to the intellectual and moral life of particular moral communities.

On the other hand, we will need to fashion a theology that draws from all communities and is a part of none. One will need to rethink again, as a serious aesthetic and moral task, the meanings that the transcendent can have for our general understandings of suffering and death. Such a theology may need to have a polytheistic dimension in drawing from traditions that include belief in many gods. We will need to learn to think as did the ancients, without embarrassment, about the possibilities for Deity. Even if Kant is correct that such a task is not an undertaking in knowledge *in sensu stricto*, it is an important intellectual and aesthetic challenge for our culture. This is, of course, to recast Kant. One assumes an aesthetic task of portraying the possibilities for meaning and purpose in suffering and death, in order to direct both moral and teleological concerns. One will be asking what visions of Deity can satisfy the human quest for meaning, while realizing there will be no definitive arguments to establish any particular portrayal. This understanding of theology, as an aesthetic undertaking completing certain moral and teleological concerns, will have implications for the kinds of contributions one would hope from theology to bioethics.

## WHAT THEOLOGY HAS TO TEACH BIOETHICS

Bioethics, where it succeeds, shows where it does not need theology. Secular bioethics, on the basis of philosophical arguments and analysis, not theological reflections, has fashioned its central notions of justice, rights, and obligations. Though theologians have contributed to bioethics, theology has not supplied its intellectual framework. There does not appear to be a special contribution to be made by theology to bioethics through ethical analysis and argument. Such generally defensible analysis and argument would just be so much more philosophy. We need to ask from religion precisely what philosophy cannot contribute: a meaning for life, for suffering, and for death. In this sense the Christian obsession with sexuality shows some merit. Aside from indicating that individuals ought not to lie, break promises, or use force, philosophy can do little to disclose the ways in which individuals *should* engage their sexuality. Even where traditional religious reflections have been misguided with regard to their particular understandings of sexuality, they have at least had the merit of directing us to the importance of goals for such central human endeavors.

Here theology will be offering precisely what philosophy cannot give, indications of ultimate meaning and purpose. Philosophy succeeds in not being a collection of traditions only insofar as it achieves generality for its arguments. As a result, the focus of philosophy tends to be upon the rational assumptions integral to such general human practices as ethics and science. However, generality is purchased at the price of content. Yet it is precisely content that is sought in questions regarding the meaning of human suffering, pain, and death.

To see theology contributing through non-doctrinal speculations and carefully articulated visions of the meaning of life, such as those of John Findlay [5], is to place theology, against tradition, close to the endeavors of literature and belle lettres. Theology would become in this general mode, insofar as it operates outside of particular traditions, a disciplined rational musing about the human condition. Unlike literature in general, it would place its accent upon the search for transcendent visions. Such a theology cannot give (as it was once thought to be able to give) meaning with sufficient surety to justify taking the lives of others or perhaps even losing one's own. One will have instead a theology encircled by hesitations and doubts, though moved by the deep human need for meaning and purpose. Such a theology will not satisfy those who have a need to lay down their lives and take the lives of others, as is the case with the passion of Khomeini's Islam, or as was the case with medieval Christian Europe. One must suspect that the middle-class children who join Marxist or terrorist groups are in search of such meaning. They hunger for a dangerous past that has been overcome in the West and that, for good moral reasons, must be set aside everywhere. The virtue of sophrosyne must be restored and such extremism in the preservation of belief must generally be seen as a vice. We must endeavor peacefully together in our attempt to understand the meaning of suffering and death, so that the search itself does not occasion more suffering and death.

Whether we can find the innocence that will allow us to return to ingenuous reflections on the ultimate meanings of our lives and the universe remains to be seen. Theology has been understood for so long as the extension either of particular religious traditions, or as a special adversary of philosophy and science, that one may doubt whether theology can assume this non-contentious understanding of its task. We will need to achieve a reflective innocence, one recaptured after a serious misadventure. We will need to take possession of a benign sense of speculation, a conjuring of ultimate images through reasoned reflection.

This is a puzzling universe. It is puzzling because of the enormity of innocent suffering. It is an intellectual puzzle, for there appears no ready account in principle for existence itself. Medicine raises ultimate questions regarding the meaning of suffering and of existence because it tends to the suffering and death of persons. Theological questions join those of bioethics. Even if theology cannot make a contribution of moral theory to the endeavors of bioethics, it can contribute aesthetic suggestions of meaning and purpose. We all face the abyss of death and of ultimate purposes obscured. Against this abyss, theology can offer images and visions and conjure meaning from a faceless night.

*Center for Ethics, Medicine and Public Issues,*
*Baylor College of Medicine, Houston, Texas, U.S.A.*

## BIBLIOGRAPHY

[1] Bergman, I.: 1976, 'Cries and Whispers', in *Four Stories by Ingmar Bergman*, Anchor Doubleday, Garden City, New York, p. 75.
[2] Bettenson, H., ed.: 1963, *Documents of the Christian Church*, 2nd ed., Oxford University Press, London.
[3] Conciliorum Oecumenicorum Decreta: 1962, ed. altera, Herder, Basel.
[4] Exodus 3:14. See *Torah: The Five Books of Moses*: 1961, Jewish Publications' Society of America, Philadelphia, p. 102.
[5] Findlay, J. N.: 1970, *Ascent to the Absolute*, George Allen and Unwin, London.
[6] Heidegger, M.: 1949, *Was ist Metaphysik?*, Vittorio Kostermann, Frankfurt/Main, F.R.G.
[7] Julianus, F. C.: 1969, *The Works of the Emperor Julian*, tr. W. C. Wright, Harvard University Press, Cambridge, Mass., Vol. 3, p. 135.
[8] Kant, I.: *Critique of Practical Reason*, Preussische Akademie der Wissenschaften edition, V. 122–134.

[9] Laforgue J.: 1958, 'L'Impossible', trans. by W. J. Smith, in *An Anthology of French Poetry from Nevral to Valery, An English Translation*, ed. A. Flores, Doubleday Anchor, New York, p. 203.
[10] Leibniz, G. W.: *The Principles of Nature and of Grace*, 7, from *Duncan Philosophical Works of Leibniz*, Vol. 32, pp. 209–217.
[11] Stapledon, O.: 1968, *Last and First Men* and *Star Maker*, Dover, New York.
[12] Vonnegut, K., Jr.: 1959, *The Sirens of Titan*, Delacorte Press, New York, p. 271.
[13] Williamson, G. A., ed.: 1965, *Foxe's Book of Martyrs*, Little, Brown, Boston.

SECTION II

# FOUNDATIONS AND FRONTIERS IN RELIGIOUS BIOETHICS

RICHARD A. McCORMICK

# THEOLOGY AND BIOETHICS: CHRISTIAN FOUNDATIONS

The title of this essay encompasses a subject so huge that sharp delimitation is necessary. On the one hand, 'bioethics' legitimately refers to the ethical dimensions of all attitudes, actions, and policies touching life (e.g., its begetting, protection, manipulation, improvement, etc.). Thus it includes the ethics of health care, of sexuality, of experimentation, of birth, of dying, etc., and public policies affecting all of these. Thus the *Encyclopedia of Bioethics* is a huge, four-volume work.

Theology, too, is an intractably sweeping term. *Webster's New Collegiate Dictionary* (1977) defines it as "the study of God and his relation to the world, especially by analysis of the origins and teachings of an organized religious community." God is, of course, the ultimate mystery. We are forced into analogical language even to achieve what Vatican I called *aliqualis intelligentia*. Furthermore, there are many organized religious communities. Thus even from this definitional point of view there will be many theologies. When one further specifies theology into theological ethics, the specification creates even more possibilities. Three types of normative judgments are fundamental to theological ethics: judgments of obligation, virtue, and value [4]. Add to this the fact that such judgments can be viewed from a variety of points of view (historical, logical, psychological, sociological, linguistic, etc.) and it becomes clear that a narrowing is required before the subject can be approached.

Let me state the problem from a slightly different point of view. It has been said (Anselm of Canterbury) that theology is "faith seeking understanding." Theology starts when faith begins reflecting on itself. There are two characteristics of this reflection that should be noted. First, it is ecclesial for the simple reason that the faith theology seeks to understand comes to us as a gift in and through a community, as Richard McBrien has noted [5]. Therefore, it is both personal and corporate. Second, the faith community exists in history and therefore must continually appropriate its inheritance in changing times, and diverse circumstances – and often with different purposes in view. From this it is clear that there can be many theologies within an individual faith community, a fact reflected in the varying theologies of the New Testament authors.



*E. E. Shelp (ed.), Theology and bioethics*, 95–113.

To render the subject 'Theology and Bioethics' somewhat more manageable, I want to introduce two qualifications and caveats. First, I shall approach the topic as a Catholic theologian and will restrict myself to viewing the relationship from this perspective. And, as noted above, even within this narrowing, there can be a variety of options.

Second, I want to note the distinction between the pairs morally right-wrong, morally good-bad. This distinction was foreshadowed in the terms 'obligations' and 'virtue'. Morally right-wrong refers to those characteristics of an action which render it either in accord with or opposed to the moral order or, in another wording, promotive or destructive to self or others or the community. Morally good-bad refers primarily to the stance, character, intentions, etc., of the agent [9]. Most discussions of bioethics are heavily concerned with the rightness-wrongness of actions – what Edmund Pincoffs has called 'quandary ethics', or problem-solving [18]. Correspondingly less attention is devoted to those considerations touching moral goodness-badness, for instance, the character of the problem-solver and the community in which it is formed. The writings of Stanley Hauerwas represent an attempt to redress this imbalance.

It is because of this imbalance that recent writings relating theology to bioethics have often examined the subject in terms of theology's contribution to decision-making. For instance, Franz Böckle of the University of Bonn argues that faith and its sources have a direct influence on 'morally relevant insights' not on 'concrete moral judgments' [2]. Thus the focus is narrowed to distilling action guides from theological sources.

Let me use my own writing as an example of this focus. I have argued that the sources of faith should be viewed above all as narratives, as a story [15]. From a story come perspectives, themes, insights. The story is the source from which the Christian construes the world theologically. The very meaning, purpose, and values of a person are grounded and ultimately explained in this story. Since that is the case, the story itself is the overarching foundation and criterion of morality. It stands in judgment of all human meaning and actions. Actions which are incompatible with this story are thereby morally wrong.

Vatican II put it as follows: "Faith throws a new light on everything, manifests God's design for man's total vocation, and thus directs the mind to solutions which are fully human" ([1], p. 209). The Catholic tradition, in dealing with concrete moral problems, has encapsulated the way faith 'directs the mind to solutions' in the phrase 'reason informed by faith'. 'Reason informed by faith' is neither reason replaced by faith, nor reason without faith. It is reason shaped by faith, and this shaping takes the form of perspectives, themes, insights associated with the story.

I have further identified some of the themes that give shape to our ethical deliberations in biomedicine as the following: (1) Life as a basic but not absolute value. (2) The extension of this evaluation to nascent life. (3) The potential for human relationships as that aspect of physical life to be valued. (4) The radical sociality of the human person. (5) The inseparability of the unitive and procreative goods in human sexuality. (6) Permanent heterosexual union as normative. There are probably many more such themes that are woven into the Christian story. But I am confident that these are dimensions of our being pilgrims created in the image and likeness of God.

Let me use the first theme (life as a basic but not absolute value) as an illustration. The fact that we are pilgrims, that Christ has overcome death and lives, that we will also live with him, yields a general value judgment on the meaning and value of life as we now live it. It can be formulated as follows: life is a basic good but not an absolute one. It is basic because, as the Congregation for the Doctrine of Faith worded it, it is the "necessary source and condition of every human activity and of all society." It is not absolute because there are higher goods for which life can be sacrificed (glory of God, salvation of souls, service of one's brethren, etc.). Thus, in John 15:13: "There is no greater love than this: to lay down one's life for one's friends." Therefore laying down one's life for another cannot be contrary to the faith or story or meaning of humankind. It is, after Jesus' example, life's greatest fulfillment, even though it is the end of life as we now know it. Negatively, we could word this value judgment as follows: death is an evil but not an absolute or unconditioned one.

This value judgment has immediate relevance for care for the ill and dying. It issues in a basic attitude or policy: not all means must be used to preserve life. Why? Pius XII, in a 1952 address to the International Congress of Anesthesiologists, stated: "A more strict obligation would be too burdensome for most men and would render the attainment of the higher, more important good too difficult. Life, health, all temporal activities are in fact subordinated to spiritual ends" ([19], pp. 1031–1032). In other words, there are higher values than life in the living of it. There are also higher values in the dying of it.

What Pius XII was saying, then, is that forcing (morally) one to take all means is tantamount to forcing attention and energies on a subordinate good in a way that prejudices a higher good, even eventually making it unrecognizable as a good. Excessive concern for the temporal is at some point neglect of the eternal. An obligation to use all means to preserve life would be a devaluation of human life, since it would remove life from the context or story that is the source of its ultimate value.

Thus the Catholic tradition has moved between two extremes: medico-moral optimism or vitalism (which preserves life with all means, at any cost, no matter what its condition) and medico-moral pessimism (which actively kills when life becomes onerous, dysfunctional, boring). Merely technological judgments could easily fall into either of these two traps.

Theology can take us this far but no farther. It yields a value judgment and a general policy or attitude. It provides the framework for subsequent moral reasoning. It tells us that life is a gift with a purpose and destiny. Dying is the last or waning moment of this 'new creature'. At this point moral reasoning (reason informed by faith) must assume its proper responsibilities to questions like: (1) what means ought to be used, what need not be? (2) What shall we call such means? (3) Who enjoys the prerogative and/or duty of decision-making? (4) What is to be done with now incompetent and always incompetent patients in critical illness? The sources of faith, in my judgment, do not provide direct answers to these questions.

Several things are notable about these themes that can be disengaged from the Christian story. First, as noted, they do not yield concrete answers to particular problems. That is the task of moral reason when faced with desperate conflicts – but moral reason *so informed.* These themes tell us how we ought to *look* on or *approach* the problem.

Second, such morally relevant insights are not mysterious, that is, utterly impervious to human insight without the story. In the Catholic reading of the Christian story such themes are thought to be inherently intelligible and recommendable – difficult as it might be practically for a sinful people to maintain a sure grasp on these perspectives, without the nourishing support of the story. Thus, for example, the Christian story is not the only cognitive source for the radical sociability of persons, for the immorality of infanticide and abortion, etc., even though historically these insights may be strongly attached to the story. In this epistemological sense, these insights are not specific to Christians. They can be and are shared by others.

Roger Shinn is very close to what I am attempting to formulate when he notes that the ethical awareness given to Christians in Christ "meets some similar intimations or signs of confirmation in wider human experience." Christians believe, as Shinn notes, that the Logos made flesh in Christ is the identical Logos through which the world was created. He concludes:

They (Christians) do not expect the Christian faith and insight to be confirmed by unanimous agreement of all people, even all decent and idealistic people. But they do expect the fundamental Christian motifs to have some persuasiveness in general practice ([21], p. 52).

Since these insights can be shared by others, I would judge that the Christian warrants are confirmatory rather than originating. I have suggested elsewhere (on abortion) that "these evaluations can be and have been shared by others than Christians of course. But Christians have particular warrants for resisting any cultural callousing of them" ([14], pp. 197–198).

Particular warrants might be the most accurate and acceptable way of specifying the meaning of 'reason informed by faith'. If it is, it makes it possible for the Christian to share fully in discussions in the public forum without annexing non-Christians into a story not their own.

The third aspect of such themes – and the one I want to highlight here – is that they have been attended to with an eye to problem-solving, that is, to ethics understood as the moral rightfulness and wrongfulness of individual decisions or policies. This is, of course, an entirely legitimate and even necessary undertaking for those concerned about the impact of theology on bioethics. But it does not exhaust that impact any more than rightfulness-wrongfulness considerations exhaust the notion of ethics. It leaves relatively untouched the inner dynamics of the faith experience and how this might relate to bioethics. Indeed, to limit the impact of theology on bioethics to such themes and perspectives is to collapse the significance of Jesus Christ into triviality. Therefore, in the remainder of this essay I want to sketch out a somewhat different approach to the problem suggested in my title. It will focus generally on what I earlier referred to as the goodness-badness distinction, that is, on the personal transformative influences of the faith experience and what this means for bioethics. In doing this, I will call to witness several theological colleagues, but especially Joseph Sittler, William F. May, and Gerard Gilleman, S. J., and Enrico Chiavacci.

Let me begin with a statement of Johannes B. Metz: "Christ must always be thought of in such a way that he is never merely thought of" ([17], pp. 39–40). Merely to 'think of' Christ is to trivialize Him, to reduce Him to one more (among many) observable historical event, to an example of humane benevolence. For the person of Christian faith, Jesus Christ is God's immanent presence, His love in the flesh. As William F. May puts it: "Jesus himself is the event in which the promises of God are fulfilled. He is the terrain, temple, and king, if you will, in which men may encounter God in his own person" ([16], p. 50).

Joseph Sittler has noted that the theme of the biblical narratives is God's "going out from Himself in creative and redemptive action toward men" ([22], p. 25). Sittler refers to "God's relentless gift of himself," "the undeviating self-giving God," "the total self-giving God," "God's undeviating will to restoration," the "history-involved assault of God upon man's sin," "the gracious assault of his deed in Christ." Jesus Christ is no less than God's self-giving deed.

The response of the believer to this person-revelation is the total commitment of the person known as faith. The term 'faith' has had an uneven history in the hands of Christians. Too often it has been tied to a pale propositional understanding of God's deed in Christ. Once again, Sittler:

It is not possible to state too strongly that the life of the believer is for Paul the actual invasion of the total personality by the Christ-life. So pervasive and revolutionary is this displacement and bestowal that terms like influence, example, command, value are utterly incapable of even suggesting its power and its vitally recreating force ([22], p. 45).

The believer's response to this specific, momentous and supreme event of God's love is total and radical commitment. For the believer, Jesus Christ, the concrete enfleshment of God's love, becomes the meaning and *telos* of the world and of the self. God's self-disclosure in Jesus is at once the self-disclosure of ourselves and our world. 'All things are made through him, and without him was not anything made that was made' (John 1:3). Nothing is intelligible without reference to God's deed in Christ. The response to this personal divine outpouring is not a dead and outside-observer 'amen'. It is a faith-response empowered by the very God who did the redemptive and restorative deed in Jesus Christ and is utterly and totally transforming – so much so that St. Paul must craft a new metaphor to articulate it. We are 'new creatures', plain and simple. Faith is the empowered reception of God's stunning and aggressive love in Jesus. As theologian Walter Kasper summarizes it,

Faith is not simply an intellectual act or an act of the will. It includes the whole man and every aspect of the human reality .... It embraces the whole of Christian existence, including hope and love, which can be seen as two ways in which faith is realized ([13], p. 82).

This same point is underlined by Sittler when he notes that faith is the proper term "to point to the total commitment of the whole person which is required by the character of the revelation"([22], p. 46).

Sittler has noted that "to be a Christian is to accept what God gives" ([22], p. 25). And what God gives is the going-out from himself in Jesus Christ. Something *has been done* to and for us and that something is Jesus. There is a prior action of God at once revelatory and response-engendering. Sittler correctly insists that the passive verb dominates the New Testament. "I love because I am loved; I know because I am known; I am of the Church, the body of Christ, because this body became my body; I can and must forgive because I have been forgiven" ([22], p. 11). This prior action of God is reflected in the Pauline 'therefore' (*ουν*) which states the entire grounding and meaning of the Christian ethic.

The Italian theologian Enrico Chiavacci puts it this way:

> In the New Testament the unique obligation of charity, which is the giving of self to God who is seen in one's neighbor, is grounded on the unique fact that God is charity .... 'Walk in love *as* Christ has loved us and given himself.' (Eph. 5:2). '*Therefore*, I exhort you brethren, through the mercy of God to offer yourselves ...' (Romans 12:1). The fact that God – in his manifestation as philanthropy – is love does not refer to further justification; it is the ultimate fact. The obligation to love is based only on God's love for us.... It is true ... that in the 'therefore' of Romans 12:1 we find the entire New Testament ethic ([7], pp. 291–292).

Here I want to make six points in a systematic way. First, as already noted in Christian ethics, God's self-disclosure in Jesus Christ as self-giving love allows of no further justification. It is the absolutely ultimate fact. The acceptance of this fact into one's life (*fides qua*) is an absolutely originating and grounding experience.

Second, this belief in the God of Jesus Christ means that "Christ, perfect image of the Father, is already law and not only law-giver. He is already the categorical imperative and not just the font of ulterior and detailed imperatives" ([7], p. 288).

Third, this ultimate fact reveals a new basis or context for understanding the world. It gives it a new (Christocentric) meaning. As a result of God's concrete act in the incarnation, "human life has available a new relation to God, a new light for seeing, a new fact and center for thinking, a new ground for giving and loving, a new context for acting in this world" ([22], p. 18).

Fourth, this 'new fact and center for thinking' that is Jesus Christ finds its deepest meaning in the absoluteness and ultimacy of the God-relationship. The person of Jesus is testimony to the fact that "no effort of man to know himself, find himself, be himself, is a viable possibility outside the God-relationship' ([22], p. 33).

Fifth, this God-relationship is already shaped by God's prior act in Jesus (self-giving). "To believe in Jesus Christ, Son of God, is identical with believing that God – the absolute, the meaning – is total gift of self." Therefore, the "active moment of faith takes place in the recognition that meaning is to give oneself, spend oneself, and live for others" ([22], p.

288). There is a German axiom that states: 'jede Gabe ist eine Aufgabe' ('Every gift constitutes an obligation'). That is profoundly true here. The very gift of God in Jesus constitutes or shapes the response, thus it is proper to refer at once to "God's love-gift and command". Thus the Christian moral life must be viewed as "a re-enactment from below on the part of men of the shape of the revelatory drama of God's holy will in Jesus Christ" ([22], p. 36). In this sense, it is the 'following of Christ'. It is a recapitulation in the life of the believer of the 'shape of the engendering deed', to use Sittler's language.

Finally, the empowered acceptance of this engendering deed (faith), totally transforms the human person. It creates new operative vitalities that constitute the very possibility and the heart of the Christian moral life.

I mention and stress these points because there has been, and still is, a tendency to conceive of Christian ethics in terms of norms and principles that may be derived from Jesus' pronouncements. That there are such sayings recorded in the New Testament is beyond question. But to reduce Christian ethics to such sayings is, I believe, to trivialize it. In this sense I agree completely with Sittler when he states:

> He (Jesus) did not, after the manner proper to philosophers of the good, attempt to articulate general principles which, once stated, have then only to be beaten out in corollaries applicable to the variety of human life.... His works and deeds belong together. Both are signs which seek to fasten our attention upon the single vitality which was the ground and purpose of his life – his God relationship ([22], pp. 50-51).

And in and through Jesus we know what that God-relationship is: *total* self-gift. For that is what God is and we are created in His image. To miss this is to leave the realm of Christian ethics.

In the remainder of this essay, I should like to cover three points: (1) an interpretative understanding of the Christian moral life; (2) the relativizing influence of the new love command; (3) the relation of these reflections to bioethics.

## AN INTERPRETATIVE UNDERSTANDING OF THE CHRISTIAN MORAL LIFE

In the New Testament charity holds a unique place, which will make it appear as the principle of the moral life and even the substance, in a true sense, of Christian revelation (Romans 13:10). Christ says: "God is love" (1 John 4:8 and 16). This love manifests in this that God loved us gratuitously (1 John 4:10), when we were as yet sinners (Romans 5:8; Mt. 5:45). The best image of God's love is Christ himself (John 1:49). Christ's love manifested itself in self-donation to blood (John 15:13), out of love for sinners and to manifest his affection for the Father (John 15:10). Life, therefore, in the eyes of Christ, consists in donation of self from love to the Father and to sinners.

The Christ-event revealed that we are destined to be one with him in the Mystical Body (Romans 12:4–5). Clothed with Christ in baptism we must take on his ways and mature to his stature (Eph. 4:13–15), even to the point where we can say "it is no longer I who live, but Christ lives in me" (Gal. 2:20). The moral life, therefore, reproduces in the Christian the moral attitudes of Christ himself. This will be a following of Christ (Luke 9:23) which means: Love God by keeping his commandments (John 4:14; 14:21), and love one's neighbor as Jesus loved us (John 13:33–35), even to death (John 15:12–13).

There is no greater commandment (Mt. 22:37–38): This love is the epitome of the entire law (Romans 13:8–9; Gal. 5:14) and is a way more elevated than all charisms and is a beginning of eternal life (I Cor. 13:1–3). It does not suppress other precepts but is the source from which they flow (Eph. 3:17) and is the bond of perfection. This is a new law (John 13:34) since it is internal, 'natural' to the new creature (II Cor. 5:17), and so characteristic of Christians that one is to recognize them by it (John 13:35) and see in them, through this love, a continuing revelation of the unity of the Divine Persons and the presence of the Spirit of Christ (John 17:20–23). From these quotes it is easy to conclude with Cerfaux: "Charity is the normal occupation of the Christian..." ([6], p. 326).

This simple, glorious but demanding morality, where individual external acts flow from and express charity like rays of the sun, was couched by St. Paul as being 'in the Lord'. St. Thomas Aquinas summarized it with the sweeping dictum: Charity is the form of the virtues.

The meaning of the Thomistic assertion will be clearer when we show how Thomas develops the thought [10]. First of all, he shows how charity is the form of all virtuous *acts*. His reasoning can be put as follows. That which ordains an act to its end gives it its form. But charity is the virtue that ordains to its end any virtuous act. Therefore charity gives form to all virtuous acts. The form of which Thomas speaks is no intruder which would collapse and destroy the differences of acts and make them all uniquely and identically acts of charity only (a kind of 'love monism'). Rather, the gist of his thought is that actions possess ultimate meaning (we used to say 'supernatural perfection') only insofar as they are caught up in the conferred divine life. And in the new creatures it is the term 'charity' that expresses this 'being caught up in', this 'being grasped by God in Christ', this ordering to the end, this animating or forming.

Thomas moves on to the virtues themselves, the origins of these acts, and says that charity is the form of these virtues. Charity does not destroy or replace the virtues; but the virtues are rooted in and depend on charity, and in such a way that there is no true virtue in the fullest sense without charity. Thus the virtues, while retaining their identity, are participations in charity so that they and their acts are in some sense also emanations of and acts of charity. Or as Thomas puts it, at a single battle command from their leader, one soldier draws his sword, another prepares his horse, and so on. All are alive in their own way with this single stimulus. This is, in brief and impoverishing summary, the Thomistic turning of the New Law, what Sittler constantly refers to as the God-relationship revealed to us in Christ.

One of the most stimulating modern reworkings of this analysis is that of Gerard Gilleman [10]. Human persons, Gilleman reasons, are at their profoundest depth a tending, a teleological drive. All that the person wills and does is a manifestation of this tending, which even at the natural level has the character of love. Since the person is in space and time, this radical tending does not obtain its object by a single leap. The person moves by halting, free steps to the goal of this tending. This tending, therefore, must exteriorize itself through moral actions. It struggles toward its goal through a variety of material combinations, choices, situations.

It is these exteriorizations of which we are clearly and reflexively conscious; but this explicit consciousness is only a momentary fixation of a more profound total movement. If we fix our attention only on the exterior, representative. static elements of human action, and ignore the profound dynamism, we fail to grasp the moral act.

In the New Covenant in Jesus, in the order of grace, our *tendance foncière* (Gilleman) has been transformed and divinized by the interior grasping by God of our beings. All that was noted above about the source and origin of moral activity must be understood in a transformed context.

The transformed-divinized person must express him/herself. Thus to meet the indefinite possibilities of choice in our complex life, the person takes on the nuances of patience, temperance, justice – all instruments of connection with the material. These are the virtues. Charity is eminently each of these, just as the light of the sun is eminently red, violet and so on.

That is, charity is now the dynamic depth from which both the virtues and their acts emanate. Virtues and acts are partial mediations of charity because charity is the source from which they flow.

We might put it this way. Because of God's gracious grasp of us in Christ, charity is now the faculty of the end. It is not an adventitious thing, like a clove jammed into a flank of ham. It is the person finalized toward the God of Jesus Christ. The virtues that adorn the person so encountered and grasped by God will enjoy this finalization, as do all the acts that proceed from these virtues. Every virtue, every virtuous act is a mediation of this profound and gracious dynamism.

Take justice as an example. If one must view justice as transparent of charity – as one must in the Christian view where charity is the form of the virtues – then it follows that charity must somehow enter the very definition of justice. If we define justice simply as that habit which inclines one to render to another his/her due, we have disengaged it from the subject and from that which confers its complete Christian intelligibility. We have conceptualized it with no reference to the Christian context. Gilleman argues that if we build a treatise of justice on this definition alone, we run the risk of allowing the life itself that justice regulates to slip away. Rather, Gilleman suggests that we should regard the function of justice as a realization, in the area of goods capable of being objects of possession and rights, of a normal climate wherein the Christian communion among persons can blossom and mature. Justice is the mediation of charity in this particular area.

In summary, then, if in viewing human persons and their actions we restrict ourselves to the representative content of the act or virtue, it is as if we were viewing a dead body. It is as if we were viewing the beautiful stained glass of the Sainte Chapelle from outside, where it appears gray and drab. The soul is missing. The same windows burn with beauty when one sees the sun through them from the inside. Our being and actions are alive with full meaning only insofar as they are viewed 'from the inside', with charity flowing.

This is, I submit, a defensible account of the Pauline crypticism 'in the Lord'. Contemporary theologians often refer to the depth of our being and activity as a fundamental option, as the vertical depth of our horizontal activity, as the ultimate meaning of the working out of the God-relationship.

THE RELATIVIZING INFLUENCE OF THE LOVE COMMAND

No one in our time has catalogued and discussed the nature of human failure more persuasively and profoundly than William F. May ([16], p. 50). May notes that our root sin is impurity of heart, a form of idolatry where "men deny God by turning away from him toward some creaturely power, whether it is the glitter of gold, the fertility of the soil, the excitement of a career, the fascination of a woman or the claim of a great public cause" ([16], p. 28). Impurity of heart or idolatry is elevating a fragment of the world into the position of God. It is to take something out of its place and invest it with divinity. Purity of heart, therefore, is not a sexual term. "It refers primarily to singlemindedness, wholeness, integrity or unity of heart" ([16], p. 33). It is what Sittler calls 'righteousness', the living out of our lives under the absoluteness and ultimacy of our structural God-relationship.

Jesus Christ is the primary analogate here. As May puts it: "God made himself savingly present to men by uniting himself with a man of absolute purity of heart. He presented himself to men only through the medium of that purity of heart to which man himself is summoned" ([16], p. 35).

Christ stated: "I am giving you a new commandment: Love one another. As I have loved you, so you too must love one another" (John 13:34). I want to concentrate on the "as I have loved you." Raymond Brown points out that this phrase emphasizes that Jesus is the source of the Christian's love for one another. In this sense it is effective, sc., "it brings about their salvation" ([3], p. 612). Only secondarily does it refer to Jesus as the standard of Christian love.

I want to attend to this secondary sense, Jesus as the standard of our love.[1] In this sense I believe it can and must be said that Jesus' love was that of absolute righteousness or purity of heart. That is, it was a love shaped by the absoluteness and ultimacy of the God-relationship. The human goods that define our flourishing (life and health, mating and raising children, knowledge, friendship, enjoyment of the arts, and play), while desirable and attractive in themselves, are *subordinate* to this structural God-relationship.

As May has so brilliantly pointed out, the characteristic of the redeemed but still messy human condition is to make idols, to pursue these basic goods *as ends in themselves*. This is the radical theological meaning of secularization: the loss of the context which subordinates and relativizes these basic human goods and which prevents our divinizing them. The goods are so attractive that our constant temptation (our continuing enslavement, our bondage to the world, our constant need for liberation and deepening conversion) is to center our being on them as ultimate ends, to cling to them with our whole being.

Jesus' love for us is, of course, primarily empowerment. But it is also, in its purity and righteousness, the standard against this type of collapse, this idolatry. Whatever he willed for us and did for us, he did within the primacy and ultimacy of the God-relationship. Since this relationship is our very being and destiny, his love took the form of a constant reminder of this momentous dignity to people hellbent on their idolatries.

His love, as standard, suggests the shape of our Christian love for each other. It is conduct which reminds others of their true dignity, which reminds them of their being and destiny, and therefore which pursues, supports, protects the basic human goods *as subordinate.* In this sense I believe it is possible to say that Jesus is the norm above all of our self-perception, of what we do in a novel sense. That sense is: all that we are must be the ultimate judge of what we do. And 'being in the Lord' is what we are. That radically relativizes all human goods. Accordingly, 'following Christ', imitating Christ means, negatively, never pursuing human goods as *final ends,* and positively, pursuing them *as subordinate.* In this perspective, the Christian community (which the Lord described as a community of love) is a community that celebrates and shares its profoundest being and life, that pursues, therefore, the basic human goods in interpersonal life *as subordinate.*

St. Ignatius Loyola, in his *Spiritual Exercises,* proposes to the exercitant that he/she "desire and choose poverty with Christ poor, rather than riches; insults with Christ loaded with them, rather than honors" ([20], p. 69). This should not be understood as a flattening and spurning of human goods, but as a relativizing of them to a world always prone to absolutize them. Christ's suffering and cross were both a symbolic contextualizing of human goods *and* the profoundest act of love. Therefore, those who strive to follow Him ('as I have loved you') are performing the profoundest act of love for the world by pursuing in interpersonal life the basic goods *within their context, as subordinate.* In this sense, both Christ's love for us as standard, and therefore our love for each other, constitute a profound relativizing of basic human values.

This can easily be missed in the culture of Western capitalistic societies, just as it was folly to the Greeks. In such cultures we are frequently enslaved by our own self-conceptions and therefore our conception of the 'good life'. To many, this term means *having things* (beautiful property, leisure, education, health, wealth, pleasures). From this follows the notion of ethics as creating the conditions that make such a life possible, especially expanding the freedom that allows the agent (a solitary, morally autonomous individual) to pursue and maximize whichever of these goals appeal to the individual.

For the believing Christian, 'the good life' does not deny these goods but radically relativizes them. They are instrumental to that which truly defines the 'good life' – becoming what we are. We are with our whole being 'in the Lord'. Thus the opening prayer for the mass of the seventeenth Sunday in ordinary time reads: "God our Father and protector, without you nothing is holy, nothing has value. Guide us to everlasting life by helping us to use wisely the blessings you have given to the world." 'Wise use' was specified in Jesus' love for us. It is non-idolatrous use, use within the primacy and ultimacy of the God-relationship.

Christian theological ethics – faith reflecting on itself and its behavioral implications – must talk like this. Otherwise it fails to be Christian and ultimately theological. Johannes Metz notes that "Christological knowledge is formed and handed on not primarily in the form of concepts but in accounts of following Christ" ([17], p. 40). The saints are its exemplars. That is why the history of Christian theological ethics is the history of the practice of following Christ, and must assume a primarily narrative form. It is also why the character of our moral agency as Christians should have its most fundamental formative ground in Christian public worship. It is above all in liturgy where we are exposed to the narratives that ought to profoundly shape our lives. In liturgy, for the Catholic Christian, we make Christ present by remembering 'the shape of God's engendering deed'.

## THE RELATION OF THESE REFLECTIONS TO BIOETHICS

I have already indicated that the Christian story generates general themes and perspectives relevant to bioethics. I have futher indicated that these relate above all to quandary ethics. Here I want to relate what has been said to the health care practitioner, to the goodness-badness distinction mentioned earlier. I shall do so briefly under two headings: (1) how to view the health care profession; (2) how to support and motivate this view.

### *How to View the Health Care Profession*

From the Christian point of view, the field of health care is a privileged context in which to encounter another person – and hence Christ. "The same Lord who meets, judges, heals and forgives, in the solitary and naked aloneness of the self, plunges that self into the actuality of the world as its proper place for faithful activity in love" ([22], p. 69). It is to be viewed as an apostolate, as the 'Decree on the Apostolate of the Laity' of Vatican II makes repeatedly clear.[2]

In the Christian understanding, the encounter of persons has a certain structure. It is the categorical moment for the faithful activity in love that describes the very being of the 'new creature'. It is literally our way of loving God in this context. It is the vertical in the horizontal, or, as Sittler words it, "Love is the function of faith horizontally just as prayer is the function of faith vertically" ([22], p. 64). This is true of both the curing and caring dimension of health care. If we do not view health care in this way, we interpret and restrict its reality short of the depths of faith. The laity decree cited above, after noting that Christ made the commandment of love his own and endowed it with a new meaning, continues: "For he wanted to identify himself with his brethren as the object of this love when he said, 'as long as you did it for one of these, the least of my brethren, you did it for me' " ([1], p. 498). But if this profound structure of the health care encounter is to be lived, "certain fundamentals must be observed. Thus attention is to be paid to the image of God in which our neighbor has been created, and also to Christ the Lord to whom is really offered whatever is given to a needy person" ([1], p. 499).

If the medical encounter is viewed and lived in this way, it would be both guided by and further generative of the moral dispositions and perspectives implied in Christ's phrase 'as I have loved you'.

I am not speaking here of a psychological 'immediacy' or 'breakthrough'. Certainly William May is correct when he notes that physicians among others "must often accept without false dismay the incompleteness of their contacts with those over whom they exercise authority" ([16], p. 63). The same is true of patients. Only a rather inflated religious romanticism would expect a direct, immediate I-thou encounter in every human relation. I am rather speaking of the profoundest ontological structure of the encounter, fully disclosed in and by Christ, a structure we perceive only dimly (in faith), but one that ought to be the organizing shape and power of our responses. That is nothing more or less than Christ enjoined upon His followers. This is exactly what 'The Decree on the Apostolate of the Laity' meant when it urged professional people to remember that in fulfilling their secular duties of daily life, "they do not disassociate union with Christ from that life." It further urged professionals to "see Christ in all men whether they be close to us or strangers" ([1], p. 493).

If health care personnel view their profession in this way, I believe we can reasonably predict three important results. First, there will be the growth of those dispositions that nourish, protect, support the medical encounter as a truly human (not merely technological) one: compassion, honesty, self-denial, generosity. Furthermore, it can be expected that such dispositions will powerfully support and implement the very themes I lifted out at the beginning.

Second, we may reasonably expect that a profession deeply penetrated with persons of such faith and such dispositions will be transformed. 'The Decree on the Apostolate of the Laity' states of persons of faith that "their behavior will have a penetrating impact, little by little, on the whole circle of their life and labors" ([1], p. 505). It regards this as the "penetrating and perfecting of the temporal sphere". I take it as indisputable that, in a time of hugeness, high technology, and deepening impersonality in health care, this would be a truly desirable leavening influence.

In this respect, Thomas Clarke, S. J., has stated:

> The Christian deeply committed to health ministry who approaches Jesus with a view to deepening discipleship to him brings something distinctive to the relationship, a gaze and a listening ear made sensitive to certain accents by engagement in the healing experience. Similarly, the Christian comes from a contemplative deepening of discipleship back to dealing with health technology enriched by distinctive habits of perception and response ([8], p. 252).

Faith creates sensitivities in the believer beyond the reach of natural vitalities. It bestows sensitivity to 'dimensions of possibility' not otherwise suspected.

Third, this 'penetrating and perfecting of the temporal sphere' will be shaped off the phrase 'as I have loved you'. Therefore it will be permeated with the ultimacy and absoluteness of the God-relationship, and the corresponding relativizing of other human goods. In health care delivery this can be very important.

The characteristic temptation of the ethos of the medical profession is to make life and the profession's ability to preserve it an idol. The symbol of this is patient-abandonment when cure is no longer possible and death is imminent. ('I can do no more.') For many physicians death is defeat. ('No one dies on my shift.') This can skewer and distort the ministry of health care, decontextualize its instrumentalities, technologize its value judgments, and bloat its practitioners – to say nothing of limitlessly expanding its cost.

Let me again cite 'The Decree on the Apostolate of the Laity'. At a key point it states that "only by the light of faith and by meditation on the word of God can one... make correct judgments about the true meaning and value of temporal things, both in themselves and in their relation to man's final goal" ([1], p. 493). In the bioethical context I take this to mean that it is precisely connection with and commitment to God's deed in Christ ('as I have loved you') that is the best guarantor against absolutizing the relative. Specifically, it is corrective to the judgment that death is ultimate defeat. This is no signal for the profession to relax its vigorous pursuit of the preservation of life. It is simply an insistence that its ministry is to serve our best interests. And for the Christian, accumulation of added minutes is not always the measure of best interests.

In a speech he composed but never gave because death intervened, the late André E. Hellegers, founder and first director of the Kennedy Institute of Ethics, noted:

> As the caring branches of medicine were gradually pushed aside by the curing ones, there seemed to be less use for the Christian virtues. I think that shortly the need for those old Christian virtues will return and once again be at a premium. Our patients will need a helping hand and not a helping knife. This is no time to dismantle the low-technology care model of medicine.... We must either recapture the Christian virtues of care or we shall be screaming to be induced into death to reach the 'discomfort free society' [11].

That strikes me as a vivid application of the ordering of values implied by 'as I have loved you'.

*How to Support and Motivate This View*

It is often said that health care personnel 'need basic ethics'. As noted, this is frequently understood as 'answers to cases'. A symbol of this is a recent book by Albert R. Jonsen, Mark Siegler and William J. Winslade [12]. It is physically designed to fit in the pocket of white-coated practitioners. Just reach in for the answer. I do not mean to belittle such work. It is vitally important. But it is not all of ethics.

The theological reflections outlined above suggest that what is no less important, and even more so, is a spirituality for health care personnel, or ethics in that broader sense. By a 'spirituality' I refer to a personal and corporate life-climate designed to foster and deepen belief in and insight into the basic structure of our lives as this is revealed in God's self-disclosure in Jesus, and particularly as this is encountered in the medical context. If such a spiritually is not developed, then a gap will exist between personal belief and professional life.

The results of such a gap can be both personally and professionally disintegrating. I refer to the distinction and eventual opposition that can arise between the role and the person (true self). Roles, of course, are essential for social structure. They delimit behavior and make it predictable. We expect a physician or a nurse to act in a certain way. When each carries a role as a response to societal needs, there is harmony in the system.

But this role is not necessarily the true self. If one is constantly relating to others through a role, that person can be alienated from the true self. The role may grow, but not the true self. That is, the qualities that nourish human exchange (patience, other-concern, communication, compassion, listening, caring) will be restricted to the role. They will become an 'assumed manner', an adopted etiquette that will not hold up very long. The true self will remain anemic, infantile, immature. At this point, health care ceases to be a personal apostolate; it becomes a job. When that happens, patients become increasingly mere sources of income and manipulable.

This is a particularly intense trap for health care personnel, especially the physician, because the role is in itself so exacting and kaleidoscopic. The physician must be concerned and care; yet a certain detachment is in order for patient best interests. The physician is a father or mother hearing secrets, often of abuse and spiritual malaise. At times he/she must give gentle orders. Professional expertise is mandatory. Too many mistakes cannot be tolerated. The physician must relate coordinatingly with a whole network of health care personnel, especially nurses. The physician is a government employee (Medicare, Medicaid), a consoler and comforter of the dying and their families. Many patients who consume his/her time are not sick, or not as sick as they think.

In sum, the physician is under great pressure, a pressure that can exhaust the qualities of patience, compassion, other-concern, communication in the role, and leave the personal self stunted. A spirituality should aid enormously in preventing such a collapse.

No one can devise a spiritually adequate to everyone. The challenge upon each of us is to establish a daily climate that reveals and reinforces the contours of our Christian faith. There is not space here to enter into the particulars of this type of program (what we might call a continuing examination of consciousness). Suffice it to say that no Christian health care deliverer who would remain precisely that is exempted from the implications of Sittler's statement that "prayer is the function of faith vertically" ([22], p. 64).

*Kennedy Institute of Ethics, Georgetown University,*
*Washington, D. C., U.S.A.*

## NOTES

[1] It is not clear to me how we can love one another *as* Jesus did in the first sense, so as to 'bring about their salvation', unless it is that our loving faith is a channel for *His* love (saving grace) to others.

[2] Cf. [1], pp. 489–521. An apostolate is there defined as all activity that brings "all men to share in Christ's saving redemption and that through them the whole world might in actual fact be brought into fellowship with him" (p. 491).

## BIBLIOGRAPHY


[1] Abbott, W. (ed.): 1966, *The Documents of Vatican II*, America Press, New York.
[2] Böckle, F.: 1976, 'Glaube und Handeln', *Concilium* **120**, 641–647.
[3] Brown, R.: 1970, *The Gospel According to John*, Doubleday, Garden City, New Jersey.
[4] Carney, F. S.: 1978, 'Theological Ethics', *Encyclopedia of Bioethics*, Vol. 1, Macmillan, New York, pp. 429–437.
[5] *Catholic Chronicle*: 1982, July 30, Toledo, Ohio.
[6] Cerfaux, L.: 1958, 'La Charité fraternelle et le retour du Christ', *Ephemerides Theologicae Lovanienses* **24**, 321–332.
[7] Chiavacci, E. (ed.): 1980, 'The Grounding for Moral Norms in Contemporary Theological Reflection', in C. E. Curran and R. A. McCormick (eds.), *Readings in Moral Theology*, Vol. 2, Paulist Press, New Jersey, pp. 270–304.
[8] Clarke, T.: 1980, *Above Every Name*, Paulist Press, New Jersey.
[9] Frankena, W. K.: 1973, *Ethics*, Prentice Hall, Englewood Cliffs, New Jersey.
[10] Gilleman, G.: 1959, *The Primacy of Charity in Moral Theology*, Newman Press, Westminster.
[11] Hellegers, A.: 1979, 'Reflections on Health Care and its Possible Future', unpublished manuscript.
[12] Jonsen, A. R. *et al.*: 1982, *Clinical Ethics*, Macmillan, New York.
[13] Kasper, W.: 1980, *An Introduction to Christian Faith*, Paulist Press, New Jersey.
[14] McCormick, R. A.: 1981, *How Brave a New World*? Doubleday, New York.
[15] McCormick, R. A.: 1982, 'Theology and Biomedical Ethics', *Logos* **3**, 25–43.
[16] May, W. F.: 1967, *A Catalogue of Sins*, Holt, Rinehart & Winston, New York.
[17] Metz, J. B.: 1978, *Followers of Christ*, Paulist Press, New Jersey.
[18] Pincoffs, E.: 1971, 'Quandary Ethics', *Mind* **80**, 552–571.
[19] Pius XII: 1957, *Acta Apostolicae Sedis* **49**, 1027–1033.
[20] Puhl, L. J. (ed.): 1953, *The Spiritual Exercises of St. Ignatius*, Newman Press, Westminster, London.
[21] Shinn, R. L.: 1969, 'Homosexuality: Christian Conviction and Inquiry', in R. W. Weltge (ed.), *The Same Sex*, Pilgrim Press, Philadelphia, pp. 43–54.
[22] Sittler, J.: 1958, *The Structure of Christian Ethics*, Louisiana State University Press, New Orleans.

LANGDON GILKEY

# THEOLOGICAL FRONTIERS: IMPLICATIONS FOR BIOETHICS

Fortunately, I have been asked to concentrate on 'the current frontiers of theology' – or at least that is how I will interpret my assigned title. On this matter I have some reasonable or at least reasoned opinions, though everyone's 'frontiers' are his own and will be disputed by the next theologian called upon. I shall try – and here I may not do so well – to relate these theological frontiers, as I view them, to issues in bioethics. One assumption behind my title, if I understand it aright, is that theology changes as well as do biology, genetics, and medicine. (Often it is assumed that theology and ethics, deriving from old traditions, remain stable and that it is in science, technology, and medicine that important changes reside.) Further, it seems assumed that these changes within theology and theological ethics will affect the relation of theology to bioethics as much as will developments in the theories and practices of the relevant scientific and medical disciplines. Even though these assumptions seem relatively obvious, frequently they are overlooked. Actually, however, that point being granted, I will suggest that the most fundamental 'frontier' of theology (and this is part of the image of a frontier) arises in *extra*-theological areas; that is, it appears to or confronts theology not as a result of its own labors but from outside, from its wider historical and cultural context. Thereby does it force on theology – as in the end it will on every other intellectual discipline – a renewed self-scrutiny, self-criticism, and self-overhauling. If I am right, therefore, this frontier will also confront bioethics, though perhaps further down the historical road. Thus, this is not so much a frontier *in* theology as a frontier *for* theology, a frontier established by deep, pervasive and fundamentally transformative trends in contemporary culture, or better, in contemporary world culture.

The image of the frontier assures us among other things that frontiers must be *new*. Clearly this is strange or at least parochial: China and India have some very old and stable frontiers. This is, therefore, peculiar to America where the frontier, while it lasted, kept moving; and thus did this *spatial* image become appropriate for use in a *historical* and so a *tem-*



*E. E. Shelp (ed.), Theology and Bioethics*, 115–133.

*poral* context: frontiers as the new edges of 'advance' of a discipline, tomorrow's issues that are already challenging the latest inquiries today. I shall mostly use it in this temporal sense as a new frontier in this piece; but first let me remind us of our *older* frontier of theology, the one that still dominates most academic theology and through that a great amount of theological ethics. That this is not a new but an old frontier does not mean that presently it is either unreal or insignificant, as our spatial image reminds us. It is also worthy of note that this long dominant frontier was, like the 'new' one I will here propose, originally presented to theology *not* as a result of theology's labors themselves, nor even out of special disciplines in the university. Rather, it 'arose' out of various changes in the larger cultural-historical context of theology and of the special disciplines. The impact of this frontier, to be sure, was mediated through special disciplines – but it presented itself *to* them before it presented itself *through* them to theology.

I refer – in this image of the old frontier – to the challenge issued to theology by the Enlightenment: that whole cultural-historical development of new this-worldly and optimistic attitudes, new methods of inquiry, new critiques of traditional authorities, new moral and social aims and standards, and, perhaps most important, new criteria for truth. This was a historical 'sea change' that characterized roughly the 17th and 18th centuries in Europe and that has continued – at least in the academic life that it 'fathered' – in further developed forms since; it represents the cultural Gestalt that we refer to when we speak of 'modern culture'. Natural science was fundamental for this deep intellectual transformation, and so this frontier was first mediated to theology by the sharp challenge of the new sciences and the even more important scientific method. But it brought with it also changes in ethics and social theory characterized by a new emphasis upon humanitarianism, individual rights and democracy. And subsequent to its rise it resulted in the development of a knowledge of and interest in history that led to the significant historical consciousness and the new interface of historical inquiry with every aspect of theology.

Since the start of the 19th century, then, the frontier of theology, the point where its major constructive work was done (or more candidly, its most important *defensive* operations were carried out) was in relation to this multifaceted cultural transformation. Most academic theology has been dominated, and on every level, by this interchange or interrelation with modern culture; and this is still the main concern of the work of theologians and of theological books and courses: (1) the reinterpretation of the *cognitive* claims of theology and of the status and character of religious truth in the face of natural science, social science, and modern philosophy; (2) the reinterpretation of theology in the face of religious and cultural *pluralism*; (3) the reinterpretation of theology in the face of the *socio-historical-psychological interpretation* of religion; (4) the reinterpretation of theology in the face of historical change, and hence of the historical *relativity* of every theological expression whether in scripture or in tradition – evident in current theology in the vast interest in hermeneutics.

This has been a 'dangerous' frontier. On each of these issues a deep secularistic critique of the very possibility of theology intellectually and so of religion existentially (*vide* Marx and Freud) has surfaced, seeming to require of all those who call themselves 'modern' not just the *revision* of theology but its *abandonment.* And with that 'secular' possibility has appeared the relegation of theology, of theological ethics – and finally of ethics itself as a discipline! – to the status of at best a subjective 'preference' and at worst a subjective neurosis and/or ideology. Theology and theological ethics thus have had plenty to do defending themselves against this implied or open secularistic onslaught along the entire length of this frontier. Perhaps their only comfort has been that metaphysics and systematic ethics, certainly also any form of prescriptive ethics, have also (to their own surprise) found themselves a bit later similarly assaulted.

Neither the sciences nor medicine have experienced this frontier – in fact, having come from the Enlightenment, they could hardly have conceived of it! Except from strange 'outsiders' (Nietzsche, Kierkegaard, Dostoiyevski and so on) and from waning traditional forces, representing premodern religious and ethical dogmas, no essential critique of natural science, of scientific method, of medical science or technology appeared on the 19th- and early 20th-century intellectual scenes, and certainly not in academia. Thus, again as writings of theological ethicists and ethicists in bioethics illustrate, because of this their lonely and dangerous 'March' (a frontier under continual attack), theology and ethics experienced a somewhat one-sided set of relations to biology, genetics, psychotherapy, and medicine. They were deeply conscious of the plurality of viewpoints, of methods, of criteria and of conclusions that characterize all modern theology and ethics. They, therefore, felt an inner uncertainty that was complemented externally by the vast indifference felt in much of the professional and academic communities to their contributions – if they could ever decide what that contribution was to be! In other words, they had busily to get their own disarrayed house in order before they could speak: get their sources, methods, and categories clear and precise. But even after they had done that, in fact the better they did it, the more their results seemed – over against the massive agreement, objectivity and relative certainty of scientific theory and praxis – to reflect a set of *special* viewpoints and so, in a universal scientific culture, to be so infinitely *subjective* that these results lacked credibility, persuasive power, and *a fortiori* any prescriptive authority.

I have brought us up and down the length of this still active 'frontier' because I think the character of this frontier explains a lot about the uncertain and often unbalanced character of contemporary bioethics discussion; and because this is still an important, indeed basic locus for crucial theological and ethical work. After all, the intellectual culture of the West, at least its academic culture, still reflects Enlightenment and post-Enlightenment perspectives on all these issues, especially on the unambiguous status of science and its praxis. To get clear on the cognitive status of theology and of ethical judgments; to reduce the confusion consequent on religious and ethical pluralism; to interpret the authority and relevance of religious-ethical judgments in the face of sociological and psychological interpretations of religion, and of the issue of historical and cultural relativity – all this is *fundamental* to this dialogue. Thus, new metaphysical and epistemological interpretations, new non-nihilistic ways of interpreting sociological and historical relativism, new forms of theological anthropology to encompass modern psychological and sociological theories of religion, of conscience and of moral obligation – are all very important conquests within contemporary theology for its dialogue with the human sciences and medical praxis. There are new and significant answers to old and well-soldiered frontiers; and they represent the current and necessary life-blood of relevant theology. It seems to me, however, that activity is now beginning (or has been begun) on quite a *new* frontier, and that this will (1) shift and redress noticeably the balance between theological ethics on the one hand and the sciences on the other, and (2) it will force a number of new elements on both sides into this refashioned dialogue.

## 'TIME OF TROUBLES'

Let me begin with the most important new frontier, that deep and vast cultural shift which, I believe, is fundamental for its two flanks which I shall mention and that, if I am right, will force revisions of one sort or another for all of our disciplines. This 'frontier' I shall call (after Arnold

Toynbee) the 'Time of Troubles' of the modern culture we have just described; it means, I think, almost certainly the shrinkage and gradual transformation of the culture and the self-understanding of the modern West. Whether it means a 'decline' in the stronger sense, I have no idea. It is impossible to argue this thesis effectively in a short paper on something else.[1]

Suffice it here to point to the following items: (1) the vast shrinkage of *power* of the West after 400 years of unchallenged dominance, since in 1945 its empires were disbanded; and corresponding to this loss the appearance since of a whole array of non-Western powers and blocs representing quite different cultural, political, and ideological traditions; those that are of roughly equal power with the West are explicitly anti-Western. (2) The relative loss of the *luring* or *grasping* power of Western (Enlightenment and post-Enlightenment) political, moral, social, and religious ideals, customs, standards, and spiritual attitudes in the world as a whole since roughly 1945, and the corresponding luring and grasping power of those of other cultures (not merely of Marxism but also, e.g., of Eastern religions). (3) The undeniable fact (cf., [3]) that the creative centers of Western 'advance' and domination: science, technology, and industrialism have become precisely those aspects of its life that have generated and are generating its most serious, even lethal problems (e.g., atomic weapons, dehumanization, population growth, reduction of supplies, and ecology). These problems, arising out of the creative power of Western culture, present a potential menace to that culture's life utterly undreamed of before and in stark contrast to the optimism on which the culture had previously lived. This represents, so to speak, a radical change of 'futures', an important point for a culture whose confidence lay precisely in the future. Toynbee defines a 'Time of Troubles' as the period when the creative answers of a culture to its initial problems turn and become destructive – and it can find no *new* answers. (4) The erosion of the ideology ('Civil Religions' or the fundamental beliefs) of *both* sides of post-Enlightenment culture: the belief, characteristic of the liberal democratic West, in Progress, dependent on scientific, technological, and industrial development, on the one hand, and the Marxist vision characteristic of socialist countries on the other. In neither place do these older 'myths' have much persuasive power; elsewhere both appear as masks of political domination and economic exploitation. In both cultures many if not most of these crucial symbols (e.g., progress or the material dialectic) remain unheeded, ignored, or if possible, severely criticized.

Much is new here. What is especially new viz-a-viz the earlier frontier and so new to each of the two partners to the bioethics dialogue is a rather definite 'reversal of roles' that follows from this new situation. Before, as a result of the first frontier, the secular culture was overwhelmingly strong, confident, and irresistible – and religion and ethics unneeded, in theoretical disarray and internally confused. Now it is the secular culture that primarily finds itself threatened, by internal uncertainty and self-doubt and by forceful external criticism. And its need in turn for ethical guidance, for inner renewal and even for deeper, more fundamental spiritual grounding is beginning to be very much more apparent. In this way *this* frontier is a potential frontier for both theological ethics *and* the life and medical sciences.

In a 'Time of Trouble' – for example, in the late Hellenistic period – certain unexpected things happen to a culture; clearly many of these have already begun to happen in our post-Enlightenment culture. First of all, quite against the expectations of this secular culture, a wide variety of forms of religion have begun to reappear or to appear anew. This is not just that old traditions (Judaism and mainline Christianity) have refused to die. On the contrary, it is the *newer* forms that have been arising with power, and they have arisen on both the private, personal and the public levels. Fundamentalist Christianity, Judaism, and Islam, imported cults from other cultures, newly developed cults within our own society, and the 'ascent' of political, economic, and nationalistic ideologies to a 'religious' level, all of this has been happening during our century. Correspondingly, with the recent awareness of the potential menace of the culture's new and awesome scientific and technological power (in weapons, genetics, medical treatments) the consciousness of the need for ethical guidance has grown. It is beginning to be evident that a technological culture can destroy itself if its humanistic values and its ethical will are not finely tuned, and if its fundamental convictions are either shaky or warped. Science and technology, in such an age, *need* religion and ethics – and all of them are terrified at the many new religious forms and ethical standards appearing all around them.

Corresponding to this, and the clear results in part of the baleful results of scientific technology (e.g., E.P.A.) or of medicine (take the exceedingly low public rating of the medical profession and of hospitals compared to 75 years ago), is a growing and often angry skepticism about science, even about the 'scientific consciousness'.as a fundamental attitude toward truth and the world. This has not yet reached significant proportions in the culture as a whole, but it well might. It reminds one of the uncomfortable status of the Church from about 1400 on: still utterly needed, publicly admired, powerful, queen of the sciences – and yet in many ways harmful and corrupt, and beginning to be destructive and even scorned. The trust in 'science' and the scientist as objective and reliable, in scientific knowledge as an *exhaustive* knowledge, and the scientific enterprise as a purely beneficent social force creating good – all of this has drastically declined in my own lifetime. Much of this is overdone, much is silly; scientific knowledge and technical reliability are vast and admired and have been immensely creative in our common life. But *something* has happened here that is culturally important: not least when one hears public comment on our patterns of medical healing, of psychiatric cure, or on the spectre of genetic manipulation.

Ours is an *advanced* culture. Consequently it is experiencing its 'Time of Troubles' and with that all the harrowing uncertainties and anxieties that such a time of breakup and disarray inevitably brings forth. It is not surprising, therefore, that, as did the Hellenistic or the late medieval peoples, many in our age seek renewed religion, ethical guidance and renewal, and that many more tend to lose their faith in our established forms of cognitive inquiry and of social authority. Ours is, however, as those other two epochs were not, an *advanced scientific* and *technological* culture. That is, science and technology have become 'established' in our cultural life. By established here I mean three things, the same three connoted by the category when we say 'in 324 Christianity became established.' Science and technology are now essential to the contemporary life and well-being of the culture; they *must* be there or the culture will die. Agriculture, industry, hospitals, defense, and so on are now impossible without them. Thus whatever they cost, the bill will be footed; wealth and influence pour into their enterprises; and intellectual life is dominated by their standards and goals. When a religion or ideology is established, precisely all these characteristics become true as well.

Now, two results accrue relevant to our subject with this establishment of science and technology in our national life.

(1) The community that is established becomes now subject to new problems centering about its own integrity: actual corruption and dishonesty begin to appear (after all, it is now profitable to be in that community, and so all types sign up); unclarity about what science (or Christianity) in fact represents begins to characterize members, even important ones, of the movement; commitment to the ideals of science begins to be exceptional – and so on. Thus the long assumed conception of a 'superior, intellectual, and moral character' as *intrinsic* to the scientist qua sci-

entist is now dubious in the extreme – as the same optimistic assumption about the integrity and dedication of a Christian leader after Constantine became infinitely questionable. Needless to say, if this be so, the question of the *ground* for ethical judgments when scientific and technological capacities are now used – for example, in genetics or in medicine – becomes suddenly a much more serious and complex issue.

(2) With the establishment of science in a culture – and in fact in all of modern cultures – the appearance of what we may call *deviant forms*, or at least variant forms, of science is now to be expected. One thing is relatively certain in modern life: *every* form of contemporary culture will, if given the chance, become scientific, technological, and industrialized (see, e.g., the incredible speed with which Japan accomplished this after 1857). As a corollary, it is evident that the scientific, technological, and industrial process will in each case be shaped by the fundamental cultural *ethos* basic to the culture in question. Thus there appeared a *fascist* and a *Nazi* science, technology, and industrial process, a *Stalinist* one, and (as we discovered) a *Shinto* one. Each of these 'ideologies' promoted, shaped, and used for its own political and moral purposes the scientific and technological forces within their common cultural life. Had his ideological fervor continued on in China, Mao would have created in the end a *Maoist* form of science; if he lives long enough, Khomeini will in time develop a Shiite version. While each of these may at certain points resemble *our* ('normative') form of science and technology, nevertheless, in certain crucial respects each of them is very dissimilar to what we know as science and technology. In this sense they represent *deviant* forms of scientific and technological culture. Creation Science in this country is an early indication of what a 'home-grown' form of deviant science might look like, namely a deviant mode of science shaped by American ideological forces, namely literalistic Protestant fundamentalism with more than a dose of right-wing politics. Moreover, in each case it has been and is trained scientists: in Germany, Tokyo, Moscow, Peking and Tehran – and San Diego! – who necessarily fashion and continue to fashion these deviant forms, and it was and is persons trained in scientific technology who implement them.

What these experiences, almost unknown in the 18th and 19th centuries but repeated throughout the course of the 20th century, show is that an established science, like an established religion, can be taken over, reshaped, and redirected by powerful cultural and spiritual forces, perhaps those of political, economic, and nationalistic ideologies or even those of powerful traditional religions – but in each case they are spiritual forces of a 'religious' sort, forms of ultimate commitments with their own set of global convictions, of authoritative moral standards and binding moral obligations. This pattern also indicates the potential *vulnerability* of every aspect of the wider community, including its academic, scientific, and technical spheres (for these in Japan, Germany, Italy, and Russia were taken over almost without a whimper), for a spiritual takeover. This occurs, or can occur, when through loss of direction, uncertainty, and deep anxiety, a community finds itself deeply unsettled; and this occurs when a preceding ethos with its 'orderly world' becomes shaken and enfeebled (see the loss of our 'civil religion' of Progress mentioned earlier). In such situations any population, including its intelligentsia (as these cases show), finding itself spiritually empty, its world askew and its anxieties unbearable, becomes 'ripe' for a new spiritual center, a new spiritual identity binding the community together once more and empowering it for the future.

Although they have traditionally found such a notion incredible, the scientific and the academic communities themselves are thus deeply dependent on a frequently unconscious and unheeded spiritual ethos, a 'civil religion', some deeply intuited and validated system of symbols which organizes their wider spiritual horizon and provides them with assumed moral standards and guidance. For most of our current intellectual community this ethos was inherited – with some changes – from the Enlightenment. This 'religious substance' is now seriously weakened: much of its optimism is counterfactual to the experience of the 20th century, and so it lacks credibility. At the present, much the way traditional religions were once vulnerable in relation to the new forces of the Enlightenment, this modern ethos finds itself vulnerable in the face of our current 'Time of Troubles'. As the 20th century has shown, this creates a dangerous situation: a culture whose spiritual ethos has dissipated is emptied of confidence, guidance, and commitment. But this becomes unbearable in difficult times – and thus is it vulnerable to new and powerful, and frequently demonic, forms of faith. In and of themselves, science and technology have no guards against this process. And in the hands of some new faith, the power of modern culture can destroy itself. It is mainly because of this suddenly relevant new possibility – that science and technology can be guided by deviant religious and moral frameworks – that the *balance* between theological and secular ethics on the one hand, scientific and technological expertise on the other, has been partly redressed. In our present *both* sides are of vast importance if scientific knowledge and technical power are to be creative and not destructive.

The new frontier we have labelled our 'Time of Troubles' opens up, therefore, immense new tasks both for theology and theological ethics and for the biological and medical sciences. For theology it means, while continuing to understand itself in the light of the present culture (frontier #1), it must also provide new bases for courage, serenity, hope, and normative obligation that can help preserve the best in the cultural inheritance, along with a note of judgment and warning that can help prevent the worst – in itself and in the culture as a whole. For the sciences it means a new and critical look at their own points of corruption and deviance; a new and humble self-understanding of their dependence on other aspects of culture (legal, moral, and religious) if they are to be whole; and a new sense of the ambiguous social consequences of what they do.

## A NEW DEPENDENCE

Granting, then, the fundamental new frontier of the 'Time of Troubles' and its general effects of redressing the balance and the dialogue represented by bioethics, in what more concrete ways in this new situation can there be 'theological implications' for this dialogue, ways that theology and theological ethics can in this new situation contribute to bioethics? The key point of this new situation is, as we have shown, the new *interdependence* in an advanced scientific culture of the religious and scientific, of the theological/ethical dimension of our existence with the cognitive and the technical. Neither one can now function creatively without the other's participation. Or, put better, they will unite in some way or another: as Marxist ideology and science unite in Moscow. Neither one, we now know, will be able in history to 'go it alone' or to be self-sufficient. The point to be striven for, therefore, is their creative union, and that the dialogue can help to bring about.

The first point where the new dependence on theology of scientific inquiry and technology has recently manifested itself is with regard to what we can call 'general attitudes'. By general attitudes I mean *how* we regard persons and nature, what fundamental attitudes we take up towards them. Heretofore our popular academic piety has considered such attitudes as, like religious doctrines, 'subjective preferences', part of the 'belief systems' that characterize individuals, to which (however silly they be) they have a right, but which have little or no bearing on the common, objective reality with which the community, through its science, technology, and industry, has to deal. Obviously, this view that fundamental attitudes are merely subjective is quite wrong, a strange 'myth' bred out of Enlightenment and post-Enlightenment biases.

First of all, we know that attitudes toward persons and nature are also social and historical phenomena, not individual 'preferences'; they vary with cultures and so are held in common by members of a given culture. Any good sociologist can tell us this. Secondly, the 20th century (our new frontier) has shown us over and over that these fundamental attitudes are desperately important: certain persons (and so in principle everyone) *can* be seen as objects, and then a scientific and technological culture (e.g., Germany) can run amok in dealing with these persons. Here fundamental attitudes, not scientific knowledge or technical advance, make all the difference. And these attitudes, if they are to help, cannot be based on our ordinary, common-sense ways of assessing people by their obvious worth or charm, their usefulness to us, their merit according to the standards of the community. For masses of humans flunk these tests; and it was precisely because they were regarded as 'useless', 'alien from the community', of only 'negative value', that whole populations were dealt with as unwanted objects – and still are. Only some religious or semi-religious valuation of the person as *intrinsically* of worth, as *inalienably* a subject and not an object; as an 'end in him/herself' (but how Kant could *establish* that rationally, is an interesting question) can guard us here from great sin. The symbol of the 'image of God' in *each* human, however apparently worthless to us or even clearly guilty according to our laws and standards, best expresses (for me) this intrinsic, elusive and yet absolutely crucial integrity of the human. In any case, here the career of bioethics is deeply dependent on a moral and 'religious' ethos and so on its theological and ethical partner. Such fundamental attitudes don't tell us *what to do*; but they open us to the possibility of uncovering what might be the moral alternative in a given situation; they enable us to deal responsibly with concrete questions of genetic manipulation, of euthanasia and so on.

The question of the attitude toward nature in the West is a much trickier issue, as innumerable commentators have noted. Again, it is absurd to call such attitudes forms of 'individual belief', matters of personal opinion or preference – as we generally label most 'religious attitudes'. A given attitude or stance towards nature is a common, quite objective aspect of modern culture, visibly different from the attitudes of other cultures past and present. And as with attitudes toward persons in general, we have recently – with the appearance of the crisis of the environment – discovered to our surprise how important such attitudes are and how wrong, even destructive, our traditional attitudes have been. Here, moreover, even the best elements of our religious and humanistic traditions have been at fault. While the biblical heritage grounded the value of the person, its mainline message was to subordinate nature to the dominance and oversight of men and women. This did no great harm until these humans, through the development of modern science and technology, amassed power enough effectively to implement this dominance. And then this traditional approach joined hands with a non-biblical humanism, developed out of the implications of the new science, to establish the modern attitude in the West toward nature. Here nature represents an objective, necessitated, mechanical system, a system of 'objects, with neither inwardness nor integrity' ('vacuous', as Whitehead said), and thus does nature become a fit object for human domination and use. Since nature has no inherent purposes, said John Dewey, it is appropriate that it become of worth through subordination to *our* purposes. In this way humans bring purposes into nature by using her. The ultimate results of these attitudes are, as we now know, lethal to nature and so in the end to ourselves. In order for science, technology, and industry to be creative and not destructive, radical changes here are necessary – and changes in our traditional religious as well as in our traditional secular attitudes toward nature.

What is doubly tricky here is that changes *from* an anthropomorphic attitude toward nature *towards* a 'supra-anthropomorphic attitude' bear with themselves, as many have noted, their own perils: in order to free nature from subordination to our own frequently evil purposes, we do not wish to subvert the high evaluation that traditional religion has given to the human. At this point it seems reasonable to maintain that a positive evalution of *both* the human and the natural world must perforce be an evaluation with a transcendent religious base, a base beyond both nature and human kind because creative of both. On a humanistic basis, human being is set over against blind nature; the former alone has intrinsic value, for here the human is the only valuer, and nature has worth only as it serves men and women. On such a view, in order to protect the value of the human over against nature, the value of nature in itself must be reduced. In a religious framework more sensitive to this issue, on the other hand, where the locus of value and of valuation lie outside as well as inside the human, the cosmos can bear its own intrinsic value along with that of human kind. In our tradition, this is expressed well in many psalms which portray nature as also 'witnessing to the glory of God', that is to say, being in its own way an 'image of God'. Since the seat of value and valuation lies *outside* of both and is creative and supportive of both, each can possess its own integrity, value, and mode of self-fulfillment – a far better base for a creative attitude toward nature than the objective, mechanical, and vacuous nature of modern scientific and technological culture. It is by no means irrelevant that other religious traditions than our Western theistic ones have had their own, in many ways far more impressive and effective, mode of grounding a positive, creative, and participatory attitude towards nature. In any case, clearly as a result of this new frontier, a supremely important issue has appeared, namely, that of our fundamental attitudes toward persons and nature. Moreover, it is an issue that predominates on the 'theological' side of the dialogue. And finally, it is evident that the way this issue is resolved will be immensely important for any future interface between religion, science, and technological implementation in our day.

The second large area where the new dependence of scientific inquiry and technical implementation on the general area represented by theology, theological ethics, and ethics has manifested itself is in a new awareness of the importance of the *doer*, that is, of the character, attitudes, beliefs, and purposes of those who know and those who wield the power generated by knowledge. Francis Bacon was right; knowledge of causes leads to power over those causes. This was taken by him and by subsequent generations as not only true but hopeful, confidence-building. When we know how nature (or society or persons) works, we can control it; then we can enact *our* human purposes – and by effecting our purposes, we can thereby increase our common well-being. Knowledge leads to human power, and, just as naturally, human power leads to greater well-being. From Bacon through Dewey to the present, this confidence in the benevolence and wisdom of the trained doer – if he or she be trained by scientific method, what Dewey liked to call 'informed intelligence' – has been the largest basis of our culture's confidence in progress. More especially, it has provided the major portion of the scientific community with *its* 'faith' in our culture's future, in the beneficence of science as a human enterprise, and in the human worthwhileness of what they were severally about. Innumerable quotations from Julian Huxley, Theodosius Dobzhansky, George Gaylord Simpson, Jacob Bronowski, Harlow Shapley, Glenn Seeburg, and countless others will verify the importance alike for scientific inquiry and for scientists of this 'faith', a faith not only in the cognitive effectiveness of scientific inquiry itself but in the ethical and moral effectiveness of *being a scientist*. 'Being a scientist' has represented for traditional modern culture a kind of spiritual metanoia almost akin to the transformation of self envisioned for the 'true philosopher' in much of Greek thought. The objective of the scientist would result in a rare and enviable 'wisdom' in moral matters, and his or her freedom from bias would issue in an utter reliability on the part of the trained inquirer and implementer to use that new power 'for the well being of human kind'.

If, as has happened more and more, the baneful consequences of much of the recent use of our new technology and medical power are pointed out, the answer has been (and still is, as I found at Iowa State two days ago): 'Oh, that is not science but the political use of scientific knowledge, the way the public and the politicians choose to use what science has discovered. Scientific man has progressed, political man lags behind – you can't blame science for that.' True – you can't *blame* science for that; but correspondingly, science cannot abstract itself from the problem of the implementation of its knowledge and locate that problem somewhere else in some other aspect of culture as if *others* were doing it. It is, after all, scientists and technologists, not politicians or lawyers, in industry and in the Pentagon who directly implement this knowledge; and it is the scientific community as a whole, as an integral and essential part of any modern civilization, that develops and applies this knowledge so important to the entire community. To divide up a cultural whole, or its individuals, into separate segments (political man, economic man, scientific man, technological man) and to blame one part and leave the other spotless is to think in vast mythical abstractions. In this way medieval theologians, had they so wished, could have blamed practical churchmen, secular rulers, and a fanatical public for religious persecution.

The establishment of science in modern culture has resulted in the support, funding, and encouragement of science by the culture as a whole – which all of science welcomes and fosters. But it also results in the *use* of science, its knowledge, and the power both bring for the uses of the culture – and inevitably those who implement that use are members of the scientific community. A special community cannot become an essential aspect – the intellectual center – for the life of an entire culture and so enjoy generous funding and prestige, and yet claim to be 'outside' the culture when what it knows and does is used. This is clear enough. Science and technology have been immensely creative in our common life, in all its many dimensions. But in its development in modern culture science has not so much resolved, dissipated or transcended the culture's moral dilemmas (as was surely hoped) as it has exacerbated many of them by providing groups in the culture with a gigantic increment of power. The power to control has become as well the power to destroy.

Again as a consequence, our concentration is turned to the *doer*: to the question of the character of the human who wields the power as well as to the question of the amount of knowledge or information that human possesses. Correspondingly, no longer are we able to believe that knowledge, information, and know-how by themselves represent the optimum needs of a modern society and so the main thrust of its creative education. Just as important, if not more important, are wisdom, empathy, self-criticism, responsibility, and obligation to the common good – and the possibility even of self-sacrifice – not to mention a legal and political structure directed toward freedom, justice, and equality. For it is clear that without these 'moral' attributes of a society, a scientific and technological culture can easily destroy itself. Indeed it would be ironical if technological intelligence, once thought the key to our survival, revealed itself when socially embodied to be the 'fatal flaw' presenting continuing human adaptability. One can only wonder that in such a social and historical situation – so different in its self-awareness from the confident pre-World War I world – scientific and technological education ('how-to' education) should in college after college be crowding out education in the humanities: 'We don't have time for that in a scientific culture'. One might say we will have no time at all unless there is a concentration of thought and of will on these issues uniting about the human 'doer', the subject, her or his attitudes, character, aims, norms, obligations, and convictions, issues that are moral and religious in the broadest sense.

As these remarks – by means new or even idiosyncratic – make clear, the 'Time of Troubles' that has dominated the 20th century have made us all much more wary and skeptical about the 'subject', or at least about the civilized, rational, scientific, and trained human subject. After all, the 20th century's largest and most lethal problems have developed *out of* modern civilization, not in spite of it; and they have arisen especially out of its most creative elements: its science, technology and its industrial organization. It is not the Third World that has oppressed us, ignited our conflicts or that threatens our common future – though, as we realize, had they preponderant knowledge, power, and 'civilization', they too might well do so!

This wariness about the subject, even the civilized, rational, and scientific subject, has its roots in 20th-century experience, not in dogma – for there was no such dogma in the post-Enlightenment world. It is based on the experience of what can happen to 'advanced' cultures: Germany was the most advanced scientific and technological culture, and it became demonic, captured by a primitive ideology, with very little resistance from the advanced circles and institutions of that culture. This wariness has also been based on the awareness that 'objective' boards or panels of trained minds can well enact genetic or medical policies favorable to their own race, class, and ideological bias. Our natural, common, conventional, even 'liberal' moral judgments have become suspect as their partiality has become evident. This is, let us note, *not* a technical problem or even of one scientific understanding; nor will a competent psychoanalyst solve it for us. It represents the deepest problem of ethics: the problem at once of unbiased ethical *judgment* and of creative ethical *enactment*, a problem both of mind and of will, a problem, therefore, deeply of 'character'. For what has also become evident is that an objective mind is more the result of a cleansed will, a will free of the constraints of self-concern, than the reverse is the case (i.e., that the health of the will depends on the intellectual training of the mind). It is obvious that here we encounter one of the deepest concerns of all profound religion.

That the depth of this problem and its utter relevance to the dilemmas of a scientific culture have begun to surface in our consciousness does not indicate at all that a religious conversion has taken place – though it might abet it. Nor does it mean that the serious moral problems of our common life will as a consequence be resolved; only technical and minor scientific problems are resolved simply because we are aware of them and seek deliberately to solve them. As Socrates reiterated, virtue can hardly be taught; and as Kierkegaard echoed, becoming a self, that is, an ethical self, cannot be communicated from one person to another through objective knowledge or information. Still, careful reflection on the ethical issues raised by science, on the bases and consequences of various alternative courses of action, and on the frequently distorted attitudes and biases characteristic of those (i.e., of *us*) who make the relevant judgments and decisions, all of this is extremely important. And it is even important to share reflection on our most fundamental attitudes: towards nature, towards persons, towards reality itself, and towards ourselves – for it is *here*, in these areas, that the frameworks for moral judgments and moral action are shaped and reshaped. Surely such rational dialogue, inadequate as it is, is better than assuming, against every bit of evidence, that American scientists and technologists (any more than American preachers!) have, because of their training, a secure grasp of values and so will judge each issue in bioethics objectively and morally. The real implications of the new frontier – the 'Time of Troubles' for a scientific and technological culture – are very great indeed for the dialogue that is represented by bioethics.

## RELIGIOUS DIALOGUE

Until my pen ran away with me I had planned to outline one other 'new frontier' for theology which has, so I believe, implications for bioethics. Because this is already more than long enough, I will only mention it briefly. This new frontier is represented by the quite new relation which now exists among or between the religions of the world. Not only are they no longer isolated from one another; much more important, no longer are Western religions considered, explicitly by themselves and frequently implicitly by others, to be superior. There now exists a kind of strange, paradoxical, and yet largely uninterpreted sense of *rough equality* among the religions, at least in the sense that each recognizes, or has to, the presence of truth and of grace in the others.

The main cause of this new sense of equality is, I think, not dogmatic but historical, namely, the impingement of our first *new* frontier, the new travail of Western culture on every level, its pervasive 'Time of Troubles', its inner uncertainties and loss of confidence in itself. Since the Enlightenment, Western religions (especially Christianity) had assumed their own superiority, or, as they said in the 19th century, their own 'absoluteness'. And they had assumed this not only on internal dogmatic grounds, but probably more importantly as the religions of a clearly superior civilization, intellectually, morally, politically, and socially. Correspondingly, they regarded other religions, while surely quaint and fascinating in nature, as 'relatively primitive', representing, as was taken for granted by all, relatively backward civilizations. All of this is now gone for the consciousness of Western religious traditions – not least because of the many conversions of Western sons and daughters to Eastern religious practice and to Eastern religious convictions about reality. The tide of spiritual influence, and so spiritual power, is now flowing in the other direction. As a consequence, theology is faced with an entirely new set of issues – as it had been by the shaking up itself of Western culture. How are Christianity or Judaism to understand themselves, their respective revelations, their law or their gospel, if other religions are *also* true and *also* saving? These issues represent – to change the metaphor – an as yet uncharted theological sea – and only a few exploratory trips have so far been made.

My present point, however, is that this new frontier, faced first (I think) by theology (not because it wanted to) has, or will have, important implications for bioethics. The absoluteness of our religion is not the only thing the West is now not as sure about as it once was. It is not so sure as it was of the exhaustive or ultimate status of its medical science, of its forms of therapy and healing, or, as we noted, of its fundamental views (largely scientific) of the organic and inorganic realities (including our own bodies) that make up the natural world. As the clear superiority of Christian doctrines has on many scores been seen to be dubious in the face of the relative profundity and attraction of other religious views, so the innate 'Cartesianism' (to use a useful label; poor Descartes expressed and did not invent Western dualism) of our culture: a mechanical body and the rational consciousness strangely joined together, has revealed itself to be woefully inadequate in innumerable ways and for innumerable purposes. Certainly *healing* cannot be understood in its terms anymore than can meditation, self-identity, or self-centering. And it is dubious that either body or self are illumined by the mechanical model with which we think about patients and the intellectual model with which we think about doctors. Here the surprisingly nondualistic cultures of Asia (we always thought *they* were the dualists!) have a lot to teach us, about the interpenetration of body and spirit together, about the energy animating both, about the unified healing of bodily and spiritual illnesses, in other words about a deeper understanding of bodily health, of ordered selfhood, and of centered identity.

So far our academic and scientific cultures have remained on the whole intellectually confined within traditional Western viewpoints – as once Western religious missionaries did. Theology can no longer do so even if it wished; the penetration of our world by Eastern religions is far too unavoidable. One theological implication for bioethics of theology's new frontier may well be this mediatorial role of bringing to that dialogue – and to the attention of the academic community generally – perspectives quite different both from those of our Christian tradition *and* from those of our scientific *and* medical traditions. And least expected of all, I suspect that what they will mediate that will be most important to that dialogue is *not* so much new insight into the Divine or Spirit, as on the contrary new awareness into the reality, unity, and effectiveness of our bodily, organic, and spiritual life on the one hand and new ways of feeling our participation as human beings in the whole systematic order of nature on the other – issues very relevant to the future health in turn of the biological and the life sciences.

*Divinity School,*
*University of Chicago,*
*Chicago, Illinois, U.S.A.*

## NOTE

[1] Cf., for a fuller treatment of this theme: [1]; [2], Ch. 1; and [4], Vol. 1, p. 53.

## BIBLIOGRAPHY

[1] Gilkey, L.: 1981, 'The New Watershed in Theology', *Soundings* **LXIV** (Summer), 118–131.
[2] Gilkey, L.: 1981, *Society and the Sacred*, Crossroads Press, New York.
[3] Heilbroner, R. E.: 1974, *An Inquiry into the Human Prospect*, W. W. Norton, New York.
[4] Toynbee, A. J.: 1945, *A Study of History*, Oxford Press, London.

DOUGLAS STURM

# CONTEXTUALITY AND CONVENANT: THE PERTINENCE OF SOCIAL THEORY AND THEOLOGY TO BIOETHICS

Reason which is methodic is content to limit itself within the bound of a successful method. It works in the secure daylight of traditional practical activity. It is the discipline of shrewdness. Reason which is speculative questions the methods, refusing to let them rest. The passionate demand for freedom of thought is a tribute to the deep connection of speculative Reason with religious intuition ([53], pp. 65–66).

In the exercise of any discipline, methodological questions may be ignored. But they do not disappear. The practitioners of the discipline merely assume the questions have been answered, permitting them to get on with the immediate tasks at hand. Yet it is wise, now and then, to pose methodological questions deliberately with the prospect that new light will be shed on the discipline and its shape and direction modified or transformed in some creative and fruitful manner.

In this essay, I pose three methodological questions – questions of scope (what are the boundaries of the discipline?), focus (what should be the point of concentration of the discipline?), and grounding (what is the foundation of the discipline?). All three questions bear on the general issue of the meaning and character of bioethics. With respect to each question I shall present a claim. In its scope, bioethics should become less strictly case-oriented and more contextually oriented. In its focus, bioethics should be linked with social theory – social interpretation and social ethics. In its grounding, bioethics should be allied with theology.

My intention is not to prove these claims, whatever proof might mean, but to make them intelligible. Overall, I shall suggest the conjunction of a contextual bioethics, a critical social theory, and a process theology. From critical social theory, the principle of structuration provides a means of understanding the intrinsic connection between individuality and sociality. Given that principle, bioethics must be a form of social ethics. From process theology, the reformed subjectivist principle provides a means of understanding the intrinsic connection between God and the world. Given that principle, bioethics must become a form of religious ethics.

There are two reasons I am brought to present these claims. One of them is practical. The second is theoretical.



*E. E. Shelp (ed.), Theology and Bioethics,* 135–161.

On the practical side, the professions are under siege. Because of their special role within the social order, the professions have long claimed and been granted autonomy. The alleged aim of the professions, each with its own particular focus, is to serve the well-being of the human community. Under conditions of modern society, however, suspicions have been roused. It is not altogether clear the professions can be trusted. The crisis of confidence that is expressed generally in attitudes of skepticism about large organizations and about persons in high office is expressed more particularly in attitudes of doubt about the professions. Serious questions are raised about the motivations of professionals, about the actual beneficiaries of professional practice, and about the social and political implications of professional conduct. The overarching question is whether the professions, given their current form and character, genuinely serve the common good.

On the theoretical side, a distinction may be drawn between two principles of understanding, an analytic principle and a relational principle. Each promotes its own method of interpretation. The former is characteristic of the cosmology of scientific materialism. The latter is characteristic of the philosophy of organism. Analysis separates, distinguishes, isolates. An analytic principle presses one toward simplification and precision. It drives toward careful definition and the delineation of narrow boundaries. In its social model, it is atomistic and individualistic. A relational principle, on the other hand, leads one to discern connections, associations, relationships. A relational approach is less concerned with sharp definition than with potential relevance. It is open to new pursuits, new possibilities, and is tolerant of complexity. In its social model, it is holistic and communal. Each of these principles has merit, yet the employment of each also runs a risk. The merit of the analytic principle is precision; its risk is loss of realism. The merit of the relational principle is its attention to full-bodied context; its risk is vagueness and uncertainty. This essay is dedicated to the promotion of the relational principle as the more appropriate in our time, given trends in cosmological thought and social reality.

About a decade ago, Daniel Callahan, among the creative pioneers in modern development in bioethics, confessed to the fluidity of the boundaries of the field.

Bioethics is not yet a full discipline. Most of its practitioners have wandered into the field from somewhere else, more or less inventing it as they go. Its vague and problematic status in philosophy and theology is matched by its even more shaky standing in the life sciences. The lack of general acceptance, disciplinary standards, criteria of excellence and clear pedagogical and evaluative norms provides, however, some unparalleled opportunities. It is a discipline not yet burdened by encrusted traditions and domineering figures. Its saving grace is that it is not yet a genuine discipline as that concept is usually understood in the academic and scientific communities. One has always to explain oneself and that leaves room for creativity and constant re-definition; there are many advantages in being a moving target ([3], p. 68).

Bioethics is thus in search of an identity, although, in Callahan's judgment as expressed above, there is some virtue in that identity not being too carefully prescribed or rigidly defined. More recently, one of the practitioners of medical ethics (whose relation to bioethics is an unsettled question) asserted that medical ethics "has come of age once again" ([50] p. 3). Yet the place of bioethics (and medical ethics) in the organization of knowledge is a point of contention. Some claim it as a subfield of philosophical ethics, the application of general principles of rational morality to a particular subject matter [5]. Others would join it to the practice of medicine as an integral part of the healing professions [40]. Still others conceive it as a new multidisciplinary science grounded in biology and needed at the present time for the survival of the human species [41].

In this search for an identity for bioethics, I would, given the claims already indicated, offer three suggestions.

First, among its current practitioners, bioethics tends toward one or the other of two approaches. For ease of reference, I shall designate these approaches personalist and contextual. The personalist approach is the narrower and the contextual is the broader of the two. The virtue of the personalist approach is its seemingly immediate practicality. But it runs the risk of committing the 'Fallacy of Misplaced Concreteness' (A. N. Whitehead). It attends so closely to the interaction between physician and patient, it neglects the institutional and cultural context within which the interactive process makes sense and finds its meaning. What a contextual approach may lose in immediate practicality, it gains in historical and social realism.

Second, a contextual bioethics, given its character, must be allied with social theory. It must deal with the fundamental problem of social theory, the relationship between individuality and sociality. In doing so, it must look to and take account of investigations in the history, sociology, and politics of medicine for, even in one's individuality, one is a historical, social, and political being. Health and well-being, disease and illness are experienced by individuals but they are, at the same time, expressive of the relational context of an individual life. Critical social theory, moreover, is both interpretive and normative. In its concern with what is the case, it poses the question of what ought to be the case. Thus a contextual bioethics, in its alliance with social theory, must be a form of social ethics.

Third, a contextual bioethics must, at least implicitly, enter the 'theological circle' (Paul Tillich). To be sure, theology in the modern world is very much in flux: "that the contemporary theological scene has become chaotic is evident to everyone who attempts to work in theology. There appears to be no consensus on what the task of theology is or how theology is to be pursued" ([25], p. ix). The chaos of the contemporary theological scene, however, may be no more characteristic of theology than it is of philosophy or the social sciences. These disciplines, too, are in flux. Besides, uniformity and single-mindedness are no assurance of profundity or validity. Yet theology, in and through its many forms, brings to the fore the issue of the nature and destiny of human life. In theological terms, this is the issue of the relation between God and the world, for God is the objective ground of confidence in the ultimate significance of the world and, in particular, of human activity. This is the elemental issue that resides at the foundation of all human practice, including the practice of medicine and of the life sciences. In Max Horkheimer's phrase, "behind all authentic human activity stands theology" ([19], p. 60). By theology, Horkheimer means the consciousness that the world as it exists is not the last word. Theology is expressive of the hope for well-being, the yearning for freedom, the aspiration for justice. Without that kind of consciousness as its grounding, it is difficult to conceive of a contextual bioethics.

## BIOETHICS: PERSONALIST AND CONTEXTUAL

The question of the scope of bioethics may be posed various ways. What is bioethics for? What are its purposes? To whom is bioethics addressed? For whom are its principles, policies, and prescriptions meant? What kinds of issues are most central to its task? Is there a difference between medical ethics and bioethics? What is the source from which the principles and insights of bioethics derive? The question of scope is a function of one's perception of social reality and the level at which one interprets the meaning of the biomedical field.

In a remark made initially for another purpose, Paul U. Unschuld suggests a distinction that bears on this question: "Standards governing practice of private and public health care have arisen as a result of continued interaction on at least two distinct, but not exclusive, levels of cultural and social activity. These levels include (1) the perception and propagation of *world views* by groups in society intending to maintain or establish social order, and (2) the *interactions* between various medical practitioners and their clients" ([49], p. 901, italics added).

Both levels are present in the current literature of bioethics. Each promotes its own response to the question of the scope of the field. The former tends toward a more macrocosmic and the latter toward a more microcosmic form of bioethics. A macrocosmic bioethics places issues of private and public health care within a broad theoretical or practical (political and historical) context within which the practice of a physician or scientist is comprehended as manifesting meanings that transcend the practice itself. A microcosmic bioethics is more limited in its concern with individual cases and the relationship between particular persons, professionals and clients or subjects. The latter appears to be the more concrete of the two levels, but in reality may be the more abstract and may, in certain circumstances, be deceptive.

In Daniel Callahan's sketch of 'bioethics as a discipline' of about a decade ago, there are evidences of a more microcosmic approach. Among the central tasks of bioethics, Callahan specifies "that of helping scientists and physicians to make the right decisions; and that requires willingness to accept the realities of most medical and scientific life, that is, that at some discrete point in time all the talk has to end and a choice must be made, a choice which had best be right rather than wrong" ([3], p. 68). What Callahan means by choice is particularized by the criterion he presents to test any proposed method in bioethics: "that it enables those who employ it to reach reasonably specific, clear decisions in those instances which require them – in the case of what is to be done about Mrs. Jones by four o'clock tomorrow afternoon, after which she will either live or die depending upon the decision made" ([3], p. 72). More recently, Callahan has urged that the discipline adopt a somewhat broader scope in order to deal with emergent conflicts and tensions between its traditional patient-centered concern and legitimate interests of the public [4].

Yet even Callahan's revised version of the scope of bioethics stands in sharp contrast with a proposal of Van Rensselaer Potter. Potter, a cancer research scientist, claims to have coined the term 'bioethics' to designate a new discipline made necessary by the ecological crisis of our times ([42], p. 201). Bioethics, in Potter's conception, is a multidisciplinary science formed to concern itself with the survival of the ecosystem. Biology is at its core, but a biology in collaboration with humanities and social sciences. We shall need such a corps of persons "who respect the fragile web of life and who can broaden their knowledge to include the nature of man and his relation to the biological and physical worlds... who can tell us what we can and must do to survive and what we cannot and must not do if we hope to maintain and improve the quality of life during the next three decades" ([41], 2). Out of such a discipline would emerge prescriptions for individuals (see Potter's "Bioethical Creed for Individuals", [41], p. 196; [42], p. 204) and for public policy (see Potter's proposal for a "Council on the Future" as a fourth branch of government, [41], pp. 75–82).

At first blush, it appears that Callahan and Potter, despite the common rubric, bioethics, are concerned with wholly different disciplines. Yet both are concerned with the life sciences and both cite the issue of death and survival as paradigmatic. The difference is, in part, a matter of focus. But it is also a matter of perspective on the character of social and historical reality.

A comparison of the approaches of Paul Ramsey and Ivan Illich to the ethics of medical practice illustrates the point. Ramsey's is a more microcosmic approach. In his understanding, bioethics is casuistic. It should attend to particular cases of medical care. There is, however, nothing unique about bioethics: "The moral requirements governing the relations of physician to patient and researcher to subjects are only a special case of the moral requirements governing any relations between man and man" ([43], p. xii). Bioethics is addressed to professionals and patients simply as persons who are obligated to conduct themselves as persons in their mutual relations. To Ramsey, as a Christian ethicist, this means to honor the covenant that binds them to each other.

> We are born within covenants of life with life. By nature, choice, or need we live with our fellowmen in roles or relations. Therefore we must ask, what is the meaning of the *faithfulness* of one human being to another in every one of these relations? This is the ethical question ([43], p. xiii).

However compelling Ramsey's Christian personalism is in its application to particular cases of professional interaction with subjects, it is, from the perspective presented in Ivan Illich's controversial text, *Medical Nemesis*, woefully deficient. To Illich, the most crucial problems of health and illness in contemporary industrial civilization are systemic, not personal. Physicians and researchers, given their inherited professional mind-set, may be faithful to their clients but at the same time exacerbate patterns of ill-health and disease which, presumably, they are committed to combat. Illich's concern is iatrogenesis, illness induced by the medical profession itself, which, he argues, assumes three forms.

Iatrogenesis is *clinical* when pain, sickness, and death result from medical care; it is *social* when health policies reinforce an industrial organization that generates ill-health; it is *cultural* and symbolic when medically sponsored behavior and delusions restrict the vital autonomy of people by undermining their competence in growing up, caring for each other, and aging, or when medical intervention cripples personal responses ([21], p. 270–271, italics added).

Even Illich's critics at times accept his basic thesis, that "the medical establishment has become a major threat to health" ([20], pp. 1, 115), without adopting the full measure of his radical anti-technological and anti-industrial populism.

Other perspectives furnish illuminating angles of vision on the question of the scope and character of bioethics. Richard Zaner, from the perspective of the phenomenological tradition, insists that bioethical issues are 'context-specific'. They cannot be settled by the application of traditional principles. They concern the totality of life of both patient and practitioner. Zaner derives this conclusion from an exploration of the phenomenon of embodiment. One's body is one's place of being and center of action; it is expressive of who one is and what one intends. Medical practice and research are thus interventions into one's self which is why bioethical issues have such a special character. Each self is a gathering together of a unique set of values, commitments, and concerns. Each patient and each practitioner bring to their interaction some understanding of the world and some vision of the good. In effect, this means to Zaner each issue must be settled in its own way ([54], [55]). As observed in a commentary on Zaner's work:

Each choice that is made must attend to the values already found in the situation, values brought to the situation by the life-history of the patient and intertwined with the expressive character of the patient's body, and by the goals and purposes the doctor seeks to actualize in the practice of his or her profession. The values found in the situation as concretely lived must be recognized and described before discussion of ethical universals in abstract form can be at all useful or informative ([45], p. 6).

Zaner's position signals, inter alia, potential conflicts in world view between practitioner and patient which, in instances of racial, sexual, and class differences between the parties, may be vividly present.

More generally, Zaner's principle of embodiment may be translated, *per analogiam*, into the sphere of sociality. Michael Gordy, for instance, has asserted, "the majority of bioethical issues presuppose a social nexus and have to be discussed in terms of the real relations between people" ([18], p. 1605). Gordy contrasts two views of sociality, the contractual and the non-contractual or organic. Each casts its own light on the character of issues in bioethics and establishes a setting for their resolution. In the former case, medical care is a commodity to be bought and sold in the market and is susceptible to the laws of exchange. In the latter case, however, medical care is a particular set of relationships established by a community to attend to the quality of life and to promote the well-being of its members. In a complex way, views of sociality may, of course, be related to and expressive of historical circumstance. So, it can be argued, the contractual view conforms to the dynamics of early capitalism, whereas the non-contractual view is more appropriate to a post-industrial or post-capitalist civilization. If this argument has merit, then bioethics must extend its scope and sights to incorporate considerations of social and historical analyses; it must concern itself with changing views of society and with historical circumstance.

Such an extension would be in keeping with René Dubos' insistence that

> physicians must learn to work with engineers, architects, and general biologists, as well as with city planners, lawyers, and politicians responsible for the management of our social life. Only through such collaboration can they help society ward off, insofar as possible, dangers to physical and mental health inherent in all technological and social change, especially when these occur as rapidly as they do now. From urban renewal to safety measures in industry, from environmental pollution to the trial of new drugs and therapeutic procedures, the sociomedical problems are countless and require technical, legal, and ethical considerations ([6], p. 91).

Each of the two general forms of bioethics distinguished above – the microcosmic and the macrocosmic, the personalist and the contextual – is appealing. Each has its own integrity. Some bioethicists have urged that the discipline as a whole should incorporate both forms [1]. But that begs the question of how the two forms should be related and which should be paramount. The two forms exist in some tension with each other. On the one hand, particular cases present clinician and researcher with ethical dilemmas that are urgent; their resolution cannot wait upon broad historical change or new directions in public policy. The day-to-day necessities of medical practice seem to dictate giving priority to the personalist form of bioethics. On the other hand, to deal exclusively or primarily with individual cases as they emerge in clinic or research would mean ignoring the structural conditions that gave rise to the cases in the first instance. More pointedly, it would mean ignoring the full contextual meaning of the cases themselves. Moreover, it would mean ignoring a trend within the medical world. Recently, Renée Fox, in a searching interpretation of the historical development of medicine through three stages – archaic, modern, and post-modern – argued that the inner dynamics of the medical profession are moving toward a more contextual approach including but going beyond the societal toward the religious.

> Not only has the development of a more sociologically oriented theory of health and illness increased social activism in this domain, but also age-old philosophical and religious questions are revitalized by what is felt to be the mystery-laden relationship between what now seems to be alterable and what not. Who and what is man? What are the meaning and purpose of his existence? What is life, what is death, and wherein lies the essence of the distinction between them? Why do men fall ill and suffer and die? How should we understand these experiences and believe toward them? This sort of querying is characteristic of post-modern medicine and its practitioners, as well as of its patients. If one accepts the sociological view that preoccupation with such questions of meaning constitutes a religious act, then one might say that in this sense post-modern medicine is less secular and more sacred than modern medicine ([13], p. 525).

## SOCIAL THEORY: INDIVIDUALITY AND SOCIALITY

The question of the focus of bioethics is related to, yet distinguishable from, that of scope. The question of scope is a matter of the range of issues appropriate to the discipline. Some issues are seemingly not pertinent to the field, although one must be cautious in fixing boundaries too absolutely, for relevance has a seasonable quality about it. The question of focus, on the other hand, is a matter of concentration. It is a question of determining importance and centrality. It is concerned with defining priorities. The focus of a discipline may change over time without necessarily distorting its boundaries in any appreciable manner.

In a recent criticism of the prevailing concentration in bioethics, Roy Branson argues that the discipline should have a double focus [1]. The dominant tendency is to center on questions of direct professional interaction because these constitute the kind of question posed by medical practitioners. Such a focus, Branson asserts, is too narrow. It ignores crucial dilemmas of an institutional character and neglects issues of public policy. He cites, as examples, the financing of medical services, the development of a nuclear powered artificial heart, and the distribution of health care. He concludes that bioethics should be both individual *and* social in its focus (cf. [5], pp. 121–123).

The proposal that bioethics has two branches, that it deals with issues of both personal-professional interaction *and* social-institutional policy, is appealing. But it evades the question of the conjunctive, namely, of how individuality and sociality are related. It seems to assume that individuality and sociality, personal agency and social structure, are separate and distinct. But that assumption is currently under question in social theory. As a clue to the position I would propose, I take the following statement from Anthony Giddens: "Just as every sentence in English expresses within itself the total which is the 'language' as a whole, so every interaction bears the imprint of the global society" ([15], p. 22).

Giddens uses the term 'structuration' to indicate the connectedness of agency and structure [16]. Structuration is the process of presentation (or, better, representation) of structural forms in cases of interaction. Society, or social structure, is only through an abstraction independent of individuals acting. The actuality, the embodiment of social structure is in and through the agency of particular persons engaged in interchange. There is, to be sure, a relative disjunction between social order and individual agency which accounts, in part, for the dynamics of social conflict and the possibility of social change. But social structure and personal agency presuppose each other. To distinguish too radically between individual ethics and social ethics or between questions of personal-professional interaction and social-institutional policy is to falsify the congruity or, more accurately, the relative coincidence of sociality and individuality in their concreteness. The radical disjunction between sociality and individuality is an inheritance from the liberal tradition and its rejection of the relational principle.

In the biomedical field as in other professional fields each instance of interaction is part of a practice, and each practice is part of a social order and a social history. In more active language, each case participates in and re-presents a practice, and each practice participates in and re-presents a social process. A practice is a specialized set of interactions governed more or less consistently by certain conventions or rules whether or not those conventions and rules are deliberately acknowledged ([12], pp. 11–15). Yet a practice, while specialized, produces and re-produces the more encompassing social totality in which it is located. It is engaged in the making of social history.

From this perspective, precisely as bioethics attends to particular cases of professional or personal interaction, it must, *to encompass the full significance of those cases*, be a form of social ethics. As such, bioethics must incorporate more directly than it has in the past, the insights, theories, principles, and problems of the sociology and history of biomedical practice, yet it must do so critically since forms of sociology and history are conditioned by theoretical and practical assumptions and presuppositions. Bioethics, in other words, must attend to the synchronic and diachronic dimensions of professional interaction within biomedical practice and within the whole social process in which the practice participates. The distinction between synchronic and diachronic dimensions, it should be noted, is only analytic and, in any adequate interpretation, will not hold. The theory of structuration is an effort to overcome the false dualisms of both individuality/sociality and statics/dynamics.

The focus I would propose is exactly at the conjunction of agency and structure in part because of its central importance given the theory of structuration, in part because a double focus tends to misrepresent the character of action, and in part to demonstrate the full burden of responsibility of professional practice.

There is a wide range of literature from the sociology and history of medicine and the life sciences pertinent to this focus for bioethics. The following survey is presented only as suggestive of the possibilities.

DIMENSIONS OF A CONTEXTUAL BIOETHICS

| | synchronic (structural context) | diachronic (historical context) |
|---|---|---|
| internal (practice): | (1) Talcott Parsons<br>Eliot Freidson<br>John & Barbara Ehrenreich<br>David Mechanic | (3) John & Barbara Ehrenreich<br>Tom Levin<br>Renée Fox |
| external (social totality): | (2) Elliott Krause | (4) Henry Sigerist<br>René Dubos |

### *Sociological Interpretations of Biomedical Practice*

Among the more significant contributions to the sociology of medicine are those of Talcott Parsons and Eliot Freidson. From a sociological perspective, medical practice is a set of interacting roles fulfilling a specialized function within an encompassing social system. Among other things, it is an instrument of social control. From the standpoint of a contextual bioethics, it would be important to determine who is being controlled, how and for what purposes.

In Parsons' judgment, health and illness are not merely conditions of individuals. They are defined and institutionalized through the social structure, since whether one is officially designated healthy or ill bears on the performance of social tasks: "Health may be defined as the state of optimum capacity of an individual for the effective performance of roles and tasks for which he has been socialized. It is thus defined with reference to the individual's participation in the social system" ([39], p. 69).

This means physician and patient occupy special roles. That of physician is the more obvious. One of Parsons' unique contributions to the sociology of medicine is his construction of the 'sick role'. It is composed of four features. The sick are (1) exempt from normal social responsibilities, (2) understood to be subject to forces beyond personal control, hence exempt from responsibility for their condition, (3) obliged to consider their condition as undesirable, and (4) obliged to seek professional assistance and to cooperate in the process of recovery ([38], pp. 436–437; [39], p. 70). Where its exemptions make the sick role attractive as a form of social deviance, its obligations constitute a balancing factor. The sick role is a mechanism of social control intended to reintegrate those who are ill back into the normal operation of the social system. The meaning of the process is given a peculiar twist by Sigerist's quip: "To immunize colonial people against disease with the one hand and exploit them into starvation with the other is a grim joke" ([47], p. 236).

Eliot Freidson presents another sociological slant on medical practice. Given the Berger-Luckman thesis – 'reality is socially defined' – Freidson is particularly concerned with the 'social construction of illness' ([14], Part II). Medicine, like law and religion, is a profession authorized to distinguish normal and abnormal, proper and improper. An illness may have roots in biological causes, but what gets labelled illness depends upon social perceptions or professional judgment and is susceptible to placement under forms of institutional control. Increasingly, over the past several decades, the medical model has been employed to deal with perceived forms of deviancy: "In our day, what has been called crime, lunacy, degeneracy, sin, and even poverty in the past is now being called illness, and social policy has been moving toward adopting a perspective appropriate to the imputation of illness" ([14], p. 249). Some physicians assume the role of 'moral entrepreneurs' in extending or altering the definition of what is considered illness. Once lay persons enter the domain of illness, they are subject to professional management; they are forced to serve the social identity implied by the diagnosed illness. Freidson himself poses one of the central ethical questions provoked by discernment of the consequence of illness as a construct of social meaning: "How desirable is that consequence? Is it in the public interest for society to allow the profession the autonomy to define both the need and problem and to control their management?" ([14], p. 331, cf. Part IV).

Barbara and John Ehrenreich interpret Parsons and Freidson as both construing medical practice as a form of social control, but in two seemingly contradictory ways – disciplinary (Parsons) and cooptative

(Freidson). To Parsons, medical practice is exclusionary: it delimits those seeking the advantages of the social role and is oriented to return the ill as quickly as possible to normal occupations in the social order. To Freidson, medical practice is expansionist: it moves to extend the sick role to include ever more classes of people in order to manage their lives.

The Ehrenreichs resolve the seeming contradiction by suggesting the orientations pertain to different socio-economic classes. The cooptative orientation characterizes the relation of the medical system to the 'medical poor' – the non-white and the elderly. The disciplinary orientation typifies treatment of the 'medical middle class'. Moreover, the forms are promoted by different ideological groups: liberals, out of humanitarian impulse, are expansionist; conservatives, out of desire to curb costs, are exclusionary.

The Ehrenreichs, dissatisfied with both forms, propose a new beginning:

If the medical system is understood as something more than a system for distributing a 'commodity', if it is understood as a system of direct *social relationships*, then the question becomes: what kinds of social relationships do we want a medical system to foster?....The problems which our society relegates to the medical system – the care of the disabled and dependent, the management of reproduction, individual suffering, and death – are no less than some of the central problems which confront any human society. Medicine has allowed us to evade them too long ([9], p. 73).

David Mechanic's studies present several points where the organized structure of modern medical practice militates against direct social relationships. There is, for instance, a tendency, especially in a hospital setting, to neglect the life circumstances of patients and how that circumstance bears on the patient's condition even though

the physician must. . .be sensitive to patients as individuals within a family and community context, since life situations are a major source of ill health and may be a barrier to convalesence. If he treats the presentation of the patient solely at the manifest level, his treatment may be only symptomatic, and he frequently will fail to deal with the basic underlying difficulties that brought the patient to him ([32], p. 116, cf. pp. 59–67).

Such neglect leads to complaints that the medical profession is inhumane. But,

[T]he institutions and the personnel who carry out medical functions behave as they do not because they are more inhumane than others but rather because the pressures and constraints of work, the priorities health personnel have been taught, and the reward structures of which they are a part direct their attention to other goals and needs ([33], p. 1056).

The problem is a matter of structural design, not personal attitude. It is, in part, a function of compensation structure ([33], p. 1057) and, in part, an expression of the elitist organization ([33], p. 1057; [27], pp. 37–67) of medical practice.

*Political-Economic Context of Biomedical Practice*

The structural problems of health care are not limited to the internal organization of medical practice. It is virtually a truism, for example, that those living in poverty are more susceptible to disease and have less access to quality health care than those who are economically comfortable ([26], [30], [2], [29]). Environmental conditions, malnutrition, lack of income, cultural differences from medical personnel, geographic locations of hospitals, and priorities of the health system are all factors contributing to the equation ([32], pp. 55–56; [33], p. 1056).

The practice of medicine does not and cannot exist independently of its cultural surroundings. Where there is racism [22], sexism ([10], Part 2), or imperialism ([10], Part 2), cultural ethos will have its effect on the character of health care. One should not forget that the Tuskegee syphilis experiments on black males were concluded barely a decade ago.

Eliot Krause generalizes the point by insisting that "if those who study it [the health care system] refuse to explore systematically the myriad ways it is tied to power and control systems of the wider society, then a disservice is done to those whose basic interests are at stake" ([27], p. 1). Krause's basic thesis is that the health care system of the United States is part and parcel of a political and economic structure that is ridden with profound social inequalities and that is itself productive of disease.

A dominant theme in Krause's study is that the political and economic arrangements of modern society constitute a 'sickening environment': "At present, for far too many people, home is a place that makes you ill, work is a place that maims your body and spirit, and the environment is a place that functions to finish the job started by home and work" ([27], p. 296). Malnutrition, crowded housing, unenforced building codes, occupational hazards, toxic wastes, automobile exhaust constitute a limited listing of the problems. Among Krause's examples: "sulfur dioxide pollution...has caused the rate of lung cancer on that part of Staten Island downwind of the Jersey refineries to be thirty-three percent higher than that on other parts of the island" ([27], p. 319). In every instance of concerted effort to meliorate these conditions, political forces have been mobilized in opposition.

*Historical Development of Biomedical Practice*

The structure of the American health system has changed significantly over the past few decades. While family doctors are still to be found, the dominant form of medical practice is now highly institutionalized. Its features are those of modern organization: concentration, bureaucracy, specialization, and complex technology. The American health system has become, in effect, industrialized both internally and in its connection with the manufacturing and financial sectors of the economy. While the benefits of this historical development should not be discounted, neither should the problems it has introduced. The Ehrenreichs, for instance, studied the impact of this change in New York City and cite a range of problems that especially low income groups in an urban setting began to confront as the new system emerged ([8], pp. 4–17).

Tom Levin, for purposes of reform, has formulated a systematic interpretation of the development of modern medicine. On the ideological level,

> the image the present medical practitioner would like to project to the American health consumer today is that of the dedicated, horse and buggy country doctor, the family practitioner whose deep human investment in his patients and in his community has brought about a totally warranted loyalty and loving social acceptance from the community he serves ([28], pp. 11–12).

But the actual structure of health care under conditions of advanced capitalism is an uneasy amalgamation of a long-time heritage with modern social trends.

> The social organization of medicine has its roots in the medieval and Renaissance guild structure that was organized around a hierarchy of skills and knowledge. Such organization is essentially a craft social structure in which position is determined by the 'possession' of skills and information...The American health system combines the worst of both social organizations; it maintains the elitist hierarchical feudal guild system of interpersonal social organization around a master craftsman while its economic, service, and technological base is more akin to General Motors than Marcus Welby ([28], pp. 13–14).

More precisely, the social organization of medicine consists of four principles, each of which, Levin argues, contributes to the pathology of the system: (1) service for profit (which leads to "the existence of pluralistic service systems with vast disparities and inequities of service quality and availability to the poor, rural populations, and ghetto residents"); (2) a symbiosis between vendor and provider (which results in "service delivery...determined by facility site rather than patient site"); (3) a guild caste system (which causes "the continued escalation of designated health specialities and overspecialization of skills both in medicine and in allied health occupations") and (4) a distorted relationship between doctor and patient (which means "the suppression of self-help in health procedures in order to reinforce the hegemony of the professional") ([28], pp.28–31).

As Giddens comments in his theory of structuration, action is a continuous flow of conduct. It is a movement in the present from a past into a future. The historical development of a practice is always underway. That is the genius of Renée Fox's interpretation of the evolution of medical practice. As already mentioned, Fox discerns a movement through three phases, from the archaic, into the modern, and toward a post-modern. In her projection, two features are particularly characteristic of the emerging medical practice: a social orientation (the social order will be blamed for the genesis and persistence of certain kinds of illness and held responsible for their alleviation) and an existential awareness (there will be increased awareness of the mysteries of life and death and the limitations of the medical profession) ([13], pp. 499–529).

*Historical Development of the Social Totality*

During the 1930s and 1940s, Henry E. Sigerist, the most prominent historian of medicine in the United States, brought to that discipline a broad perspective leading to investigations of virtually all aspects of civilization. Repeatedly he asserted that medicine was more a social science than a natural science ([46], p. 26; [47], pp. 66, 241). To understand illness, its origins and treatment, one should focus not merely on biological and clinical research, but equally as much on the mundane features of every-day life: "Food, clothing, housing, occupations, social relations – these factors have always played a considerable role, both in health and disease" ([47], p. 7). More particularly, "In any given society the incidence of illness is largely determined by economic factors... A low living standard, lack of food, clothing and fuel, poor housing conditions, and other symptoms of poverty have always been major causes of disease" ([47], p. 55). Following a survey of housing conditions over several centuries, Sigerist pronounces

> If health is vital for human welfare, then housing is certainly an issue of major importance. And if this is the case, housing should not be the object of speculation or source of profit which it is today. Competitive business can neither honestly face nor properly solve the housing problem of contemporary society ([47], p. 40).

At one juncture in his history of the connections between civilization and disease, Sigerist concludes on a point of social ethics and social ontology.

In every country disease must be attacked with all available means and where it is most prevalent, in the low income groups. And since the world has become very small as a result of the present means of communication, we must think and plan not merely on a national but on an international scale. There is a human solidarity in health matters that cannot be disregarded with impunity ([47], p. 59).

The interconnections between personal welfare and physical and social environments are also central in René Dubos' approach to the biomedical field:

Clinical and epidemiological studies show that the inextricably interrelated body, mind, and environment must be considered together in any medical situation whether it involves a single patient or a whole community... Whatever the complaints of the patient and the signs or symptoms he manifests, whatever the medical problems of the community, disease cannot be understood or successfully controlled without considering man in his total environment ([6], pp. 65–66).

This means attending to processes of social change, for attributes of body and mind appropriate to one age may be causative agents in illness in a subsequent age ([6], pp. 52–53) and, moreover, times of social upheaval have an impact on patterns and intensity of disease ([6], p. 72). While some diseases seem to recur throughout all periods of history, nonetheless, "each civilization, each type of culture, has its own particular illnesses" ([7], pp. 10–11).

In summary, given the focus I have proposed, a central task of contextual bioethics is to demonstrate how particular cases of biomedical practice constitute conjunctions of social structure and historical process. They are, to be sure, instances of personal interaction and must be treated as such. But they are much more than that. They are, simultaneously, moments in the reproduction of social forms and in the passage of time.

## THEOLOGY: GOD AND THE WORLD

Two questions in the methodology of bioethics have so far been raised – a question of scope and a question of focus. In response to the first question, I have suggested that, of two general tendencies in the field, personal and contextual, the latter approach manifests a profounder social and historical realism than the former. In response to the second question, I have suggested that a contextual bioethics, focusing on the conjunction of individuality and sociality and appropriating resources from the sociology and history of medical practice and its social setting, should attend to synchronic and diachronic dimensions of the cases and subjects into which it inquires.

There remains a more difficult yet more fundamental methodological question, namely, a question of the grounding of bioethics. What are the foundations of bioethics? What basic understandings or beliefs underlie its working? Where there appear to be irreconcilable procedures or principles among different forms of bioethics, how might one adjudicate the conflict to determine the more justifiable alternative? Why should one engage in bioethics at all? What is its ultimate purpose and point? What in human experience constitutes the impetus behind bioethics? What gives warrant to its premises and first principles?

There are varying levels of approach to this set of issues. Adapting a distinction from Anthony Giddens, I would propose at least two levels: practical consciousness and discursive consciousness. In Giddens' usage, the former includes those "tacit stocks of knowledge which actors draw upon in the constitution of social activity," whereas the latter involves "knowledge which actors are able to express on the level of discourse" ([16], p. 5). I shall use them in this context to refer, respectively, to popular understanding and philosophical (and theological) reasoning.

According to William Glaser, "Since every society has the fundamental problems of explaining illness to the sick and their families and of guiding practical action, each society develops a body of medical theory from its basic ideas about the universe, life, and man" ([17], p. 95). Such ideas may, in David Mechanic's judgment, have an ideological cast.

> The character and distribution of health care, in many respects, reflects the ideological preferences of society...A health care plan must encompass a particular political philosophy concerning both the patients' and the providers' roles and obligations. These philosophies usually arise from the political and economic orientations prevailing in the community ([32], pp. 1–2).

To illustrate his point, Mechanic contrasts health care in the United States and the U.S.S.R. In the former, health care is "a manifestation of a capitalist entrepreneurial viewpoint emphasizing self-help, the maintenance of the private sector, and a pluralist system. Barriers to access are prevalent and are legitimized on the basis that free goods are unappreciated and exploited." Contrariwise, the U.S.S.R. "conceives of the health of workers as an important economic resource and the health system as an aspect of the government structure concerned with the public welfare"; thus access to primary services is open and encouraged, and industrial medicine is stressed ([32], p. 2, cf. [39], pp. 79–81, [27], pp. 123–155).

The ideologies of political communities and socio-economic classes constitute an everyday justification for institutional forms and practice. They exercise a powerful influence in the continuous production and reproduction of patterns of social relationship. They serve, on a popular level, as a grounding for group interest and as an apparently sufficient reason for supporting (or criticizing) social policy. Yet in an age imbued with historical consciousness, cognizant of cultural and political difference, and suspicious of self-aggrandizing rationalization, ideologies are strongly discounted as well. When they are discounted, there is an impetus to move from practical consciousness to discursive consciousness, from popular morality to critical morality. This is the point at which an explicit move may be made to philosophy or theology. In a provocative 'essay on theological method', Gordon Kaufman has asserted

> the ideas of God and the world are constructed by the human imagination for essentially practical purposes: in order to live and act it is necessary to have some conception or picture of the overall context, the fundamental order, within which human life falls. The ideas and images of God and the world supply this. Thus, they are created primarily to provide orientation in life ([25], p. 30).

While accepting Kaufman's assertion that ideas of God and the world constitute a conception of the overall context within which human life is lived and in accordance with which the processes and forms of human existence are ultimately grounded, I shall not pursue the basic thesis of Kaufman's argument, that theology in the final analysis is nothing but an imaginative construction of the human mind. I shall instead pursue a clue taken from process thought, a clue which forms the basis of the philosophy of organism (A. N. Whitehead) and the metaphysics of internal relations (B. E. Meland). The clue is the reformed subjectivist principle: "The starting point for a genuinely new theistic conception is what Whitehead speaks of as 'the reformed subjectivist principle' " ([37], p. 57).

The reformed subjectivist principle is to philosophy and theology what the theory of structuration is in social theory. The theory of structuration points to the conjunction of individuality and sociality and to the coincidence of structural forms and historical passage. The reformed subjectivist principle indicates the connection between self and world and between self and God. Moreover, the reformed subjectivist principle, I mean to suggest, gives rise to a cluster of principles of action that constitute a grounding for social ethics and thus also bioethics.

Modern philosophy and modern theology are characterized, in large part, by a turn toward the subject. What can be known about the world, it is assumed, is constructed from subjective experience. There are two versions of the subjectivist principle. In an earlier version, it is individualistic and, in an extreme form, solipsistic. In a later reformed version, it is relational and processual. In Whitehead's formulation: "The way in which one actual entity is qualified by other actual entities is the 'experience' of the actual world enjoyed by that actual entity, as subject. The subjectivist principle is that the whole universe consists of elements disclosed in the analysis of the experiences of subjects" ([52], p. 252). More concisely, according to the philosophy of organism, "Each actual entity is a throb of experience including the actual world within its scope" ([52], p. 290). Self and world, from this perspective, are engaged in a continuous process of interaction. The world is the condition for the self's realization and the self, in turn, is a contribution to the on-going world. Each is important to the other. *Per analogiam*, God and the world are engaged in constant interchange. The world is the condition for God's activity and God's valuation of and intentionality for the world constitute the impetus for its next moment of becoming.

In this metaphysics of internal relations,

> Relationship is of the essence. It is not just a connecting link forming the parts into a mechanism; it is a live and serious confrontation of created centers of dignity pursuing the intentions and ends of self in and through the drama of communal existence ([35], p. 206).

Religion, Whitehead has argued, lies at the heart of this philosophical understanding, a religion

> founded on the concurrence of three allied concepts in one moment of self-consciousness, concepts whose separate relationships to fact and whose mutual relations to each other are only to be settled jointly by some direct intuition into the ultimate character of the universe. These concepts are:
> (1) That of the value of an individual for itself.
> (2) That of the value of the diverse individuals of the world for each other.
> (3) That of the value of the objective world which is a community derivative from the interrelations of its component individuals, and also necessary for the existence of each of these individuals ([51],p. 59).

The self thus exists within a matrix of relationships whose dominant member is God. What matters is the quality of those relationships. In Meland's terms, the self confronts

> a threefold demand, which is what gives complexity to his existence. He is made for God, he is made for other people [and, I would add, for other beings within the entire ecosphere], he is made for himself. The living out of these relationships becomes man's daily burden as well as his opportunity. And it is his ultimate hope ([35], p. 207).

This threefold demand may be translated into three principles, each of which has its ethical import and each of which bears on bioethics: the principles of autonomy (the value of an individual for itself), relationality (the value of the diverse individuals of the world for each other), and community (the value of the objective world). These may be construed in traditional ethical language as principles of liberty, equality, and the common good.

The principle of liberty, which pertains to the self, honors the individuality of experience. The principle of equality, which belongs to relations between beings, promotes the contribution each life can make to all future lives. The principle of the common good, which, in its most encompassing form, rests in God's apprehension of the universe as a whole and becomes a condition for each instance of self-realization, respects the ever-evolving quality of the connectedness of life with life as drawn together in every new moment in the passage of time.

Self, world, and God are thus conjoined in what Meland has called the 'structure of experience', that is,

> a depth in our natures that connects all that we are with all that has been within the context of actuality that defines our culture. It is a depth in our nature that relates us as events to all existent events. It is a depth in our nature that relates us to God, a sensitive nature within the vast context of nature, winning the creative passage for qualitative attainment ([34], pp. 111–112).

Out of that structure of experience are born the principles of liberty, equality, and the common good which, in turn, constitute a foundation for the assessment and direction of personal agency and social structure. They therefore constitute, in particular, a grounding for biomedical practice and bioethics. But I would add, out of the Hebraic-Christian traditions, the principle of covenant. On the one hand, the principle of covenant conveys a 'myth of identity' between the world and God and, as such, it points to an ideal conformity of the principles of liberty, equality, and common good. It embraces them in a single vision. On the other hand, the principle of covenant, in its presentation in the narrative of the Hebraic-Christian traditions, conveys a 'note of dissonance' in relations between world and God and, thus, it points to tensions and conflicts among considerations of liberty, equality, and common good (cf. [36], pp. 96–101).

In the literature of bioethics and, more generally, of health and disease, various forms of these four principles have been invoked as fundamental. Selected instances are summarized in the following table.

*Fundamental Principles of Biomedical Practice and Bioethics*

| | |
|---|---|
| (1) Principle of Liberty | T. Engelhardt<br>W. Reich |
| (2) Principle of Equality | L. Kass<br>H. Sigerist |
| (3) Principle of Common Good | H. Sigerist<br>T. Parsons<br>V. R. Potter<br>A. Hellegers / A. Jonsen |
| (4) Principle of Covenant | P. Ramsey<br>R. Veatch<br>W. May |

(1) In a searching discussion of meanings of health and disease, H. Tristram Engelhardt ultimately defines health as "a state of freedom from the compulsion of psychological and physiological forces." Treatment, therefore, is to focus "on securing the autonomy of the individual from a particular class of restrictions" ([11], p. 42). Throughout his wide-ranging discussion, Engelhardt debates several important issues of philosophical anthropology pertinent to biomedical practice – e.g., the contention between dualists and monists on the body-mind problem and the difference between Platonic realism and contextualism on the etiology of disease. In its final conclusion, however, his position is centered in a strong affirmation of the liberty of the individual. That is the reason for biomedical practice and, presumably, a governing principle of bioethics (although he claims the principle is a non-moral regulative ideal): "While there are many diseases, there is in a sense only one health – a regulative ideal of autonomy directing the physician to the patient as person, the sufferer of the illness, and the reason for all the concern and activity" ([11], p. 43).

Warren Reich, acknowledging the central importance of the principle of autonomy in bioethics particularly as it applies to the matter of informed consent, argues that the principle tends to be construed much too narrowly as 'non-interference'. Respect for autonomy requires that persons be provided with those conditions – physical, mental, and social – that enable them authentically and positively to make determinations about their future. In some instances, therefore, active intervention by medical professionals even against the expressed wishes of the patient may be a means of honoring the principle of positive autonomy ([44], pp. 191–215).

(2) In a now classic essay on the goal of medicine, Leon Kass argues that the governing purpose of medicine is not, as with Engelhardt, to free persons from restraints, thus enhancing their autonomy. It is rather to promote the actualization of specifically human potentialities. The underlying anthropology is Aristotelian. It assumes a kind of equality that underlies all differences among human beings and it assumes equivalence of treatment relative to the standard of human need. Health, the goal of medicine, is grounded in the nature of being human. Thus, in a summary definition, health is 'the well-working of the organism as a whole' or, again, 'an activity of the living body in accordance with its specific excellences' " ([24], p. 29).

(3) Versions of both equality and the common good are discernible in the writings of Henry E. Sigerist. As mentioned already, Sigerist attacks the unequal distribution of wealth as a dominant factor in the incidence and apportionment of disease. But Sigerist postulates, without argument, "The goal of medicine is not simply to cure diseases; it is rather to keep men adjusted to their environment as useful members of society, or to readjust them when illness has taken hold of them" ([47], p. 66). In short, the goal of medicine is the common good of society. Sigerist, however, would have society understood inclusively: "the health and welfare of every individual is the concern of society, and *human solidarity beyond the boundaries of nationality, race, and creed is a true criterion of civilization*" ([47], p. 242, italics added).

Talcott Parsons' rendition of the common good as the central principle of medical practice is much narrower. He defines health "as the state of optimum capacity of an individual for the effective performance of the roles and tasks for which he has been socialized" ([39], p. 69, italics removed). By contrast, Van Rensselaer Potter's approach to the field of bioethics is cosmic in scope, for, in his construction, the fundamental purpose of the field is the "survival of the total ecosystem" ([41], p. viii).

The position of Albert Jonsen and André Hellegers stands between the extremes of Parsons and Potter. Jonsen and Hellegers propose that an adequate ethics of health care would be comprised of three moral concerns: virtue, duty, and common good. Virtue pertains to the character of the practitioner. Duty pertains to the quality of particular actions in the relationship between professional and patient. Common good pertains to the structure of the institutions through which and in which health care occurs.

> A theory of the common good seeks to elucidate the nature of human communities. These are the institutional forms that human actions create and human virtues sustain and, in their turn, should become the objective conditions nurturing virtue and sustaining action ...Properly conceived, the theory of the common good is a third dimension in which virtues and actions take on depth and tone that they do not have in isolation. The very meaning of a virtue or an action depends on its social or institutional setting ([23], p. 133).

By invoking the principle of the common good, Jonsen and Hellegers are calling for an institutional ethic, that is, a theory of how the institution of modern medicine should be formed to serve the general cause of social justice for the whole community of humankind.

(4) Diverse understandings of the principle of covenant are manifest in the bioethics of Paul Ramsey, Robert Veatch, and William F. May. Ramsey's construction is *confessional* in its derivation and *personalistic* in its result. As a Christian ethicist in the Barthian tradition, Ramsey holds "that covenant-fidelity is the inner meaning and purpose of our creation as human beings." The basic requirement of fidelity for the medical practitioner as for anyone is to respect the sanctity of human life which, where effective,

> prevents ultimate trespass upon him even for the sake of treating his bodily life, or for the sake of others who are also only a sacredness in their bodily lives. Only a being who is a sacredness in the social order can withstand complete dominion by 'society' for the sake of engineering civilizational goals – withstand, in the sense that the engineering of civilizational goals cannot be accomplished without denying the sacredness of the human being. So also in the use of medical or scientific technics ([43], p. xiii).

Veatch, however, proposes a principle of covenant that is *contractualist* in its preferred origin for, with John Rawls, Veatch believes that a form of contractualism accessible to all rational persons will more nearly satisfy the need for a universally acceptable basis for bioethics than a theologically based convenantalism. In result, the principle of covenant entails the rejection of utilitarianism and the adoption of a set of prima facie duties in human relationships (e.g., promise-keeping, truthfulness, respect for autonomy) ([50], pp. 110–126; [48], pp. 38–29).

William F. May's principle of covenant appears to be *ontological* in its derivation although expressed in and through religious traditions of Judaism and Christianity. In its results it enjoins a reciprocity of giving and receiving between medical professional and patient as also between the institution of medical practice and social order [31].

Of these three understandings of the meaning of covenant, May's most closely approximates what I intend by the reformed subjectivist principle as a theological means of grounding the ethics of biomedical practice.

Moreover, the reformed subjectivist principle constitutes a framework for incorporating the principles of liberty, equality, and common good into a single conception, a conception whose full methodological and institutional implications remain to be developed. But that is a task that lies beyond the scope of this essay.

## CONCLUSION

In sum, I have posed three questions about the meaning of bioethics – its scope, its focus, and its grounding. On the first question, I have proposed the need for a contextual bioethics. On the second, I have proposed the need to concentrate on the conjunction between sociality and individuality in bioethics. And on the third, I have proposed the reformed subjectivist principle as a theological means for grounding bioethics and for developing principles for its practice.

*Bucknell University, Lewisburg, Pennsylvania, U.S.A.*

## BIBLIOGRAPHY

[1] Branson, R.: 1975, 'Bioethics as Individual *and* Social: The Scope of a Consulting Profession *and* Academic Discipline', *The Journal of Religious Ethics* **3**, 111–139.
[2] Bryant, J. H.: 1978, 'Poverty and Health in International Perspective', in W. Reich (ed.), *Encyclopedia of Bioethics*, Vol. 3, Free Press, New York, pp. 1321–1327.
[3] Callahan, D.: 1973, 'Bioethics as a Discipline', *Hastings Center Report* **1** (November 1), 66–73.
[4] Callahan, D.: 1980, 'Shattuck Lecture – Contemporary Biomedical Ethics', *The New England Journal of Medicine* **302** (May 29), 1228–1233.
[5] Clouser, D. K.: 1978, 'Bioethics', in W. Reich (ed.), *Encyclopedia of Bioethics*, Vol. 1, The Free Press, New York, pp. 115–127.
[6] Dubos, R.: 1968, *Man, Medicine, and Environment*, Frederick A. Praeger, New York.
[7] Dubos, R. and Escande, J. P.: 1980, *Quest: Reflections on Medicine, Science, and Humanity*, P. Ranum (trans.), Harcourt Brace Jovanovich, New York.
[8] Ehrenreich, B. and Ehrenreich, J.: 1970, *The American Health Empire: Power, Profits, and Politics*, Random House, New York.
[9] Ehrenreich, B. and Ehrenreich, J.: 1978, 'Medicine and Social Control', in J. Ehrenreich (ed.), *The Cultural Crisis of Modern Medicine*, Monthly Review Press, New York, pp. 39–79.
[10] Ehrenreich, J. (ed.): 1978, *The Cultural Crisis of Modern Medicine*, Monthly Review Press, New York.
[11] Engelhardt, H. T.: 1981, 'The Concepts of Health and Disease', in A. L. Caplan, *et al.* (eds.), *Concepts of Health and Disease: Interdisciplinary Perspectives*, Addison-Wesley, Reading, Massachusetts, pp. 31–45.

[13] Fox, R.: 1979, *Essays in Medical Sociology: Journeys into the Field*, John Wiley & Sons, New York.

[14] Freidson, E.: 1979, *Profession of Medicine: A Study of the Sociology of Applied Knowledge*, Dodd, Mead & Company, New York.

[15] Giddens, A.: 1976, *New Rules for Sociological Method: A Positive Critique of Interpretive Sociology*, Basic Books, New York.

[16] Giddens, A.: 1979, *Central Problems in Social Theory: Action, Structure, and Contradiction in Social Analysis*, University of California Press, Berkeley and Los Angeles.

[17] Glaser, W. A.: 1978, 'Medical Care: Social Aspects', in David Sills (ed.), *International Encyclopedia of the Social Sciences*, Vol. 10, MacMillan and The Free Press, New York, pp. 93–100.

[18] Gordy, M.: 1978, 'Sociality', in W. Reich (ed.), *Encyclopedia of Bioethics*, Vol. 4, The Free Press, New York, pp. 1603–1606.

[19] Horkheimer, M.: 1970, *Die Sehnsucht nach dem ganz Anderen*. Furche-Verlag, Hamburg.

[20] Horrobin, D.: 1977, *Medical Hubris: A Reply to Ivan Illich*, Eden Press, Montreal and Lancaster.

[21] Illich, I.: 1976, *Medical Nemesis: The Expropriation of Health*, Pantheon Books, New York.

[22] Jones, J. H.: 1978, 'Racism and Medicine', in W. Reich (ed.), *Encyclopedia of Bioethics*, Vol. 4, Free Press, New York, pp. 1405–1410.

[23] Jonsen, A. R. and Hellegers, A. E.: 1977, 'Conceptual Foundations for an Ethics of Medical Care', in S. J. Reiser *et al.* (eds.), *Ethics in Medicine: Historical Perspectives and Contemporary Concerns*, MIT Press, Cambridge, Massachusetts, pp. 129–137.

[24] Kass, L.: 1975, 'Regarding the End of Medicine and the Pursuit of Health', *The Public Interest* **40** (Summer), 11–42.

[25] Kaufman, G. D.: 1975, *An Essay in Theological Method*, Scholars Press, Missoula, Montana.

[26] Kosa, J. *et al.*: 1969, *Poverty and Health: A Sociological Perspective*, Harvard University Press, Cambridge, Massachusetts.

[27] Krause, E.: 1977, *Power and Illness: The Political Sociology of Health and Medical Care*, Elsevier, New York.

[28] Levin, T.: 1974, *American Health: Professional Privilege vs. Public Need*, Praeger Publishers, New York.

[29] Luft, H. S.: 1978, *Poverty and Health: Economic Causes and Consequences of Health Problems*, Balinger Publishing Co., Cambridge, Massachusetts.

[30] Marshall, C. L. and Marshall, C. P.: 1978, 'Poverty and Health in the United States', in W. Reich (ed.), *Encyclopedia of Bioethics*, Vol. 3, Free Press, New York, pp. 1316–1321.

[31] May, W.: 1977, 'Code and Covenant or Philanthropy and Contract', in S. J. Reiser *et al.* (eds.), *Ethics and Medicine: Historical Perspectives and Contemporary Concerns*, MIT Press, Cambridge, Massachusetts, pp. 65–76.

[32] Mechanic, D.: 1974, *Politics, Medicine, and Social Science*, John Wiley & Sons, New York.

[33] Mechanic, D.: 1978, 'Medicine, Sociology of', in W. Reich (ed.), *Encyclopedia of Bioethics*, Vol. 3, The Free Press, New York, pp. 1054–1059.

[34] Meland, B. E.: 1953, *Faith and Culture*, Oxford University Press, New York.

[35] Meland, B. E.: 1962, *The Realities of Faith: The Revolution in Cultural Forms*, Oxford University Press, New York.

[36] Meland, B. E.: 1976, *Fallible Forms and Symbols: Discourses of Method in a Theology of Culture*, Fortress Press, Philadelphia.
[37] Ogden, S.: 1966, *The Reality of God and Other Essays*, Harper & Row, New York.
[38] Parsons, T.: 1951, *The Social System*, The Free Press, Glencoe, Illinois.
[39] Parsons, T.: 1981, 'Definitions of Health and Illness in the Light of American Values and Social Structure', in A. L. Caplan *et al.* (eds.), *Concepts of Health and Disease*, Addison-Wesley Publishing Co., Reading, Massachusetts, pp. 57–81.
[40] Pellegrino, E. D. and Thomasma, D. C.: 1981, *A Philosophical Basis of Medical Practice: Toward a Philosophy and Ethics of the Healing Professions*, Oxford University Press, New York.
[41] Potter, V. R.: 1971, *Bioethics: Bridge to the Future*, Prentice-Hall, Englewood Cliffs, New Jersey.
[42] Potter, V. R.: 1972, 'Bioethics for Whom?' in P. Siekevitz (ed.), *The Social Responsibility of Scientists: Annals of the New York Academy of Sciences*, Vol. 196, article 4, pp. 200–205.
[43] Ramsey, P.: 1970, *The Patient as Person*, Yale University Press, New Haven & London.
[44] Reich, W.: 1982, 'Toward a Theory of Autonomy and Informed Consent', in L. J. Rasmussen (ed.), *Annual of the Society of Christian Ethics*, pp. 191–215.
[45] Schenck, D. *et al.*: 1981, 'Encyclopedia of Bioethics', *Religious Studies Review* **7**/1, 5–9, 12–14.
[46] Sigerist, H. E.: 1960, *Henry E. Sigerist on the History of Medicine*, Felix Marti-Ibanez (ed.), MD Publications, New York.
[47] Sigerist, H. E.: 1962, *Civilization and Disease*, Phoenix Books, The University of Chicago Press, Chicago.
[48] Sumner, L. W.: 1982, 'Does Medical Ethics Have Its Own Theory? *The Hastings Center Report* **12** (August), 38–39.
[49] Unschuld, P. U.: 1978, 'General Historical Survey', in W. Reich (ed.), *Encyclopedia of Bioethics*, Vol. 3. The Free Press, New York, pp. 901–906.
[50] Veatch, R. M.: 1981, *A Theory of Medical Ethics*, Basic Books, New York.
[51] Whitehead, A. N.: 1962, *Religion in the Making*, MacMillan, New York.
[52] Whitehead, A. N.: 1929, *Process and Reality: An Essay in Cosmology*, MacMillan, New York.
[53] Whitehead, A. N.: 1958, *The Function of Reason*, Beacon Press, Boston.
[54] Zaner, R.: 1978, 'Embodiment', in W. Reich (ed.), *Encyclopedia of Bioethics*, Vol. 1, The Free Press, New York, pp. 361–366.
[55] Zaner, R.: 1981, *The Context of Self*, Ohio University Press, Athens, Ohio.

MARGARET A. FARLEY

# FEMINIST THEOLOGY AND BIOETHICS

The aim of this essay is to explore the connections between feminist theology and issues in the field of bioethics. I have construed the task largely as a descriptive one; that is, I shall try to indicate some basic contours of feminist theology and some ways in which the values it emphasizes bear on the vast network of ethical issues related to the biological sciences, technology, and medicine. In addition, and in order to press the question of possible contributions by feminist theology to bioethics, I shall focus on the particular implications of feminist theology for the development and use of reproductive technology.

To some extent, the connection between the concerns of feminist theology and bioethics is obvious. Whatever else feminist theology does, it proceeds from a methodological focus on the experience of women, and whatever feminist ethics does, it begins with a central concern for the well-being of women. Medical ethics (as a part of bioethics) can be expected to share in some important way this focus and this concern, if for no other reason than that women constitute the majority of those who receive and provide health care ([46], pp. 119, 125; [33]). Beyond this, however, traditional religious views of women associate them symbolically and literally with nature, with the body, with human relationships, with reproduction – all themes for feminist theological critique and reconstruction, all foci for major concerns of bioethics in its broadest dimensions. The obviousness of the connection between these two disciplines, however, does not in itself give us the present and potential lines of mutual influence.

Before beginning a closer look at the relation between feminist theology and bioethics, three caveats are perhaps in order. That is, it is helpful to identify some forms of relation which we should *not* expect to find.

First, we should not expect to find feminist theology articulating for bioethics fundamental values or moral principles which are in every way unique to a feminist theological perspective. Few contemporary theological ethicists who take seriously the task of making explicit the connection between religious beliefs and ethical action claim for their theologies



*E. E. Shelp (ed.), Theology and Bioethics,* 163–185.

exclusive access to moral insight in the formulation of commonly held norms ([19], p. 9; [18], p. 119; [14], pp. 84–90; [13], p. 26).[1] It is not only religious belief, or theology, or a particular theology that can ground, for example, a requirement to respect persons, or a principle of equality, or a rational system of distributive justice. Likewise, it is not only feminist theology that can ground a view of human persons as fundamentally interpersonal and social, or that can formulate a view of nature that requires human stewardship rather than exploitation. Still, theologies do yield ethical perspectives that are unique in some respects, moral points of view that claim hermeneutically privileged insights, even particular moral action-guides that chart frontiers for human decision. Feminist theology is no exception in this regard. Indeed, it may have a more explicit ethical entailment than many other theologies. Moreover, its critical function may provide an essentially new perspective on some issues in bioethics.

Second, there is no one definitive form of feminist theology which can be looked to as representing all of its possible implications for bioethics. Theology in general is pluralistic on many levels. Feminist theology is not just one among many options in theology; it is itself pluralistic on many of the same levels as is theology generally. Thus, there are feminist theologies that are centered in ancient forms of goddess worship, and others that locate themselves, with important distinctions, in the Jewish or Christian biblical traditions, and still others that move beyond any historical traditions at all. There are diverse perspectives within particular traditions, too – perspectives that vary as much as process theology varies from medieval scholasticism. So clear have the differences in feminist theologies become that typologies abound in a growing effort to compare and contrast them ([36], pp. 214–234; [10], pp. 7–36; [31]). This wide divergence must be kept carefully in mind while we, nonetheless, explore a rather remarkable convergence of basic ethical concerns, values, and to some extent, norms for action.

Third, while it is not difficult to identify some parameters of an ethic which derives from and/or is reinforced by feminist theology, the kind of systematic development necessary to bring basic values to bear on very specific bioethical issues remains in important respects still to be undertaken. Indeed, feminist theology as such is at beginning points in its systematic formulation. While monumental strides have been taken by feminist biblical scholars, theologians, and historians ([4], [36], [39], [10], [42]), sustained theological synthesis is new on the horizon, at least for the Christian tradition [36]. Even newer is a systematic comprehensiveness and depth on the ethical side of feminist theology [14]. The import of the still limited development of feminist theology and ethics lies in the general conviction of most theologians that there is no easy route from the sources of religious faith to the specific insights needed for many of the radically new questions generated by scientific and medical capabilities. This conviction is mirrored in the reservations, though not final condemnation or approval, which many feminist theologians and ethicists express regarding, for example, some technologies of reproduction ([14], p. 37; [36], p. 226). It is also mirrored in the recognition of the necessity of collaboration with disciplines other than theology and ethics for the gradual forging of moral perspectives on the multitude of issues which a comprehensive bioethics may address.

There are limits, then, to the connections presently discernible between feminist theology and bioethics. Within those limits, however, lie meeting points, challenges, resources, of potential critical importance to both disciplines. We turn first to the methods, sources, and relevant themes of feminist theology.

## FEMINIST THEOLOGY

Of all the themes in feminist theology which have direct bearing on issues in bioethics, three can be raised up for central consideration. These are the themes of (1) relational patterns among human persons, (2) human embodiment, and (3) human assessment of the meaning and value of the world of 'nature'. Feminist theology's development of these themes includes an articulation of basic ethical perceptions and leads to the formulation of some ethical action-guides. Moreover, attention to the emergence and treatment of these themes illuminates important methodological decisions which, as we shall see, constitute not only central commitments for feminist theology but possible warrants for ethical arguments in bioethics.

### *Patterns of Relation*

Feminism, in its most fundamental sense, is opposed to discrimination on the basis of sex. It opposes, therefore, any ideology, belief, attitude, or behavior which establishes or represents such discrimination. In terms of social structure, feminism is opposed, then, to patriarchy. This opposition has the ultimate aim of equality among persons regardless of gender. To achieve this aim, however, feminism is necessarily pro-woman. Since discrimination on the basis of sex – or sexism – has been and remains pervasively discrimination against women, feminism aims to correct this bias by a bias for women, however temporary or prolonged that bias must be. A bias for women includes a focal concern for the well-being of women and a taking account of women's experience in coming to understand what well-being demands for women and men.

Feminist theology perceives profound discrimination against women in traditions of religious patriarchy. The major work of feminist theologians to date has been the unmasking of beliefs, symbols, and religious practices which establish and foster this discrimination. What they have found are massive tendencies in religious traditions to justify patterns of relationship in which men dominate women. Within the history of Christianity, for example, the major pattern of relationship between women and men has been one of dominance and subjugation. This has been sustained through a variety of beliefs about the essential inferiority of women to men and the need for a hierarchical order in social arrangements. Theological assessments of woman's nature, like many philosophical assessments, were based on views of a fundamental dualism within humanity. In these views, women and men are distinguished as polar opposites, representing body or mind, emotion or reason, passivity or activity, dependence or autonomy. The female-identified pole is always inferior to the male. Even when men and women are considered complementary in their duality, complementarity never means equality when it comes to role-differentiation. Thus, for example, men are to be primary agents, leaders, initiators; women are helpers, followers, supporters. More than this, women are often symbolically associated with evil. They are perceived as temptresses, to be feared as the threat of chaos to order, carnality to spirituality, weakness to strength. Even when women are exalted as symbols of virtue rather than vice, they bear the liabilities of impossible expectations and the burden of mediating 'femininity' to men [35].

Feminist theology's critique of religious traditions goes further, however, to the central symbols of faith. Feminists argue, for example, that Christianity's traditional formulation of a doctrine of God is itself a sexist warrant for discrimination against women. Though the Christian God transcends gender identification, personal metaphors for God are strongly masculine. This is true in the biblical tradition as well as in theological formulations of the doctrine of the Trinity. Moreover, Christian faith is centered in a savior who is male. Hence, there is a strong tendency in this tradition to consider men more appropriate as representatives of God in the human family, society, and the church. Indeed, traditional Christian theology has often granted the fullness of the *imago Dei* to men, yielding it only derivatively and partially to women. Thus is sealed the primary role of men in the human community. But more than this, feminist theologians point to the character of the Christian God as it is frequently drawn ([2]; [4]). That is, God is portrayed as sovereign, transcendent, requiring submission from human persons. It is on this model of relationship (dominance and submission) that human relationships are then patterned. Hence, as God is to God's people, so man is to woman, husband to wife, pastor to congregation, and on and on through the many forms of human life.

Some feminists have argued that the Christian view of the human self and its ideal development is also determined by the submissive role of persons in relation to God. That is, the height of Christian virtue is thought to be often portrayed as patient suffering and self-sacrificial love, and the mode of Christian action as humble servanthood. Women are socialized into these ideals in a way that men are not, for men can imitate the autonomy and agency of God in their role as God's representatives. Doctrines of sin which stress the evil of prideful self-assertion serve as a caution to men, but they only reinforce the submissiveness which already characterizes women [25]. Nietzsche's critique of Christianity as a religion for weaklings and victims can then be applied to the effect of Christian faith on women if not on men [5].

What emerges in feminist theology (in relation to Christianity, but here a harbinger of systematic developments in relation to other historical religions as well) is an analysis of what are judged to be oppressive patterns of relationship and ideologies which foster them. These patterns of oppression are identified not only in relations between men and women but in every human relation where the pattern is one of domination and subjugation on the basis of sex or race or class or any other aspect of persons which is used to deny full humanity to all. Given the radical nature of the feminist critique of Christianity (a critique which ultimately reaches to every major doctrine – of God, creation, redemption, the human person, sin and grace, the church, eschatology), feminist theologians either move away from Christianity altogether, or they take up the task of critical reconstruction of Christian theology. In either case, they have by and large moved to develop a view of human relations characterized by equality and mutuality, in which both autonomy and relationality are respected.

Feminist theologians who take up the task of restoring and reconstructing Christian theology 'beyond the feminist critique' argue that there are fundamental resources within the tradition which are not ultimately sexist and which can be brought to bear precisely as a challenge to sexism. With feminist hermeneutical methods, for example, biblical resources are available which reveal a God who does not need to compete with human beings for sovereignty, who comes forth from freedom in order to call forth freedom from human persons; a God who is able to be imagined in feminine as well as masculine terms [42], for whom 'friend' or 'partner' are more apt metaphors than 'king' or 'logos' ([20]; [39]). Reformulations of gender-assignment within the doctrine of the Trinity free a male-identified God from some of the limits of the human imagination and, indeed, from some of the problems which parent/child metaphors retain both for the life of God in itself and the relation of God to human persons. The prophetic traditions in the Old and New Testaments provide biblical grounds for challenging religious as well as secular institutions, or in other words, "every elevation of one social group against others as image or agent of God, every use of God to justify social domination and subjugation" ([36], p. 23). Biblical and historical studies using a feminist hermeneutic yield evidence of Christian community organized not on the model of sexist hierarchy but on bases of equality and reciprocity [10].

It might be argued that what feminist theology is doing offers no new insights regarding patterns of human relationships. When it argues for equality between women and men, it simply extends to women the insights of liberalism. When it concerns itself with economic structures as well as political, it only blends a form of Marxism with liberalism. When it raises up the importance of mutuality, it follows the theoretists of sociality – George Herbert Mead, Martin Buber, John MacMurray. When it criticizes notions of Christian love as self-sacrifice, it just gets clear on what has been taught all along. When it analyzes claims for self-determination and active participation in all the spheres of human life, it only repeats the agenda of liberation theology.

Feminist theologians are drawing on all of these sources of insight. Like feminists in general, however, they conclude that none of these other traditions or movements adequately address the oppression of women. This is not just a failure of extension. Rather, it represents a fundamental need for deeper analysis of the contexts of human life, concepts of the human self, and categories of human relation. From the hermeneutical vantage point of the experience of women – of their oppression and their achievements, their needs and contributions, their freedom and their responsibilities – feminist theology assumes ground-breaking work on questions of human relationships.

It has not been open to feminist theology, for example, simply to appropriate a view of the human person which makes autonomy paramount as the ground of respect or the primary principle to be protected in social relations. The issue of relationality as equiprimordial with autonomy as a feature of human personhood has pressed itself on feminist theologians from the experience of women. It is this that has demanded continued analysis of the nature of human relations and has led to historical and biblical studies of, for example, Christian communities, and to theological studies of the very nature of God (as relational). But if feminist theology cannot ignore relationality, neither has it been able to let go of autonomy as an essential feature of personhood [40]. Romantic returns to organic notions of society where relation is all, each in her place, without regard for free agency or for personal identity and worth which transcends roles – these are options that feminists judge can only repeat forms of oppression for women. It is this conviction that prompts continued biblical and theological studies of the compatibility of autonomy with dependence on God, the coincidence of activity and receptivity in peak experiences of relation, and social models which both protect individuality and promote the common good.

In another example, feminist theology has not been able to critique and then ignore interpretations of the differences between women and men. The whole enterprise of feminist theology still has something to do with demythologizing and de-ontologizing these differences, yet taking persons seriously as woman-persons and man-persons. Faced with these issues, feminist theology has had to take account of insights from the biological and behavioral sciences, and from philosophy. It has also had to maintain a focus on the concrete experience of women in systems where roles and spheres of human life are gender-specific. Refusal to defer these issues prompts unique probing of the fundamental possibilities and requirements of human relations, whether intimate or public. There is potential universal relevance for all human relationships in a move, for example, from traditional ideas of gender 'opposites', or even gender opposite 'complements', to ideas of gender 'analogies', where the primary focus is on similarity rather than difference [45].

Pluralism in feminist theology, of course, leads to some profoundly different choices regarding historical forms of human relationships. As in feminism generally, disagreements can be on the level of strategy (is there any possibility of radically transforming existing religious traditions?); or on the analysis of the cause of oppression (whether it is most fundamentally religion, or culture in a more general sense, or the conspiracy of men, or economics, etc.); or even on important characteristics of the model of relation to be advocated (do exclusivity and separatism contradict the values of equality and mutuality?). Such disagreements are extremely serious, and it would be a mistake to underestimate them. Still, there is basic unanimity among feminist theologians on the values that are essential for nonoppressive human relationships – the values of equality, mutuality, and freedom.[2] The depth of significance given to these values is testified to, not denied, in the disagreements they entail. For feminist theologians who finally reject traditional religious traditions as irretrievably sexist, the alternative is a 'women's culture' which can incorporate these values despite the impossibility of transforming existing religions or society at large. For feminist theologians who continue to stand within their traditions, the alternative is a radical restructuring of institutions and a radical revision of religious doctrine and practice.

*Embodiment*

The second theme in feminist theology which has particular bearing on issues of bioethics is the theme of human embodiment. Less needs to be said about this theme, since an understanding of it follows directly upon many of the concerns we have already explored regarding patterns of human relationships. There is, however, a clear history of association of ideas that we must trace if we are to see the import of this theme both for feminist theology and bioethics.

Body/spirit is in many ways the basic dualism with which historical religions have struggled since late antiquity. Women, as we have already noted, have been associated with body, men with mind. Those who have speculated on the reasons for this have generally noted the tendency to locate the essence of woman in her childbearing capabilities. Women's physiology has been interpreted as 'closer to nature' than men's in that many areas and functions of a woman's body seem to serve the human species as much as or more than they serve the individual woman [24]. Moreover, woman's bodies, in this interpretation, are subject to a kind of fate – more so than men's. Women are immersed in 'matter', in an inertness which has not its own agency. This is manifest not only in the determined rhythms of their bodily functions, but in a tendency to act from emotion rather than from reason, and in women's 'natural' work which is the caring for the bodies of children and men.

Whatever the reasons women have been associated with the body, they have thereby also been associated with the going evaluations of human bodyliness and matter in general. Historical religions which have made this connection have frequently devalued the body in relation to the spirit. Despite resistance from basically world-affirming attitudes in Judaism, and despite an ongoing conflict with positive Christian doctrines of creation and incarnation, both of these traditions incorporated negative views of the human body (and especially women's bodies). In late antiquity, Judaism was influenced by world-denying attitudes of Near Eastern gnosticism and mysticism. Christianity absorbed these same influences in its very foundations, along with Greek philosophical distrust of the transitoriness of bodily being.

Integral to views of the human body have been views of human sexuality. Once again, despite traditional influences toward positive valuation (of sexuality as a part of creation, as implicated in the very covenant with God, etc.), strongly negative judgments have been brought in. From ancient blood taboos, to Stoic prescriptions for the control of sexual desire by reason, to Christian doctrines of the consequences of original sin, fear and suspicion regarding the evil potentialities of sex have reigned strong in the Western conscience. So great, in fact, has been the symbolic power of sex in relation to evil that there seems to have been "from time immemorial", as Paul Ricoeur puts it, "an indissoluble complicity between sexuality and defilement" ([30], p. 28).

Central to the association of women with bodyliness has been the interpretation of their sexuality as more 'carnal' than men's, again 'closer to nature', more animal-like, less subject to rational control. Disclosure of this historical view of women's sexuality came as a surprise to many feminists whose direct learning from religious traditions had tended to be the opposite – that is, that women are less passionate than men, and hence more responsible for setting limits to sexual activity. The reversal in this regard has its roots, too, in religious traditions, and reflects the tendency we have seen before to identify women with evil, on one hand, and place them on a pedestal, on the other [3]. In either case, women's identity remains closely tied to the way they relate to their bodies, and in either case, women have learned to devalue their bodies. For women themselves, Freud's comment on beliefs about menstruation, pregnancy, and childbirth held true: "... it might almost be said that women are altogether taboo" ([12], p. 75).

With the rise of feminist consciousness, all past interpretations of the meaning of women's bodies were called into question. Women's turn to their own experience for new interpretations of embodiment was not a simple process, however. Feminist theology, like feminism in general, has continually modified original insights, not settling once and for all a meaning for every woman's experience. Thus, a beginning feminist response to past religious and cultural associations of women with their bodies was a rejection of this association. Anatomy was *not* destiny; women were not to be identified with their bodies any more than were men; women could transcend their bodies through rational choices. Such a response paradoxically freed women, however, to take their bodies more seriously. Rather than abstract from bodyliness, reinforcing a dichotomy between body and mind, women soon moved to 'reclaim' their bodies – to claim them as their own, as integral to their selfhood and their womanhood. This entailed new practical and theoretical approaches. Reflecting on their experiences, woman shared insights and interpretation, formulated new symbols, expanded and revised understandings of human embodiment as such [43].

Struggling to move beyond the dualism of body and self that had limited them for so long, feminist philosophers and theologians used a phenomenological method to describe what it means to *be* a body as well as *have* a body, to understand their own bodies as ways of being inserted into the world, as structured centers of personal activity, as body-subjects not just body-objects [28]. From an understanding of themselves as embodied subjects, women 'reclaim' their bodies not just by taking them seriously and 'living' them integrally, but by refusing to yield control of them to men. New intimate self-understandings and new philosophical and theological anthropologies yield new personal and political decisions.

*The World of Nature*

The third theme in feminist theology which is potentially relevant to issues in bioethics is the meaning and value of the world of nature. Feminist theologians' concern for this theme is directly influenced by their concern for patterns of human relations and for the world as the place of human embodiment.

Just as women have been thought of in religious and cultural traditions as 'close to nature', so the world of nature has been symbolized as female. This is a clue to the difficulties which feminist theologians have with past beliefs and attitudes regarding nature. They find, in fact, a correlation between patterns of domination over women and efforts at domination over nature ([36], pp. 72–85; [37], pp. 57–70).

Perceptions of nature change through history, of course, and its symbolism is always to some extent ambivalent. Nature has been exalted in importance beyond the being and culture of humans, or reduced to a tool for humans; it has been viewed as the cosmic source of life and goodness, or a mysterious force to be feared and fled or controlled. All of these interpretations of nature mirror similar identifications of the essence of woman. However, especially in the history of Christian thought, a pattern emerges which raises serious questions for feminist theology.

Despite the fact that a Christian world-view and specific Christian teachings have supported 'sacramental' views of the whole of creation, sometimes especially of nature (as revelatory of the fidelity, the presence, the grandeur and the graciousness of the living God), Christianity has nonetheless also tended to trivialize the value of nature. Ascetic theologies sometimes reduced nature to a transitory illusion, a distraction from 'higher things', in the manner of some Hellenistic philosophies. Christian leaders sometimes forbade the study of nature as dangerous or a waste of time. When nature and culture were paired among traditional dualisms, nature was assigned the value of the negative pole.

Similarly, while there is a strong tradition in Christian thought requiring reverence for and stewardship of nature, there is also strong support for a way of relating to nature which sees it only as something to be used, dominated, controlled by human persons ([19], p. 7). In this latter view, because nature has no value of its own, it can be treated as the private property of humanity (or of individuals), with no limits on its exploitation or manipulation short of the limits of human persons' own self-interest. Where total possession is permitted, the concept of 'rape' does not apply.

Rosemary Ruether traces a history of Western attitudes toward nature which moves from an early ascetical 'flight' from nature to a modern 'return' to nature ([36], p. 82). The rise of scientific research in the seventeenth century helped secularize a view of nature, fostered a perception of it as intrinsically rational, penetrable, manageable. Unintended negative consequences of scientific and technological development, visible from the nineteenth century on, produced romantic reactions calling for a different sort of return to nature – a restoration of 'pure' nature, uncontaminated and unalienated by human intervention. All of these attitudes toward nature, however, represented pieces of the pattern of hierarchical domination and subjugation – domination through possession and control, whether through denigration, or exploitation, or the expectation of mediated happiness and identity through 'keeping' nature as a haven for some (despite the suffering this in turn might cause for others) ([36], p. 85; [37]).

Feminist theology argues, alternatively, for a view of nature consonant with a view of a God who takes the whole of creation seriously, and a view of creation which does not see predatory hierarchy as the basis of order. Nature, in this view, is valuable according to its own concrete reality, which includes an interdependence with embodied humanity. It is limited in its possibilities, which precludes its moral use as the battleground for the ultimate challenge to human freedom. Human intelligence and freedom are not barred from addressing nature, but measures for understanding and just use are lodged both in nature itself and in ethical requirements for relations among persons.

## FEMINIST THEOLOGICAL ETHICS

Given this overview of themes in feminist theology, it may be possible to identify some characteristics of a feminist theological ethic, moving still closer to connections with issues in bioethics. We have, for example, seen enough of feminist theology to draw some conclusions regarding the *methods* likely to characterize any ethics that derives from it. First in this regard is a sense in which feminist theology and ethics can be said to be concerned with objective reality, and hence to presuppose methodologically some access to an intelligible reality. The work of these disciplines began, after all, as a result of what was at base a new understanding of the reality of women. Like feminism in general, feminist theology had its origins in women's growing awareness of the disparity between received traditional interpretations of their identity and function within the human community and their own experience of themselves and their lives. The corresponding claim that gender role-differentiation and gender-specific limitations on opportunities for education, political participation, economic parity, etc., are discriminatory was based on the argument that past interpretations of women's reality were simply wrong. That is, past theories failed to discover the concrete reality of women and represented, in fact, distorted perceptions of that reality. Moreover, the attitudes and policies they fostered often did violence to that reality.

It would be a mistake to label feminist theology and ethics in any simple sense 'naturalistic', though the term is not wholly inappropriate. Feminist theology does not, obviously, reduce to a natural or behavioral science. Nor does it rely for its access to reality on human reason alone. And while feminist theological ethics searches for and proposes universally valid norms, it does so in a way that acknowledges the historical nature of human knowledge and the social nature of the interpretation of human experience. The fact that present insights may be superseded by future ones, and that present formulations of specific principles may change, does not contradict either the methodological requirement of attending to concrete reality or the methodological presupposition that the accuracy and adequacy of theories can be tested against that reality.

Closely aligned with all of this is the methodological commitment to begin with and continue a primary focus on the experience of women ([25]; [36], pp. 12–13). This is often coupled with the qualification that no claims are made for the universality of women's experience in relation to human experience. There is a claim made, however, that until a theology based on women's experience is developed, traditionally assumed universal claims for a theology based on men's experience will continue to render inadequate if not inaccurate the major formulations of religious belief.

A methodological commitment to the primacy of women's experience as a source for theology and ethics goes significantly beyond a simple focusing of attention, however. It yields, in addition, a feminist hermeneutical principle which functions in the selection and interpretation of all other sources. While not every feminist theologian articulates this principle in exactly the same way, it can be expressed as strongly as, "Whatever diminishes or denies the full humanity of women must be presumed not to reflect the divine or an authentic relation to the divine, or to reflect the authentic nature of things, or to be the message or work of an authentic redeemer or a community of redemption" ([36], p. 19). As is to be expected, this principle functions in importantly different ways in different feminist theologies. In some, for example, it leads to the rejection of the authority of the Bible altogether [41]; in others it allows the relativization of the authority of some texts [10]; in still others it leaves all texts standing as a part of an authoritative revelation, but renders their meaning transformed under a new feminist paradigm [42]. The same is true for theological doctrines, historical events, and for sources of theology and ethics which can range from the comparative study of religions to philosophical and scientific writings and schools of thought.

A focus on women's experience, the use of a feminist hermeneutical principle, and a concern for the lived experience of women precisely as disadvantaged, can constitute for feminist theological ethics the bias for women which is the earmark of feminism in general. If this is chosen as a strategic priority, feminist theological ethics can be methodologically oriented ultimately as an ethic whose concerns include the well-being of both women and men, both humanity and the world of nature. Its theological center will depend on its ultimate warrants for these concerns.

Finally, in regard to method, feminist theological ethics has been open to both deontological and teleological patterns of reasoning ([14], pp. 12–13). On the one hand, for example, the very notion of 'strategic priority', as well as a strongly 'ecological' view of reality, implies a concern for consequences, an ethical evaluation of means in relation to ends and parts in relation to wholes, a relativization of values in situations of conflict. On the other hand, demands of the concrete reality of persons are such that some attitudes and actions can be judged unethical precisely because they contradict values intrinsic to that reality. The sorting out of what ultimately determines a specific obligation is the task of ethics, but neither of these modes of reasoning is ruled out for feminist theological ethics.

When we turn from method to *substance* in feminist theological ethics, we perhaps need only summarize the ethical import of what we have seen in regard to feminist theological themes. Thus, an ethic derived from feminist theology understands the well-being of persons in a way that takes account of their reality as embodied subjects (and hence includes considerations of persons as historical beings, living in social and cultural contexts, identified with yet transcending systems and institutions; as beings whose actuality includes potentiality for development as well as vulnerability to diminishment; beings constituted by complex structures of freedom, physiology, intelligence, affectivity, etc.; beings which are essentially interpersonal and social; beings which are unique as well as common sharers in humanity). It is an ethic which gives important status to principles of equality and mutuality. It holds together principles of autonomy and relationality. It gives ethical priority to models of relationship characterized by collaboration rather than competition or hierarchical gradation. Finally, it does not isolate an ethic of human relations from ethical obligations to the whole of nature. These sound like a list of ideals, high rhetoric which any ethic may incorporate somehow. Some test of it can be made by turning now to issues in bioethics.

## FEMINIST THEOLOGY AND BIOETHICS

Feminist theology offers something of a distinctive perspective on many issues that we today include under the general rubric of bioethics. This is because women's lives are deeply implicated in areas of personal medical care, public health, and the development and use of biomedical technologies. Feminist theology also gains from analyses of issues arising in these areas, for here the lived experience of women reveals some of the central opportunities and limitations of the human condition. Here it is that "reflection upon the goals, practices, and theories of medicine validates philosophical reflection upon many issues that have traditionally been of concern to women," but ignored by the traditional disciplines of philosophy and theology ([46], p. 120).

We can explore the interrelation between feminist theology and bioethics in a number of ways. Thus, we can examine the perspective offered by feminist theology on the principles usually considered central to bioethics – principles of, for example, nonmaleficence and beneficence, veracity and fidelity, as well as autonomy, mutuality, and justice. We can also look to numerous specific issues for which femenist theology and ethics can be expected to have special relevance – issues such as abortion and sterilization, medical care of the elderly, psychiatric treatment of women, medical settings for childbirthing, conflict between nursing roles and moral rules, the use of amniocentesis for gender selection, models of doctor/patient relationships in a culture and in relation to a profession marked by sexism. Among these and other possibilities, however, let me select for consideration the issue of the development and use of reproductive technologies.[3] A feminist theological approach to this issue may, in even a brief attempt, show some of the implications of feminist theology for understanding both context and principles in the area of bioethics.

The potentialities of reproductive technology have for some time caught the attention of feminists. It raises issues, however, on which unanimity of view does not exist. Some feminists have argued that the ultimate source of women's oppression is their physiological capability of bearing children. While physical motherhood can constitute individual and social power, it also renders women powerless – before nature, before men, before their children, before society (which judges them, and which determines the conditions under which their children must grow). In the face of this powerlessness, and the suffering it entails, technology offers a solution. Indeed, in an extreme view, women's liberation can only be achieved with a revolution not only against forms of society, but against nature itself. Thus, Shulamith Firestone argued for the "freeing of women from the tyranny of their reproductive biology by every means available", including technology which could separate women once and for all from a gender-identified responsibility for reproduction ([11], p. 238).

This was a relatively early position, however, and strong disagreement has come from other feminists on a variety of grounds. Many consider the analysis of the causes of oppression to be wrong ([21], pp. 87–91). Others see in the development of reproductive technologies a new means of devaluing women, rendering them "expendable in the procreative process" ([14], p. 37). Still others argue that some uses of technology, such as amniocentesis for the purpose of gender selection, will pit women against themselves [26].

Feminists agree on at least two things with regard to these questions. First, the history of women's experience in relation to the power and process of reproduction is, indeed, a history of great pain. While fertility, pregnancy, and childbirth have been a source of women's happiness and fulfillment, and an occasion for powerful expressions of great human love and enduring fidelity to duty, they have also been the locus of a cumulative burden of immense oppression and suffering. The twentieth century incursion of technology into reproduction (the 'medicalization' of pregnancy and childbirth) has often added to this suffering, extended this oppression.

Second, and closely following, feminists are in agreement that the development and use of reproductive technology cannot be evaluated apart from its concrete, socio-cultural context. This context has been, and remains, an "historically specific social order in which positions of power and privilege are disproportionately occupied by men" ([7], p. 41). As long as sexism continues to characterize the lived world which women know, technology will have different consequences for women and for men. Far from freeing women from unnecessary burdens in reproduction, further technological development may result in greater bondage.

Given these agreements, however, neither feminism in general nor feminist theology renders wholly negative judgments on reproductive technology. One obvious reason for this is that such technology can take many forms. Evaluations of developments of contraceptives, childbirth procedures, methods of abortion, artificial insemination, *in vitro* fertilization, fetal diagnosis, cloning, and many other technologies can hardly be lumped together in a single comprehensive judgment. Only a total anti-technology approach could yield that. Generally, despite deep ambivalence toward reproductive technologies, feminists can affirm that

> natural-scientific breakthroughs represent genuine gains in human self-understanding. The widespread social irresponsibility of medical practice, exacerbated by male monopoly of the medical profession that is only now changing, must not be confused with the value of scientific discoveries ([14], pp. 169–170).

Science and technology have, in fact, been instruments of reform at times, even in regard to almost intractable problems of sexism ([32], pp. 22, 83, 136).

But if a single ethical judgment cannot be made for all forms of reproductive technology, then it will be helpful here to narrow our focus still more to one form. Once again, this will have implications beyond itself for reproductive technology more generally, but we cannot expect thereby to have resolved all questions. The form of technology that I will consider is *in vitro* fertilization for the purpose of producing a child (that is, not just as a procedure for the purpose of scientific research with no intention of bringing a child 'to term'). As a technology, it raises the issue of profound change in human modes of reproduction, not just the issue of improving present modes.

One place to begin a feminist analysis of *in vitro* fertilization (with embryo transfer or some other form of providing for gestation) is with women's experience to date of technology in the area of pregnancy and birth. As we have already noted, this is in many respects not a happy experience. Recent studies have helped to make visible the difficulties women have had in this regard ([44]; [22]; [29]; [17]; [5]). Recalling these difficulties can help us to formulate the questions that need to be asked of *in vitro* fertilization. If, for example, the use of medical technology in relation to childbirth has been oppressive to women, or to their children, in what way has it been so? One response to this question is that it has contributed to the alienation of women from their bodies, their partners, and their children (by, for example, moving childbirth into settings appropriate primarily for the treatment of disease, isolating mothers both from 'women's culture' and their spouses, regimenting the presence of mothers with their babies, etc.);[4] and that it has placed women in a network of professional relations which unjustifiably limit their autonomy (in the manner of a 'patient'). Does the development and use of *in vitro* fertilization hold this same potential for alienation, albeit in different ways? From a feminist theological perspective, the question can be: Does *in vitro* fertilization violate (or is it in accord with) feminist understandings of embodiment, norms for relationships, and concerns for the common good? The following considerations hold this question as their backdrop.

For many feminists the sundering of the power and process of reproduction from the bodies of women would constitute a loss of major proportions. Hence, the notion of moving the whole process to the laboratory (using not only *in vitro* fertilization but artificial placentas, *et al.*) is not one that receives much enthusiasm. On the other hand, *in vitro* fertilization is not perceived as a procedure which necessarily violates the essential embodying of reproduction. If its purpose is primarily to enable women who would otherwise be infertile to conceive a child, it becomes a means precisely to mediate embodiment. Feminists generally oppose the kind of sacralization of women's reproductive organs and functions that would prohibit all technological intervention. In fact, desacralization in this regard is seen as a necessary step in the breaking of feminine stereotypes and the falsification of anatomy as destiny. Moreover, feminist interpretations of sexuality are very clear on the validity of separating sexuality from reproduction. Without contradiction, however, they also affirm reproduction as a significant potential dimension of separating sexuality from reproduction. Without contradiction, however, they also affirm reproduction as a significant potential dimension of series of 'natural' physical connections between sexual intercourse and the fertilization of an ovum by male sperm. Indeed, it is a faulure of imagination which sees this as the only way in which integrated sexuality can be related to reproduction. All in all, then while human embodiment remains a central concern in a feminist analysis of *in vitro* fertilization, it does not thereby rule out the ethical use of this technology.

Feminists are generally clear on the need to understand and experience childbearing in an active way. Pregnancy and childbirth are not events in relation to which women should be wholly passive ([14], pp. 169, 246–247). Part of taking active control and responsibility regarding their reproductive power can include a willingness to use technology insofar as it makes childbearing more responsible, less painful, and more safe. Sometimes discernment of just these consequences for technology is difficult, but the fact that it is called for indicates, again, that *in vitro* fertilization is not ruled out in principle.

Perhaps the most troubling aspect of *in vitro* fertilization, and of other technologies which actually empower reproduction, is the question of primary agency and responsibility. This question in itself has many sides. For example, women's previous experience with reproductive technology suggests that (at least in the concrete context of present societies) women's own agency is likely to be submerged in the network of multiple experts needed to achieve *in vitro* fertilization. Far from this accomplishing a liberation of women from childbearing responsibilities, it can entail "further alienation of our life processes" ([36], p. 227). Moreover, efforts to restrict and share the agency of professionals often move only in the direction of what some feminists fear as collectivism or state control, the "total alienation of one's life to institutions external to one's own control and governed by a managerial elite" ([36], p. 226]). In any case, without a drastic change in the composition of society and the professions, widespread use of *in vitro* fertilization could, it seems, make it difficult for women to achieve or sustain control of human reproduction.

But does it matter whether women or men, parents or scientists, control reproduction? Feminists argue that those who will bear the responsibility for childrearing should have primary agency in decisions about childbearing – not just because it is their right if they are to bear the burden of such responsibility, but because this is required for the well-being of offspring. "Only those who are deeply realistic about what it takes to nourish human life *from birth onward* have the wisdom to evaluate procreative choice" ([14], p. 173).[5] Reproductive technologies that divorce decisions for childbearing from childrearing fail to take seriously the basic needs of children for not only material resources but personal relation and support, in contexts that allow the awakening of basic trust and the development of fundamental autonomy ([34], p. 258; [27], p. 65).[6] It is not only women who, in principle, can make these choices ([34], p. 262),[7] but it is 'parents', not just 'scientific facilitators' or society at large or any persons who are unprepared to take responsibility at an intimate as well as comprehensive level for our children. Such problems of agency are complex and sobering in the face of technological capabilities such as *in vitro* fertilization. They are not, in principle, intractable, perhaps not even in practice. They need not rule out the ethical use of *in vitro* fertilization, but they occasion grave moral caution.

Yet another consideration prompted by *in vitro* fertilization (and other reproductive technologies) regards the developing capability for 'selection' of offspring – from among many candidates (differentiated by gender, bodily health, intellectual capacity, etc.). The problem of 'discards' in *in vitro* fertilization is larger than the discernment of grave embryonic anomalies. For some feminists this capability can erode moral and religious obligation to accept all sorts of persons into the human community. In so doing, it undermines basic feminist principles of equality, inclusiveness, mutuality, and toleration of difference and of 'imperfection' ([7], p. 42). *In vitro* fertilization need not, of course, be used in this way. But once again, a voice of caution is raised.

Underlying all of these considerations is what might be called the need to measure *in vitro* fertilization according to norms of justice. If justice in its deepest sense can be understood as treating persons in truthful accordance with their concrete reality (a concrete reality which must be interpreted as best we can), then all the issues of embodiment, nondiscrimination, agency, responsibility, inclusive care, are issues of justice. They are not focused only on individuals, but on the human community. They converge in the fundamental question, 'How are we to reproduce ourselves as human persons?' They press us to new theories of justice which extend a requirement for 'just parenting' in relation to all human children. They include, then, too, questions of the meaning and value of *in vitro* fertilization in a world threatened by overpopulation, in countries where not every existing child is yet cared for, in communities where grave needs of children require the resources of science and technology. Questions of macroallocation of scarce goods and services may finally be unresolvable, but they cannot be ignored. At the very least, in this instance, they preclude justifications of *in vitro* fertilization on the basis of any absolute right to procreate.

A feminist analysis of *in vitro* fertilization remains provisional. It yields, however, the following position: Negatively there are not grounds for an absolute prohibition of the development and use of technology such as *in vitro* fertilization; positively, such technology may aid just and responsible human reproduction. The presence of certain circumstances, or certain conditions, sets limits to its ethical development and use – circumstances such as (a) high risk of injury to the well-being of either parent or child; (b) a context unconducive to the growth and development of any child produced (unconducive because, for example, no one is prepared to offer the child basic human personal relationship); (c) an intention to produce a child to be used as means only in relation to the producers' ends (as, for example, if the child is produced merely for the sake of the advance of scientific research, or for the duplication of one's own self without regard for the child's development into an autonomic self); (d) failure to meet criteria of distributive justice (when it is determined that other basic human needs place legitimate prior claims on the resources involved). Such conditions rule out spectres of human laboratory 'farms'. They also tell us something about the conditions for *any* ethical decisions regarding human reproduction, not just decisions made in the context of reproductive technology.

With this, then, we have one example of the relation between feminist theology and an issue in bioethics. My development of the issue must remain more suggestive than exhaustive of the particular ethical values and ultimate theological warrants that feminist theologians may offer. My suspicion is that future work in the area of bioethics will bring careful reflection on questions that I have not, within the limits of this essay, addressed at all; questions, for example, of women's interpretation not only of birth but of death, and women's evaluation of the strength of 'quality of life' claims in relation to sanctity-of-life principles. Whatever lines along which a feminist bioethics may develop, however, it will never be far from central concerns for human embodiment, for the well-being of women-persons on a par with the well-being of men-persons, for newly just patterns of relationship among all persons, and for the balanced care of the whole world of both nonpersonal and personal beings.

*The Divinity School,*
*Yale University,*
*New Haven, Connecticut, U.S.A.*

## NOTES

[1] For a contrary emphasis, see ([15], p. 48).

[2] To the argument that separatist feminist movements do indeed contradict these values (by affirming a new form of elitism, by simply 'reversing' the order in the hierarchy of men and women, etc.), the response is sometimes given that separation does not entail domination, and that elitism is no more a necessarily substantial charge against separatist feminism than it is against any religious sectarianism. It is more difficult for some separatists to answer the criticism that they are duplicating oppressive patterns of 'identifying an enemy'.

[3] Reproductive technologies include all those technologies which relate to human reproduction. They are sometimes differentiated from technologies of genetic engineering, though I do not in this essay maintain a sharp separation. For some helpful distinctions, see ([27], pp. 8–10).

[4] 'Tales of horror' are told more and more in recent sociological studies in this regard. However, it should be noted that very recently there have come significant changes – changes, for example, such as an increase in home birthing, the provision of birthing rooms in hospitals, the rise once again of the profession of midwifery, etc. Some feminists express concern that some new movements, such as natural childbirth, incorporate an alienating technology just as previous methods did ([44], pp. 183–198; [22], pp. 628–630).

[5] I am not, here, focusing on the grounds for women's right to procreative choice which are often central to feminist arguments – that is, a right to bodily integrity or a right to privacy. One reason I am not focusing on those grounds is that *in vitro* fertilization *can* be understood to prescind from women's bodies in a way that, for example, abortion cannot.

[6] This can be maintained without conflicting with contemporary concerns for 'too much mothering', etc.

[7] Nor should it be the exclusive prerogative of women. When it is this, it justifies a male dismissal of obligation regarding childbearing – something feminists have long been concerned to oppose.

## BIBLIOGRAPHY

[1] Carr, A.: 1982, 'Is A Christian Feminist Theology Possible?' *Theological Studies* **43**, 279–297.

[2] Christ, C. P.: 1977, 'The New Feminist Theology: A Review of the Literature', *Religious Studies Review* **3**, 203–212.

[3] Cott, N. F.: 1978, 'Passionlessness', *Signs* **4**, 227–228.

[4] Daly, M.: 1973, *Beyond God the Father*, Beacon, Boston.

[5] Daly, M.: 1978, *Gyn/Ecology: The Metaethics of Radical Feminism*, Beacon, Boston.

[6] Donegan, J. B.: 1978, *Women and Men Midwives: Medicine, Morality, and Misogyny in Early America*, Greenwood Press, Westport, Connecticut.

[7] Elshtain, J. B.: 1982, 'A Feminist Agenda on Reproductive Technology', *Hastings Center Report* **12**, 40–43.

[8] Farley, M. A.: 1975, 'New Patterns of Relationship: The Beginnings of a Moral Revolution', *Theological Studies* **36**, 627–646.

[9] Farley, M. A.: 1976, 'Sources of Sexual Inequality in the History of Christian Thought', *The Journal of Religion* **56**, 162–176.

[10] Fiorenza, E. S.: 1983, *In Memory of Her: A Feminist Reconstruction of Christian Origins*, Crossroad, New York.

[11] Firestone, S.: 1971, *The Dialectic of Sex: The Case for Feminist Revolution*, Bantam, New York.

[12] Freud, S.: 1918, 'The Taboo of Virginity', *Collected Papers*, Vol. 8, pp. 70–86.

[13] Gustafson, J. M.: 1975, *The Contributions of Theology to Medical Ethics*, Marquette University Press, Milwaukee.

[14] Harrison, B. W.: 1983, *Our Right to Choose: Toward A New Ethic of Abortion*, Beacon, Boston.

[15] Hauerwas, S.: 1978, 'Can Ethics Be Theological?' *The Hastings Center Report* **8**, 47–49.

[16] Heyward, C.: 1979, 'Ruether and Daly: Theologians Speaking and Sparking, Building and Burning', *Christianity and Crisis* **39**, 66–72.

[17] Lebacqz, K.: 1975, 'Reproductive Research and the Image of Woman', in Fischer, C. B. *et al.*, *Women in A Strange Land*, Fortress, Philadelphia.

[18] McCormick, R. A.: 1983, 'Bioethics in the Public Forum', *Milbank Memorial Fund Quarterly* **61**, 113–126.

[19] McCormick, R. A.: 1981, *How Brave A New World? Dilemmas in Bioethics*, Doubleday, Garden City, New York.

[20] McFague, S.: 1982, *Metaphorical Theology*, Fortress, Philadelphia.

[21] Mitchell, J.: 1971, *Woman's Estate*, Vintage, New York.

[22] Oakley, A.: 1979, 'A Case of Maternity', *Signs* **4**, 606–631.

[23] O'Brien, M.: 1981, *The Politics of Reproduction*, Routledge and Kegan Paul, London.

[24] Ortner, S. B.: 1974, 'Is Female to Male as Nature is to Culture?' in M. Z. Rosaldo and L. Lamphere (eds.), *Woman, Culture, and Society*, Stanford University Press, Stanford, pp. 67–87.

[25] Plaskow, J.: 1980, *Sex, Sin, and Grace*, University Press of America, Washington, D.C.

[26] Powledge, T. M.: 1981, 'Unnatural Selection: On Choosing Children's Sex', in H. B. Holmes *et al.* (eds.), *The Custom-Made Child? Women-Centered Perspectives*, Humana Press, Clifton, New Jersey, pp. 193–199.

[27] President's Commission for the Study of Ethical Problems in Medicine and Biomedical and Behavioral Research: 1982, *Splicing Life*, U.S. Government Printing Office, Washington, D.C.

[28] Rawlinson, M. C.: 1982, 'Psychiatric Discourse and the Feminine Voice', *The Journal of Medicine and Philosophy* **7**, 153–177.

[29] Rich, A.: 1976, *Of Woman Born: Motherhood as Experience and Institution*, Prometheus, Buffalo.
[30] Ricoeur, P.: 1967, *The Symbolism of Evil*, Harper, New York.
[31] Robb, C. S.: 1981, 'A Framework for Feminist Ethics', *The Journal of Religious Ethics* **9**, 48–68.
[32] Rosenberg, R.: 1982, *Beyond Separate Spheres: Intellectual Roots of Modern Feminism*, Yale University Press, New Haven.
[33] Rothman, B. K.: 1979, 'Women, Health, and Medicine', in J. Freeman (ed.), *Women: A Feminist Perspective*, Mayfield, Palo Alto, pp. 27–40.
[34] Ruddick, S.: 1980, 'Maternal Thinking', *Signs* **6**, 342–367.
[35] Ruether, R. R. (ed.): 1974, *Religion and Sexism: Images of Women in the Jewish and Christian Traditions*, Simon and Schuster, New York.
[36] Ruether, R. R.: 1983, *Sexism and God-Talk: Toward A Feminist Theology*, Beacon, Boston.
[37] Ruether, R. R.: 1981, *To Change the World*, Crossroad, New York.
[38] Russell, L.: 1982, *Becoming Human*, Westminster, Philadelphia.
[39] Russell, L.: 1979, *The Future of Partnership*, Westminster, Philadelphia.
[40] Saiving, V.: 1981, 'Androgynous Life: A Feminist Appropriation of Process Thought', in S. G. Davaney (ed.), *Feminism and Process Thought*, Edwin Mellen Press, New York, pp. 11–31.
[41] Stanton, E. C. (ed.): 1974, *The Original Feminist Attack on the Bible: The Woman's Bible*, Arno, New York.
[42] Trible, P.: 1978, *God and the Rhetoric of Sexuality*, Fortress, Philadelphia.
[43] Washbourn, P.: 1979, *Becoming Woman: The Quest for Spiritual Wholeness in Female Experience*, Harper, New York.
[44] Wertz, R. W., and Wertz, D. C.: 1977, *Lying-In: A History of Childbirth in America*, Free Press, New York.
[45] Whitbeck, C.: 1984, 'A Different Reality: Feminist Ontology', in: C. Gould (ed.), *Beyond Domination: New Perspectives on Women and Philosophy*, Rowman and Allenheld, Totawa, N.J., pp. 64–85.
[46] Whitbeck, C.: 1983, 'Women and Medicine: An Introduction', *The Journal of Medicine and Philosophy* **7**, 119–133.

MARK JUERGENSMEYER

# DOING ETHICS IN A PLURAL WORLD

The nearly universal moral acclaim granted to Gandhi – even more after his death than during his lifetime – may eventually make him one of history's first global saints. If so, it would be a significant phenomenon: rarely has a figure been admired as a moral exemplar simultaneously in several cultural traditions. To what can we attribute Gandhi's multicultural appeal? One explanation – and a enticing one at that – is that Gandhi manifests a set of virtues held in common by many ethical traditions.

This suggests that there is a universal model of morality, one ready to be tapped by a man like Gandhi. Perhaps there is, and it is a possibility that I want to consider later on in this essay. But before we come to that conclusion, we should take a close look at the way in which a figure like Gandhi is admired in different cultures, for it could lead to a somewhat different estimation of Gandhi and, at least initially, to a somewhat different conclusion about ethics.

Gandhi radiates such a diversity of virtues that many cultures can find something in him to admire, even though it is not always the same thing. In this sense he is more a multiple saint than a universal one. Christians, for instance, see him as Christ-like – humble and self-effacing – and Hindus regard him as the champion of *rajdharma*, the morality of political leadership that requires assertiveness, along with qualities of discipline and inner strength. Even though Gandhi is admired in several cultures at once, this fact may not demonstrate that ethics are global. In fact, it may show that ethics can be persistently plural.

One example of the distance between Indian and Western expectations about Gandhi's ethical behavior is nicely illustrated by a scene that appears in the recent cinematic portrayal of Gandhi's life. Theater audiences watch as the wife of the Mahatma, Kasturbai, explains to a Western devotee of Gandhi how difficult it was for her husband to resist his sexual urges and remain celibate. It is a poignant scene, but to many members of Western audiences it is a curious and expendable moment in the film. Why, they wonder, is it so important for Gandhi to remain celibate, and how did Kasturbai feel about his vow of abstinence? She was his wife, after all: didn't his sense of marital responsibility require such a decision to be made jointly?



*E. E. Shelp (ed.), Theology and Bioethics,* 187–201.

Indian audiences react to the scene in a different way, and these questions that trouble Westerners do not seem to concern them. They regard the scene as critical to a spiritual portrayal of the Mahatma. The film is about Gandhi, not about Kasturbai, the Indians would say, and it is a film about a saint. To be a saint in India means to gain mastery over spiritual and physical powers, including the power of sexual passion. Moreover, in the traditional Indian view, semen is thought to contain both spiritual and physical potency, and one of the first tasks in a morally disciplined life is to restrain the seminal flow. A male saint is celibate in order to keep that vital fluid to himself. The ultimate ascetic is the god Siva, but Siva is portrayed in erotic as well as ascetic forms, because the magnitude of his sexual powers is related to his spiritual potency. Sublimated and restrained, they provide the fuel for his ascetic capabilities [23]. Gandhi's spiritual potency is also seen to have sexual dimensions: he was obsessed with sex as a youth (at least in Gandhi's own estimation [7]) but became celibate later in life, after his children were born.

The differences between the Indian view of sexual continence and the Western view help explain why audiences in Delhi and San Francisco respond in opposite ways to a scene in the cinematic portrayal of Gandhi's life. Interesting though these ethical differences may be, for most of us the matter is of academic interest; it does not present a personal dilemma. But if we were to live with our spouses or our lovers in a Gandhian ashram, our views of celibacy and sexuality might quickly clash with those of Gandi. And if we wanted to adopt Gandhi as a personal saint and moral model, the aspects of Gandhian morality that do not rest easy with Western points of view would be more than cultural quirks. They would be personal problems. Our own moral confusion would mirror the clashes of ethical assumptions that occur in political and economic spheres. Such are the difficulties of taking on a universal saint in an ethically plural world.

The problems are magnified when the issues at stake are ones that are dear to us – not only sexual continence, but human rights, for instance, and the sanctity of life. When these issues seem to be regarded differently by moral authorities in cultures other than our own, we are tempted to question whether they are right – whether these spokespersons adequately express the views of their own cultures – and if so, whether the moral systems they represent should be regarded as legitimate.

In such cases our moral and intellectual task is a double one: the descriptive duty to find out what the tradition says and why its views may be different from ours; and the normative one of deciding what, if anything, we can do to bring about a conciliation between those views and our own.

## THE CASE OF ABORTION

One of the most sensitive issues of life and death is that of abortion, which provides a useful example of cross-cultural ethics. It also show how exasperating the differences can be. The induced destruction of the fetus has been practiced in virtually every culture of the world, from ancient times to the present ([2], p. 54; [4]), and in every time and place where it has been committed, the practice is the subject of ethical debate. Yet the way in which it is viewed as an ethical problem varies vastly from culture to culture.

It also varies greatly from gender to gender, and it is quite possible that these differences are even greater than those of cultural dissimilarities. In the examples that follow, however, I will concentrate on the cultural differences, and the reader should keep in mind that these are largely male perspectives. Many Catholic and Muslim women, for examples, might view the issue quite differently than have the official, usually male, religious authorities of their traditions.

The conservative Protestant and Roman Catholic positions against abortion, which are based on the inviolability of human life and the sacredness of the human soul, are often contrasted with the prevailing liberal humanistic view shared by many Protestant Christians. The liberal view is also supported by moral principles, including the right of freedom of choice and the responsibility for maintaining the quality of family and public life through limiting the growth of population. Moreover, the liberal view sees life before birth as developing incrementally, not full-blown at the point of conception. The Soviet point of view goes even beyond liberal humanism in stressing the responsibility that parents have towards the whole of society, and allowing for abortion as a matter of public policy [20].

The Jewish and Islamic points of view fall somewhere in between the conservative Christian and liberal Protestant extremes. Like conservative Christians, most Jews and Muslims are, above all, concerned about the taking of human life, but like liberal Protestants they do not accept that life begins at the moment of conception. This is often a matter of some debate.

One example of the ambivalence within the Muslim position is the judgment made by the Grand Mufti of Jordan in 1964, a position that has since been reversed. The Mufti determined that although the Qur'an has an injunction against 'murdering one's own children', the fetus cannot be defined as a child until the last 120 days of normal pregnancy. The Mufti

Mufti implied that abortion could be permitted before that time because human life had not yet been formed ([20], p. 268). Since the Mufti's decision, however, the official position of Jordan's Islamic courts has changed. Even though there is general agreement that a child cannot receive the rights accorded other humans until birth, fetal life is now regarded as human life from the time of conception, and abortion is allowed only if the mother's life is in danger or if there is a likelihood that the child will be born deformed or seriously diseased [13].

From at least one traditional Hindu point of view, the soul does not enter the body until the time of birth [*Katha Upanishad* I.ii.18], and perhaps for that reason the Indian government has been remarkably permissive about abortion. At the same time, however, other traditional Hindu texts have denounced the destruction of any form of life, including fetal life [*Atharva-Veda* xlii, 165 and the *Manusamiti*]. The Buddhist and Jain traditions express a similar revulsion against the taking of life, and even though Buddhists feel that there is no such thing as a human soul and therefore no need to protect its bodily container, they feel that the act of destruction imposes a great weight of moral demerit on those who commit it.

These Buddhists, Jains, and Hindus seem to align themselves with the conservative Christians who vigorously defend the right to life. Yet from the Asian perspective the Christian concern with the protection of the human soul seems curiously narrow. All life is sacred, in the Asian view, and in the Hindu understanding the soul may travel through various bodily habitations, including animal forms, before it reaches its ultimate abode. The same logic that would compel Hindus to disavow abortion, therefore, would require them also to reject war, capital punishment, and the killing of animals. It is a source of great puzzlement to many Indians that Christians who are exercised about abortion can at the same time advocate the death penalty, condone an aggressive military posture, and maintain a diet dependent on meat.

Gandhi abhorred all of these practices, and relied upon Hindu tradition in defending his position. For him the Hindu insistence on the protection of life, including that of animals, was an abiding concern. In fact his attitude toward the cow and its virtues would strike the Western mind as a peculiar obsession. Yet Gandhi's attitude was the logical conclusion of an ethical system that regards the lives of all sentient beings as interconnected. Hindus usually object to the destruction of the fetus, just as they object to the destruction of any other living thing. But their reasons for doing so are quite different from those of conservative Christians who, in arguing for the 'right to life', mean by that only its human form.

The Native American points of view differ from both the conservative and the liberal Christian perspectives, and those of many other traditions as well, for the Native American attitude emphasises the process of making a moral decision rather than the decision itself. In most Native American groups there is no single principle regarding abortion; moral attitudes vary according to the circumstances of each case. Harmony with nature and with the tribal society are the guiding principles for a collective decision regarding whether a wrong was committed and what should be done about it. If punishment is meted out to those involved in an abortion, it varies according to the factors surrounding the decision to abort, and the impact of the act [4].

Considering the variety of views and the depth of the disagreements, one might well be skeptical about the possibility of arriving at a global moral consensus on the issue of abortion. Skepticism might also be appropriate for similar attempts at universal agreement in the fields of social equality and human rights. In Louis Dumont's cross-cultural studies of the value systems of traditional India and the modern West, the universality – and even the moral superiority – of the notion of individual equality is put to question [5, 6]. This casts some question over the way we evaluate the situation of those in India who call for equality, especially in such cases as the treatment of Untouchables. As I have found in my own studies of this issue, the problem of making some judgment about untouchability is complicated by the fact that the concept is viewed differently from one ethical framework to another [12, 15]. Understanding how these frameworks differ, however, does not make it easy for us to accept these views as equally valid.

Those of us who observe the ethical attitudes of other cultures may at times be disturbed by what we observe. The ethical discrepancies among cultures may then become compelling personal problems. This is true not only for those of us who occasionally travel from one culture to another, but for those who are responsible for political and economic policies on an international scale, and for those who feel a sense of responsibility for the quality and moral character of the whole global community. For us these ethical inconsistencies are matters of pressing moral concern, and we have no choice but to try to think of ethics in global terms. But how should we begin?

## DESCRIPTIONS OF PARTICULAR ETHICAL TRADITIONS

One way to approach the matter is to make a distinction between the descriptive and the normative problems involved in cross-cultural ethics, and tackle them separately. The division of labor would be something like this: the task of describing how different ethical systems function would be that of the secular scholars, especially those in the fields of anthropology and comparative studies; and the normative task of discerning what to do about the differences would be left to the theologians and moral authorities, and to the policymakers and other ethical actors who have to cope with them. I am not wholly comfortable with this dichotomy, for it seems to me that every ethical description carries implications for action, and every guideline for moral behavior is based on some sort of ethical analysis, inchoate though it may be. Nonetheless, the two terms, descriptive and normative, are useful for indicating two quite different approaches to the study of ethics. Every ethical analysis or judgment will contain these two approaches in varying degrees.

In the field of cross-cultural ethics, it is the descriptive rather than the normative kind of ethical analysis that is more often done, and it is easy to understand why. The normative task is the more difficult one. And it is presumptuous. Making judgments about how to act cross-culturally may imply that the moral standards of one culture are being imposed on another. This is not necessarily the case, but the very suggestion that moral judgments should be attemped on a global scale may evoke ghosts from the past. We are reminded of those years in which Western culture, impressed with the superiority of its own tradition and buttressed with political power, was able to recommend itself with missionary zeal to the rest of the world. This era of Western moral presumptuousness is not entirely dead, even though many of us may wish it were. Avoiding making intercultural ethical judgments is one way to avoid the pretentiousness to which we have been prone in the past.

This may be one reason why many modern scholars steer away from normative judgments and limit themselves to descriptions of ethical systems. There is a more positive reason for choosing the descriptive approach, however: the conviction that such descriptions are absolutely essential and logically prior to making normative judgments. For example, descriptions of the Hindu and Islamic attitudes towards matters such as celibacy, abortion, human rights, and the like, and descriptions of the theoretical positions upon which such attitudes are based, are absolutely necessary before any sensible cross-cultural judgment about the relative value of these positions can be made.

This task – which we might call 'single culture descriptions', in order to contrast a bit later in this essay with a somewhat different sort of descriptive approach – is one well suited to the textual and social scientific training of most comparative scholars. And it has been badly neglected. If attempts at making normative ethical judgments have foundered in the past, surely one of the most pervasive causes of these failures was the lack of cultural descriptions. To fill this gap, a number of regional studies have recently emerged, and a national project on the comparative study of values has as one of its goals the compilation of a guidebook to studies of ethics in each of the major religious traditions [19]. The project regards cultural descriptions as the first step towards conceiving ethics in global terms.

Yet the description of single cultures is only a first step, and it is of limited use in the normative task of trying to frame ethical guidelines. Moreover, if ethical analysis stops with single culture descriptions, two kinds of problems are apt to arise. One is the problem of relativity. Single culture descriptions may promote the notion that all ethical positions are relative to particular cultures, and that there are no enduring moral constraints that universally apply. Those in a position to make normative judgments who accept this notion might then adopt a laissez-faire attitude to ethics. In such a view moral strictures do not extend accross cultural boundaries. Recent examples of the consequences of this view are the double standards adopted by some governmental agencies and multinational corporations: they feel morally free to adopt looser standards abroad than they do at home.

The other problem that arises is an intellectual one. For if one's ethical map of the world is a congeries of particular descriptions, the world one imagines is a myriad of cultural islands separated by impassable seas. One misses the continental cohesion that draws some traditions together, and the layers of continuity that link seemingly disparate cultures somewhere beneath the surface.

When the descriptions of particular cultures are used as the basis for comparing one ethical system with another, yet another problem arises. The problem is identifying which ethical strand within a tradition is the representative one. As we have seen, there are great differences between conservative and liberal Christian views, and as Ninian Smart has pointed out, the social ethics of Hindu *bhakti* religious communities may be as different from that of Brahmanical Hinduism as, say, Theravada Buddhist ethics are from Islamic [26]. Those who rely upon single culture descriptions must be sensitive to the diversity of religious styles and the kinds of ethical attitudes that result from them.

What is needed as a complement to the descriptions of single cultures, therefore, is a somewhat different sort of descriptive approach, one that looks at particular cultural values within a wider context, and searches for links among traditions. This approach might be called 'intercultural description'. It is, like 'single culture description', a regrettably unwieldy term, but the point of using it is to indicate that descriptions of this sort include more than one cultural tradition at a time and are concerned with their interactions.

## THE SEARCH FOR AN INTERCULTURAL ETHIC

One way to accomplish intercultural description is through historical studies. In such studies one does for ethical concepts what Wilfred Cantwell Smith does for theological ones: search for the emergence of parallel ideas in several traditions and observe the historical interactions among them [27]. Pioneering efforts at intercultural descriptions of social values include the works of Max Weber on several religious traditions [31, 32, 33] and the studies of Louis Dumont comparing the historical growth of social values in India with that in the West [5, 6]. Eventually an even wider variety of traditions may be encompassed by other scholars, and attempts may be made to categorize and compare the social teaching of the world religions in the manner of Ernst Troeltsch's studies of the social teaching of the Christian churches [29].

Another way of describing intercultural ethical relations is to report on the logical connections that tie different ethical systems together on a fundamental level. This requires a search for the missing links, the common patterns of moral reasoning that bridge the differences among ethical cultures. The Western philosophical tradition has provided resources for some interesting recent attempts to do just that. Ronald Green, for instance, employs Kant's approach to moral reasoning and finds a common logic in the religious ideas of all traditions: the motive forces of altruism and spiritual reward that allow individuals the risk of self-denying action [8]. David Little and Sumner Twiss also search for a common logic in the religious ethics of various traditions, but in a somewhat different way. Their concern is with formal structures of moral reasoning, and they attempt to supply a framework by which comparison among traditions is made possible [16].

Although Green and the team of Little and Twiss have criticized each other's approaches [9, 18], the formats of their two books are remarkably similar: a general theory is presented in the first half, and in the second half the theory is applied to case studies from various religious traditions. It is that format that has upset some historians of religion who see it as an indication of a flaw in the authors' approaches. The historians of religion would rather reverse the procedure and deduce the logical frameworks from the case studies instead of imposing them on them [28]. The point is a valid one, but regardless of the limitations of their ideas, the efforts of Green, Little, and Twiss have already made a positive impact on the field of cross-cultural ethics. The theories have offered interesting models of ways that intercultural descriptive ethics might be done, and they have provided the occasions for lively discussions about the legitimacy of the enterprise, and how intercultural ethics might be done otherwise.

The ventures of Green, Little, and Twiss might be compared with those of scholars in the field of comparative linguistics who search for common deep structures within the various systems of languages. But just as there is a debate in that field over whether there are semantic deep structures as well as syntactic ones, there is also some question in the field of comparative ethics as to whether the search for intercultural ethics may reveal areas of agreement that are substantive as well as formal. The attempt to unearth a substantive layer – a deep stratum of universal values lying underneath all religious traditions – is fraught with difficulties, but the possiblities are exhilarating: one hopes for a sort of ethical Rosetta Stone that will allow one to decode the differences among traditions and find them marvelously the same.

One way to look for this common underlying stratum is to go directly to those areas of human life that are universal. Sumner Twiss has attempted this by taking the essential facts of life and survival and proposing several fundamental ethical principles that are always associated with them [30]. One might also move beyond our species to those things that all animate life has in common, and attempt, as the sociobiologists have done, to trace the bases for all social values in animal behavior [1, 33].

Or one might invert the process and move inductively from a comparison of different traditions to find those moral ideals that they hold in common. Following this approach, David Little has attempted to find a similarity between Islamic and Christian notions of religious tolerance [17]. Other attempts have compared whole systems: Roderick Hindery's study of Buddhist and Hindu ethics [11], Frank Reynolds' comparison of Christian and Theravada Buddhist modes of action [24], and Ninian Smart's linkage of religious and moral styles in the Islamic and Theravada Buddhist traditions [26]. In each of these studies the differences have been stressed as much as the similarities, and yet points of agreement – or at least of complementarity, as Smart puts it [25] – have emerged.

## THE NORMATIVE IMPERATIVE

So some tentative steps are being taken in the direction of descriptive intercultural ethics. In the meantime, however, decisions are being taken every day on issues of global dimensions, decisions that simultaneously affect many different cultures. What kind of normative stance can be taken on the basis of these kinds of intercultural ethical description? Or to ask the question a different way, how have these scholarly attempts helped to inform those who are concerned about such issues as celibacy, abortion, and human rights, and who must confront them in intercultural settings?

It would be foolish to suggest that the field of intercultural ethics has developed to the point of providing immediate resources for decision-makers, and it is possible that a reflective discipline like descriptive ethics never will be able to satisfy such practical needs completely. Yet some forms of intercultural ethical investigations that we have considered are as relevant to the normative task as they are to the descriptive one.

One approach that is especially applicable is the descriptive scholars' search for universal areas of ethical agreement. The normative ethical actors face a similar task. Like the descriptive scholars, they attempt to look within every ethical position, including their own, to find elements that are not culturally limited, and that can apply in any cultural area. In the case of abortion, such an element might be the moral requirement to preserve life, a requirement all traditions maintain even though they may disagree over the concept of the soul, or the point at which life is thought to commence. The normative actors would seize upon this area of agreement as a place to begin in formulating an ethical opinion on the subject that would stand up cross-culturally.

Another approach that is as useful for formulating normative judgments as it is for making descriptive analyses is the one that looks for similarities in ethical patterns. Using this approach, the normative ethical actor tries to stand outside his or her own tradition and see issues from the point of view of someone within another culture. The goal is to see if there are points of similarity on which a linkage across traditions or a synthesis between them can be built. Again, turning to the case of abortion, a normative actor might find that even though two traditions disagree over when fetal life becomes human life, in certain extenuating circumstances (such as those involving the safety and care of the mother) they both might permit some acts of abortion.

Through case-by-case studies, therefore, those scholars and actors who are concerned about normative issues can begin to piece together something resembling an intercultural ethic that will relate to the issue of abortion and to many other issues as well. Such a store of ethical similarities and agreements cannot rightly be called a global ethic, however. This term should be reserved for the ethics that emerges from the whole global culture.

## GANDHI AND GLOBAL ETHICS

As we have seen, it is possible to view world history in just this way: as embodying a single, albeit complex, cultural tradition. One of the approaches to intercultural descriptive ethics that we have considered follows the approach advocated by Wilfred Cantwell Smith in viewing the history of ethics in the world uniculturally. Smith's proposal for a 'world theology' to match the unitary religious experience of humanity would be followed by a 'global ethics' that would reflect the common experience of people throughout the world who grapple with issues that transcend cultural boundaries. These shared values might be sufficiently broad to be the base for making normative judgments.

How might a normative ethical actor adopt a global ethic? To consider that question, we return to the figure with which we began this essay: Gandhi. Here, however, we are not going to consider the image of Gandhi and whether it deserves universal canonization, but rather the actions of Gandhi and whether they can be considered the fulcrum for a global ethical position.

There are two reasons for regarding Gandhi's ethical position as global. One is his attempt to view the impact of actions in which he was involved from multiple cultural perspectives. For instance, even though he encouraged Indians to spin yarn at home in order to free India from reliance on the British textile industry, Gandhi also attempted to see the British millworkers' view of the problem, and he lived with them in London for a time in order to do so. It was a gesture similar to the one in which Gandhi attempted to bridge two points of view in a labor strike at Ahmedabad. There Gandhi resided in the home of the owner of a factory at the same time he was leading the workers in a strike against the very owner with whom he was staying. The notion of 'truth', in Gandhi's view, was something that could only be determined through the parallax formed by two differing points of view. Hence Gandhi's insistence on understanding both of those points of view and allowing their similarities and differences to emerge. This process was integral to Gandhi's method of fighting, *satyagraha*, which I take to be a significant contribution to ethical theory [14]. It has special significance for inter-cultural issues, where the points of view can be disparate and strongly defended.

The other reason why Gandhi's ethics may be regarded as global has to do with the eclectic nature of his ethical views. Gandhi relied upon a variety of cultural traditions, regarding all moral traditions as different 'experiments with truth' [7]. He denied that any one moral code or set of principles had more claim to validity than any other. Gandhi's own religious background showed influences from both mainstream Hindu culture and the egalitarian religious movement to which his mother was attracted. As a lad, his best friend and alter ego was a Muslim, and his religious mentor was a Jain. As he developed his political and religious communities in South Africa, two of his closest colleagues were a Jewish architect and an Anglican missionary. The religious socialism of Tolstoy made a deep impact, as did the concerns of Theosophists, Fabian socialists and anti-vivisectionists with whom he associated in London during his school days. Gandhi loved preaching from the Sermon on the Mount, and his favorite hymn was 'Lead, Kindly Light'. It is no surprise, then, that Gandhi could claim to be a Muslim among Muslims, a Christian among Christians, and a Jew among Jews.

I do not mean to imply that Gandhi lumped all of these religious traditions together indiscriminately. He relied foremost on his own Hindu background, but it was a Hinduism informed by his encounters with other traditions and his acceptance of at least some elements as valid from each. His own moral stance was forged through a sort of ethical *satyagraha*, a contesting of many different moral points of view.

Is Gandhi the harbinger of a new style of global ethics? His actions hold more promise in that regard than his image does, for even if he cannot be taken piously as a universal saint, he can be taken seriously as someone who lived as a world citizen – perhaps inadvertently – and gleaned ethical attitudes from the several cultural traditions he encountered. Much the same can be said of many of us, and it seems to me that Gandhi's form of intercultural ethics may increasingly become the general pattern. Like it or not, as participants in a global culture we are all heirs to each of the world's cultural traditions, and we may be affected by Islamic and Buddhist ethical ideas not only because they are similar to our own but because they are, in some sense, our own.

Current writings in Christian theology explore the plural cultural milieu and its implications for thinking about religious experience. The recent works of John Cobb [3], John Hick [10], and Charles McCoy [20] are interesting cases in point. The ethical implications of this theological direction are significant. Like Gandhi, these theologians take a global stand, widening their frame of reference from the cultural traditions of Western Christianity to the larger community, and embracing the whole history of cultural interaction. Such a stance has several implications for the doing of normative ethics, especially, perhaps, in the area of medical ethics.

In the first place, it allows us to draw upon the common themes of several traditions. This gives us the assurance that our attitudes toward the maintenance of life, for example, are widely shared and are not culturally peculiar to one section of the globe.

Secondly, a global stance requires us to submit our own tradition's views to the critical reflection that results when we encounter a moral view different from our own. Asian views of the interrelationship between the body and the soul, for instance, are helpful corrections to Western dualistic views, and the Asian concepts of the healing arts and the role of the healer could add much to the mechanistic attitudes of Western medicine.

Third, a global stance allows us to be critical of other traditions' attitudes, and use as a basis for our criticism their own moral heritage as well as our own. Our attitude toward sexuality, for instance, may call into question the traditional Hindu view, and challenge whether it is being adequately interpreted for the present day.

Finally, a global stance encourages us to think of ethical traditions in a future-oriented way, and opens up the possibilities of changes in traditional ethical positions. It even opens up the possibility of a convergence between differing ethical points of view. The increasing interest in traditional forms of Asian medicine, for example, may lead not only to an appropriation of the techniques into Western practices, but also to an assimilation into Western consciousness of some of the ideas of health and healing that lie behind them.

Such outcomes will help to break down the boundaries that exist among ethical traditions, and in doing so they will create new ethical syntheses. Here again we will need the descriptive scholar to explain and interpret these new ethical patterns. And these syntheses and their interpretations will in turn provide new bases for the judgments that normative actors so desperately need to make.

*Graduate Theological Union,*
*Berkeley, California, U.S.A.*

## BIBLIOGRAPHY


[1] Caplan, A. (ed.): 1978, *The Sociobiology Debate*, Harper and Row, New York and San Francisco.

[2] Chandrasekhar, S.: 1974, *Abortion in a Crowded World: The Problem of Abortion with Special Reference to India*, University of Washington Press, Seattle.

[3] Cobb, J.: 1975, *Christ in a Pluralistic Age*, The Westminster Press, Philadelphia.

[4] Devereux, G.: 1955, *A Study of Abortion in Primitive Societies*, The Julian Press, New York.

[5] Dumont, L.: 1970, *Homo Hierarchicus: The Indian Caste System and Its Implications*, Mark Sainsbury (trans.), University of Chicago Press, Chicago.

[6] Dumont, L.: 1977, *From Mandeville to Marx: The Genesis and Triumph of Economic Ideology*, University of Chicago Press, Chicago.

[7] Gandhi, M. K.: 1925; reprint 1957, *An Autobiography: The Story of My Experiments with Truth*, Beacon Press, Boston.

[8] Green, R.: 1978, *Religious Reason: The Rational and Moral Basis of Religious Belief*, Oxford University Press, Oxford.

[9] Green, R.: 1981, 'Review of Little and Twiss, *Comparative Religious Ethics*', *Journal of Religion* **61**, 111–113.

[10] Hick, J. and B. Hebblewaite (eds.): 1980, *Christianity and Other Religions*, Collins, London.

[11] Hindery, R.: 1978, *Comparative Ethics in Hindu and Buddhist Traditions*, Motilal Banarsidass, Delhi.

[12] Juergensmeyer, M.: 1984, 'Dharma and the Rights of Untouchables', in I. Bloom and W. Proudfoot (eds.), *Religion and Human Rights*, Columbia University Press, New York.

[13] Juergensmeyer, M.: 1983, Interview with Dr. Abdul Aziz Al-Khayyat, Dean of the faculty of Sharia'h University of Jordan and former Minister of Waqf and Islamic Affairs, Government of Jordan, September 20, 1983, at Berkeley, California.

[14] Juergensmeyer, M.: 1984, *Fighting With Gandhi*, Harper and Row, New York and San Francisco.

[15] Juergensmeyer, M.: 1982, *Religion as Social Vision: The Movement Against Untouchability in 20th Century Punjab*, University of California Press, Berkeley and Los Angeles.

[16] Little, D., and S. Twiss: 1978, *Comparative Religious Ethics: A New Method*, Harper and Row, New York.

[17] Little, D.: 1985, 'Human Rights, Natural Law and Religious Liberty', in I. Bloom and W. Proudfoot (eds.), 1985, *Religion and Human Rights*, Columbia University Press, New York.

[18] Little, D.: 1981, 'The Present State of the Comparative Study of Religious Ethics', *Journal of Religious Ethics* **9**, 210–227.

[19] Mansager, D., and T. R. Sizemore (eds.): 1984, *A Guidebook to the Comparative Study of Ethics*, Berkeley-Harvard Series in Comparative Ethics, Berkeley.

[20] McCoy, C.: 1980, *When Gods Change: Hope for Theology*, Abingdon Press, Nashville.

[21] Mehlan, K.: 1970, 'Changing Patterns of Abortion in the Socialist Countries of Europe', in R. Hall (ed.), *Abortion in a Changing World*, Columbia University Press, New York, pp. 302–314.

[22] Nazer, I.: 1970, 'Abortion in the Near East', in R. Hall (ed.), *Abortion in a Changing World*, Columbia University Press, New York, pp. 267–273.

[23] O'Flaherty, W.: 1973, *Aesceticism and Eroticism in the Mythology of Siva*, Oxford University Press, London (reprinted as *Siva, The Erotic Ascetic*).

[24] Reynolds, F.: 1980, 'Contrasting Modes of Action: A Comparative Study of Buddhist and Christian Ethics', *History of Religions*/**20**, 128–146.

[25] Smart, N.: 1981, *Beyond Ideology: Religion and the Future of Western Civilization*, Harper and Row, New York and San Francisco.

[26] Smart, N.: 1981, 'Types of Religion and the Moral Strand', Paper delivered to the annual conference of the Berkeley-Harvard Program in the Comparative Study of Values, Berkeley, California.

[27] Smith, W.: 1981, *Towards a World Theology: Faith and the Comparative History of Religion*, Westminster Press, Philadelphia.

[28] Swearer, D.: 1980, 'Nirvana, No-Self, and Comparative Religious Studies', *Religious Studies Review*/**6**, 301–303.

[29] Troeltsch, E.: 1931, *The Social Teachings of the Christian Churches*, O. Wyon (trans.), George Allen and Unwin, London (German edition published in 1911).

[30] Twiss, S.: 1984, 'Conceptual Relativism and Human Rights: Reflections on Method in Comparative Ethics', in I. Bloom and W. Proudfoot (eds.), *Religion and Human Rights*, Columbia University Press, New York.

[31] Weber, M.: 1930, *The Protestant Ethic and the Spirit of Capitalism*, T. Parsons (trans.), Charles Scribner's Sons, New York.

[32] Weber, M.: 1951, *The Religion of China*, H. Gerth (trans.), The Free Press, Glencoe, Ill.

[33] Wilson, E.: 1975, *Sociobiology: The New Synthesis*, Harvard University Press, Cambridge.

# SECTION III

# RELIGIOUS REASONING ABOUT BIOETHICS AND MEDICAL PRACTICE

STANLEY HAUERWAS

# SALVATION AND HEALTH: WHY MEDICINE NEEDS THE CHURCH

## A TEXT AND A STORY

While it is not unheard of for a theologian to begin an essay with a text from scripture, it is relatively rare, especially for those who turn their attention to issues of medicine. However, I begin with a text, as almost everything I have to say is but a commentary on this passage from Job 2:11–13:

> Now when Job's friends heard of all this evil that had come upon him, then came each from his own place, Eliphaz the Temanite, Bildad the Shuhite, and Zophar the Na'amathite. They made an appointment together to come condole with him and comfort him. And when they saw him from afar, they did not recognize him; and they raised their voices and wept; and they rent their robes and sprinkled dust upon their heads toward heaven. And they sat with him on the ground seven days and seven nights, and no one spoke a word to him, for they saw that his suffering was very great.

I do not want to comment immediately upon the text but rather, like most preachers, I think it best to begin by telling you a story. The story is about me and one of my earliest friendships. When I was in my early teens I had a friend, let us call him Bob, who meant everything to me. We made our first hesitant steps toward growing up through sharing the things young boys do – i.e., double dating, athletic activities, and endless discussions on every topic. For two years we were inseparable. I was extremely appreciative of Bob's friendship as he was not only brighter and more talented than me, but also came from a family that was economically considerably better off than my own. So through Bob I was introduced to a world that otherwise I would hardly know existed. For example, we spent hours playing pool in a room in his home that was built for no other purpose; and we swam in the lake that his house was specifically built to overlook.

Then very early one Sunday morning I received a phone call from Bob requesting I come to see him immediately. He was sobbing intensely, but through his crying he was able to tell me that they had just found his mother dead. She had committed suicide by placing a shotgun in her mouth. I knew immediately I did not want to go to see him or to confront a reality like that. I had not yet learned the desperation hidden under our



*E. E. Shelp (ed.), Theology and Bioethics,* 205–224.

everyday routines and I did not want to learn of it. Moreover, I did not want to go because I knew there was nothing I could do or say to make things even appear better than they were. Finally, I did not want to go because I did not want to be close to anyone who had been touched by such a tragedy.

But I went. I felt awkward, but I went. And as I came into Bob's room we embraced, a gesture that was almost unheard of between young men raised in the Southwest, and we cried together. After that first period of shared sorrow we somehow calmed down and took a walk. For the rest of that day and that night we stayed together. I do not remember what we said, but I do remember that it was inconsequential. We never talked about his mother or what had happened. We never speculated about why she might do such a thing, even though I could not believe someone who seemed to have such a good life would want to die. We did what we always did. We talked girls, football, cars, movies, and anything else that was inconsequential enough to direct our attention from this horrible event.

As I look on that time, I now realize that it was obviously one of the most important events in my life. That it was so is at least partly indicated by how often I have thought about it subsequently as well as tried to understand its significance. For as I have reflected on what happened in that short space of time I have only marveled at how inept I was in helping Bob. I did not know what needed or could be said. I did not know how to help him start sorting out such a horrible event in a way he could go on. All I could do was be present.

But after a considerable number of years I now realize that is all he wanted – namely, my presence. For as inept as I was, my willingness to be present was a sign that this was not an event so horrible that it drew us away from all other human contact. Life could go on and in the days to follow we would again swim together, double date, and generally waste time. Thus I now think that at that time God granted me the marvelous privilege of being but a presence in the face of profound pain and suffering, even when I did not appreciate the significance of being present.

Yet the story cannot end there. For while it is true that Bob and I did go on being friends, nothing was the same. For a few months we continued to see one another often, but somehow the innocent joy of loving one another was gone. We slowly found that our lives were going in different directions and we developed new friends. No doubt the difference between our social and cultural opportunities helped explain to some extent our drifting apart. Bob finally went to Princeton and I went to Southwestern University in Georgetown, Texas.

But that kind of explanation for our growing apart is not sufficient. What was standing between us was that day and night we spent together under the burden of a profound sadness that neither of us had known existed until that time. We had shared a pain so intense that for a short period we had become closer than we knew, but now the very pain that created that sharing stood in the way of the development of our friendship. Neither of us wished to recapture that time, nor did we know how to make that night and day part of our ongoing story together. So we went our separate ways. I have no idea what became of Bob, though every once in a while I remember to ask my mother if she has heard from him.

Does medicine need the Church? How could this text and this story possibly help us understand that question, much less suggest how it might be answered? Yet that is what I am going to claim in this essay. Put briefly, what I will try to show is that if medicine can be rightly understood as an activity that trains some to know how to be present to those in pain, then something very much like a church is needed to sustain that presence day in and day out. Before I try develop that thesis, however, I need to do some conceptual groundbreaking to make clear exactly what kind of claim I am trying to make about the relation of salvation and health, medicine and the church.

## RELIGION AND MEDICINE: IS THERE OR SHOULD THERE BE A RELATION?

It is a well known fact that for most of human history there has been a close relation between religion and medicine. Indeed, that very way of putting the matter is misleading, since to claim a relation suggests that they were distinguished, which has often not been the case. Disease and illness were not seen as matters having no religious import but rather as resulting from the disfavor of God. As Darrell Amundsen and Gary Ferngren have recently reminded us, the Hebrew scriptures often depict God promising "health and prosperity for the covenant people if they are faithful to him, and of disease and other suffering if they spurn his love". This promise runs through the Old Testament. "If you will diligently harken to the voice of the Lord your God, and do that which is right in his eyes, and give heed to his commandments and keep all his statutes, I will put none of the diseases upon you which I put upon the Egyptians; for I am the Lord, your healer" (Exod. 15:26) ([2], p. 62).

This view of illness was not associated only with the community as a whole, but with individuals. Thus in Psalm 38 the lament is

> There is no soundness in my flesh because of thy indignation; there is no health in my bones because of my sin.... My wounds grow foul and fester because of my foolishness.... I am utterly spent and crushed; I groan because of the tumult of my heart.... Do not forsake me, O Lord! O my God, be not far from me! Make haste to help me, O Lord, my salvation! (vv. 3, 5, 8, 21–22).

Amundsen and Ferngren point out this view of illness was accompanied by the assumption that acknowledgement of and repentance for our sin was essential for our healing. Thus in Psalm 32:

> When I declared not my sin, my body wasted away through my groaning all day long. For day and night thy hand was heavy upon me; my strength was dried up.... I acknowledged my sin to thee, and I did not hide my iniquity; I said, 'I will confess my transgressions to the Lord'; then thou didst forgive the guilt of my sin (vv. 3–5) ([2], p. 63).

Since illness and sin were closely connected, it is not surprising that healing was also closely associated with religious practice – or put more accurately, healing was a religious discipline. Indeed, Amundsen and Ferngren make the interesting point that since the most important issue was a person's relationship with God, the chief means of healing was naturally prayer. That such was the case thus precluded magic and thus the Mosaic code excluded soothsayers, augurs, sorcerers, charmers, wizards, and other such figures who offered a means to control or avoid the primary issue of their relation to Yahweh ([2], p. 64). Amundsen and Ferngren also suggest that this may have been the reason that there seems to have been no sacerdotal medical practice develop in Israel particularly associated with the priesthood. Rather, the pattern of the Exodus tended to prevail with illness and healing more closely associated with prophetic activity.

The early Christian community seems to have done little to change these basic presuppositions. If anything, it simply intensified them by adding what Amundsen and Ferngren call the 'central paradox' in the New Testament.

> Strength comes only through weakness. This strength is Christ's strength that comes only through dependence upon him. In the Gospel of John, Christ says: 'I have said this to you, that in me you may have peace. In the world you have tribulation; but be of good cheer, I have overcome the world' (16:33). 'In the world you have tribulation.' It is simply to be expected and accepted. But for the New Testament Christian no suffering is meaningless. The ultimate purpose and meaning behind Christian suffering in the New Testament is spiritual maturity. And the ultimate goal in spiritual maturity is a close dependence upon Christ based upon a childlike trust ([2], p. 96).

Thus illness is seen as an opportunity for growth in faith and trust in God.

Because of this way of viewing both the positive and negative effects of illness, Amundsen and Ferngren note that there has always been a degree of tension in the way Christians understood the relation between theology and secular medicine, between the medicine of the soul and the medicine of the body.

According to one view, if God sends disease either to punish or to test a person, it is to God that one must turn for care and healing. If God is both the source and healer of a person's ill, the use of human medicine would circumvent the spiritual framework by resorting to worldly wisdom. On another view, if God is the source of disease, or if God permits disease and is the ultimate healer, God's will can be fulfilled through human agents, who with divine help have acquired the ability to aid in the curative process. Most Christians have asserted that the human agent of care, the physician, is an instrument of God, used by God in bringing succor to humankind. But in every age some have maintained that any use of human medicine is a manifestation of a lack of faith. This ambivalence in the Christian attitude, among both theologians and laity, has always been present to some degree ([2], p. 96).

Moreover, this issue is further complicated by the Jewish and Christian insistence that life is sacred not only in some spiritual sense but particularly as bodily life. As Paul Ramsey reminds us, for Jews and Christians a person

...is a sacredness in bodily life. He is a person who within the ambience of the flesh claims our care. He is embodied soul or ensouled body. He is therefore a sacredness in illness and in his dying. He is a sacredness in the fruits of the generative process. The sanctity of human life prevents ultimate trespass upon him even for the sake of treating his bodily life, or for the sake of others who are also only a sacredness in their bodily lives. Only a being who is a sacredness in the social order can withstand complete dominion by 'society' for the sake of engineering of civilizational goals – withstand, in the sense that the engineering of civilizational goals cannot be accomplished without denying the sacredness of the human being. So also in the use of medical or scientific techniques ([14], p. XIII).

Thus, it is not possible to separate and/or distinguish religion from medicine on the basis of a distinction between soul and body. Religion does not deal with the soul and medicine with the body, as practitioners of both are too well aware of the inseparability of soul and body – or perhaps better, they know the abstractness of both categories. Moreover, when religion too easily legitimates the independence of medical care by limiting medicine to mechanical understanding and care of the body, it has the result of making religious convictions ethereal in character. It may be that just to the extent Christianity is always tempted in gnostic and manichean directions it accepts too willingly a technological understanding of medicine. Christians, if they are to be faithful to their convictions, may not ever be able to avoid at least potential conflict between their assumptions about illness and health and how the ill should be cared for and the assumptions about medicine. One hopes for cooperation, of course, but structurally the possibility of conflict between church and medicine cannot be excluded since both entail convictions and practices concerned with that same subject.

Put differently, given Judaism and Christianity's understanding of human-kind's relation with God, that is, how we understand salvation, health can never be thought of as an autonomous sphere. Moreover, insofar as medicine is a specialized activity distinguished from religious convictions, one cannot exclude the possibility that there may well be conflict between religion and medicine. For in many ways the latter is constantly tempted to offer a form of salvation that religiously may often come close to idolatry. The ability of modern medicine to cure is at once a benefit and potential pitfall, as too often it is tempted to increase its power by offering more than care and instead offers alleviation, if not cure, of the human condition. That it does so is not the fault of medical practitioners, though often they encourage such idolatry, but rather the fault lies with those of us who pretentiously place undue expectations on medicine in the hopes of finding an earthly answer to our death. However, we can never forget that the relation between medicine and health, and especially the health of a population, is as ambiguous as the relation between the church and salvation.

In the hopes of securing peace between medicine and religion two quite different and equally unsatisfactory proposals have been suggested, if not tried. The first underwrites a strong division of labor between medicine and religion by limiting the scope of medicine to the mechanism of our body. While it is certainly true that medicine in a unique way entails the passing on of the wisdom of the body from one generation to another, there is no way that medical care can be limited to the body and be good medicine [11]. For as Ramsey has reminded us again and again, the moral commitment of the physician is not to treat diseases, or populations, or the human race, but the immediate patient before him or her ([14], pp. 36, 59). Religiously, therefore, the care offered by physicians cannot be abstracted from or given independence from the moral commitment to care based on our view that every aspect of our existence is dependent upon God.

Equally to be avoided is the correlative division between specialization based on this view that assumes a division of expertise between the physician and the clergy. The clergy, no less than the physician, is concerned about the patient's physical well-being. No assumptions about technical skills and knowledge can or should legitimate the clergy retreating into the realm of the spiritual in order to claim some continued usefulness and status. Such a retreat is as unfaithful as abandoning the natural world to the physicist on grounds that God is a God of history and not nature. For the church, and its office holders, to abandon claims over the body in the name of a lack of expertise is equivalent to reducing God to the gaps in scientific theory. Such a strategy is not only bad faith but it results in making religious convictions appear at best irrelevant and at worse foolish.

The other alternative, rather than accepting the autonomy of medicine from our religious convictions, seeks to maintain a close relationship by resacralizing medical care. Medicine requires a "holistic vision of man" ([8], p. 9), because the care it brings is but one aspect of salvation. Thus the church and its theology serve medical care by "promoting a holistic view of man" that can provide a

> comprehensive understanding of human health [that] includes the greatest possible harmony of all of man's forces and energies, the greatest possible spiritualization of man's bodily aspect and the finest embodiment of the spiritual. True health is revealed in the self-actualization of the person who has attained that freedom which marshals all available energies for the fulfillment of his total human vocation ([8], p. 154).

To which one can only respond, 'I hope not.'

For it is now a common observation that to ask medicine to try to serve and achieve health so understood is to pervert the kind of care that physicians can provide. For the physicians rightly maintain that their skill primarily has to do with the body, as medicine promises us health, not happiness.[1] When a more general understanding of health is made the goal of medicine, it only results in making medical care and physicians' claims over our lives tyrannical. It is already a difficult task in our society to control the expectations people have about modern medicine, we only compound that problem by providing religious legitimacy to this broader understanding of health. Certainly we believe that any account of salvation includes questions of our health, but that does not mean that medicine can or ever should become the agency of salvation. It may be a fundamental judgment on the church's failure to help us locate wherein our salvation lies that so many today seek a salvation through medicine.

## CAN MEDICAL ETHICS BE CHRISTIAN?

The already complex question of the relation between religion and medicine only becomes more confusing when we turn our attention to more recent developments in medical ethics. For even though religious thinkers have been at the forefront of much of the work done in the expanding field of 'medical ethics', it is not clear that they have been there as religious thinkers. Joseph Fletcher [5], Paul Ramsey [14], James Gustafson [6], Charles Curran [4], Jim Childress [3], to name just a few, have done extensive work in medical ethics, but often it is hard to tell how their religious convictions have made a difference for the methodology they employ or for their response to specific quandaries. Indeed, it is interesting to note how seldom they raise issues of the meaning or relation of salvation and health as they seem to prefer to deal with questions of death and dying, truth telling, etc.

By calling attention to this fact I by no means wish to disparage the kind of reflection that has been done concerning these issues. I feel sure we have all benefitted from their careful analysis and distinctions concerning such problems. Yet one must wonder if by letting the agenda be set in such a manner that we have already lost the theological ball game. For the very concentration on 'issues' and 'quandaries' as central for medical ethics tends to underwrite the practice of medicine as we know it, rather than challenging some of the basic presuppositions of medical practice and care. By failing to raise more fundamental questions, the kind of concerns that might provide more access for our theological claims are simply never considered.

There are at least two reasons for this I think worth mentioning. The first has to do with the character of theological ethics itself. We tend to forget that the development of 'Christian ethics' is a relatively new development [8]. It has only been in the last hundred years that some styled themselves as 'ethicists' rather than simply theologians. It is by no means clear that we know how to indicate what difference it makes conceptually and methodologically to claim our ethics as Christian in distinction from other kinds of ethical reflection. In the hopes of securing greater clarity about their own work, many who have identified their work as Christian have nonetheless assumed that the meaning and method of 'ethics' were determined fundamentally by non-Christian sources. In a sense, the very concentration on 'medical ethics' was a godsend for many 'religious ethicists', as it seemed to provide a coherent activity without having to deal with the fundamental issue of what makes Christian ethics Christian.

This can be illustrated by attending to the debate among Christian ethicists concerning whether Christian moral reasoning was primarily deontological or consequential. This debate was particularly important for medical ethics, as obviously how one thinks about non-therapeutic experimentation, truthtelling, transplants, and a host of other issues seems to turn on this debate. Thus Joseph Fletcher, who wrote one of the first books by a Protestant in medical ethics, has always argued in favor of a consequential stance, thus legitimating the overriding of the physicians' commitment to an individual patient in the name of a greater good [5]. In contrast, Paul Ramsey has emphasized that the 'covenant' of the physician with the patient is such that no amount of good to be done should override that commitment [14].

It is interesting to note how each makes theological appeals to support his position. Fletcher appeals to love as his basic norm, interpreting it in terms of the greatest good for the greatest number, but it remains unclear how his sense of love is theologically warranted or controlled. Ramsey provides a stronger theological case for his emphasis on 'covenant' as a central theological motif, but it is not clear how the many 'covenants of life with life into which we are born' are related to or require the covenant of God with a particular people we find in the scripture. Ramsey's use of covenant language thus underwrites a natural law ethic whose status is unclear both from a theological and/or philosophical perspective.[2]

What is interesting about the debate between Fletcher and Ramsey is that it could have been carried on completely separately from the theological premises that each side at least claimed were involved. For the terms of the debate – consequential and deontological – are basically borrowed from philosophical contexts and are dependent on the presuppositions of certain philosophical traditions. Of course, that in itself does not mean that such issues and concepts are irrelevant to our work as theologians, but what missing is any sense of how the issue as presented grows, is dependent on, or is informed by our distinctive commitments as theologians.

The question of the nature of theological ethics and its relation to the development of ethical reflection in and about medicine is further complicated by our current cultural situation. For as Ramsey has pointed out, we are currently trying to do the impossible – namely, "build a civilization without an agreed civil tradition and [in] the absence of a moral consensus" ([15], p. 15). That such is the case makes the practice of medicine even more morally challenging, since it is by no means clear how one can sustain a non-arbitrary medicine in a genuinely morally pluralistic society. For example, much of the debate about when someone is 'really' dead is not simply the result of our increased technological power to keep blood flowing through our bodies, but rather witnesses to our culture's lack of consensus as to what constitutes a well-lived life. In the absence of such a consensus our only recourse is to resort to claims and counterclaims about 'right to life' and 'right to die' with the result of the further impoverishment of our moral language and vision. Moreover, the only means it seems we have to create a 'safe' medicine under such conditions is to expect physicians to treat us as if death is the ultimate enemy to be put off by every means. Then we blame physicians for keeping us alive beyond all reason, but failing to note that if they did not, we would not know how to distinguish them from murderers.

Alasdair MacIntyre has raised this sort of issue directly in "Can Medicine Dispense with a Theological Perspective on Human Nature?" Rather than calling attention to what has become problematic for physicians and surgeons – issues such as when it is appropriate to let someone die – he says he wants to direct our attention to what is still taken for granted, "Namely, the unconditional and absolute character of certain of the doctor's obligations to his patients" ([13], p. 120). The difficulty, however, is that modern philosophy, according to MacIntyre, has been unable to offer a persuasive account of such an obligation.

> *Either* they distort and misrepresent it *or* they render it unintelligible. Theological moralists characteristically end up by distorting and misrepresenting. For they begin with a notion of moral rules as specifying how we are to behave if we are to achieve certain ends, perhaps *the* end for man, the *summum bonum*. If I break such rules I shall fail to achieve some human good and will thereby be frustrated and impoverished ([13], p. 122).

But MacIntyre notes that this treats moral failure as if it is an educational failure that lacks the profound guilt that should accompany moral failure. More importantly, such an account fails entirely to account for the positive evil we know certain people clearly pursue.

Moral philosophers who tend to preserve the unconditional and absolute character of the central requirements of morality, however, inevitably make those 'oughts' appear as if they are taboo. What they cannot do is show how those oughts are rationally entailed by an account of man's true end. Kant was only able to do so because he continued the presuppositions that he failed to justify within his own philosophical position, that "the life of the individual and also of that of the human race [is] a journey toward a goal" ([13], p. 127). Once that presupposition is lost, however, and MacIntyre believes that it has been lost in our culture, then we lack the resources to maintain exactly those moral presuppositions that seem essential to securing the moral integrity of medicine.

Such a situation seems ripe for a theological response, since it might at least be suggested that it thus becomes our task as theologians to serve our culture in general and medicine in particular by supplying the needed rationale. But MacIntyre advises that theologians would be ill-advised to seize this strategy too easily.

For the theologian in the modern world finds himself in the following difficulty: either he is able to present his position in terms which are intelligible within that secularized world or he is not. If he is so able, then he may find himself in one of two difficulties. For either his assertions will conflict with some of the body of secular beliefs or they will not. The former possibility is less often realized these days than formerly; theologians no longer usually claim that certain features of the natural world require explanation in terms of the divine. But the latter possibility is all too real. A Bultmann decodes Christianity and what is left – Heidegger's philosophy; a van Buren decodes it and what is left is. . . . To decode turns out to be to destroy ([13], p. 130).

The other alternative is not to decode but to insist that the secular world must accept Christianity on its own terms. But this option fails to meet the difficulty that we all,

...including contemporary theologians, inhabit an intellectual universe in which the natural sciences are at home and theology is not. The believer has not only to make theological claims about the existence and nature of God, but he also has to make special epistemological claims about the character of his theological claims, if he wants to be heard. The unbeliever equally has to decide what grounds he might have for giving credibility of any kind to the theologian's interpretations. For both parties, the question of what would settle the argument has characteristically become unclear in a way in which it was not unclear either for medieval theologians or for eighteenth-century materialists ([13], p. 131).

This kind of dilemma is particularly acute when it comes to medicine. For if the theologian attempts to underwrite the medical ethos drawing on the particular convictions of Christians just to the extent those convictions are particular, they will serve only to emphasize our lack of a common morality. Thus the theologian, in the interest of cultural consensus, often tries to downplay the distinctiveness of their theological convictions in the interest of social harmony. But in the process we only reinforce the assumption on the part of many that theological claims make little difference for how medicine itself is understood or how various issues are approached. At best theology or religion is left with justifying a concern with the 'whole patient', but it is not even clear how that concern depends on or derives from any substantive theological conviction that is distinguishable from a sentimental humanism.

Almost as if we have sensed that there is no way to resolve this dilemma, theologians and religious professionals involved in medicine have tended to associate with the patient's rights movement. For at least one of the ways of resolving our cultural dilemma is to protect the patient from medicine by restoring the patient's autonomy over against the physician. While I certainly do not want to underestimate the importance of patients recovering a sense of medicine as an activity in which they play a role as important as the physician, the emphasis on the patient's rights over against the physician's power cannot resolve our difficulty. For it is but an attempt to substitute procedural safeguards for what only substantive convictions can supply. As a result, our attention is distracted from the genuine challenge we confront for the forming of an ethos sufficient to sustain a practice of medicine that is morally worthy.

## PAIN, LONELINESS, AND BEING PRESENT: THE CHURCH AND THE CARE OF THE ILL

If I am to be true to the sermonic mode with which I began this essay, I should now return to my original text and story to show how they might offer some solution for helping us understand how Christian convictions can or should inform the practice of medicine within our current culture. Yet I am afraid I cannot promise anything so grand. Indeed, I do not know if there is any adequate response to the dilemmas we are currently facing morally in our culture and in particular in medicine. However, I at least want to return to my original text and story to suggest how they at least might provide a means to help us go on when we lack such a solution.

Before I do so, however, I think it important to note that I am not going to try to deal with the issue of the relation of medicine and religion and/or the relation of theology to medical ethics. Rather, I want to talk about the relation of the church to the practice of medicine. This may seem to be but a quibble, but I think as I proceed its significance will be apparent. For the problem with discussing the question of 'relation' in such general terms as 'medicine' and 'religion' is that each of those in its own way distorts the character of what it is meant to describe. For example, when we talk about 'religion' rather than a specific set of behavior and habits embodied by a distinct group of people, our account always tends to be reductionistic as it makes it appear that underlying what people actually believe and do is a deeper reality called 'religion'. It is as if we could talk about God abstracted from how a people have learned to pray to that God. In like manner we often tend to oversimplify the nature of medicine by trying to capture the many activities covered by that term in a definition or ideological system. What doctors do often is quite different from what they say they do.

Moreover, the question of the relation of theology to medical ethics is far too abstract. For when the issue is posed in that manner, it makes it appear that religion is primarily a set of beliefs, a world-view, that may or may not have implications for how we understand and respond to certain kinds of ethical dilemmas. While it is certainly true that Christianity involves beliefs, the character of those beliefs cannot be understood apart from their place in the formation of a community with cultic practices. By focusing on this fact, I hope to throw a different perspective on how those who are called to care for the sick can draw upon and count on the particular kind of community we call the church.

I do not intend, for example, to argue that medicine must be reclaimed as in some decisive way dependent on theology. Nor do I want to argue that the development of 'medical ethics' will ultimately require the acknowledgement of or make recourse to theological presuppositions, Rather, all I want to try to show is why, given the particular demands put on those who care for the ill, something very much like a church is necessary to sustain that care.

To develop this point I want to call attention to an obvious but often overlooked aspect of illness – namely, that when we are sick we hurt and are in pain. I realize that we often are sick and yet we are not in pain – i.e., hardening of the arteries – but that does not ultimately defeat my general point, since we know that such an illness will lead to physical and mental pain. Nor am I particularly bothered by the observation that many pains are 'psychological', having no real physiological basis. Physicians are right to insist that someone who says he has pain, even if no organic basis can be found for such pain, in fact is in pain. He is not mistaken he is in pain, though he may be mistaken about what kind of pain it is.

Moreover, I am well aware that there are many different kinds of pain, as well as intensity of pains. What may only be a minor hurt for me may be a major trauma for someone else. Pain comes in many shapes and sizes and it is never possible to separate the psychological aspects of pain from the organic. For example, suffering, which is not the same as pain since we can suffer without being in pain, is nonetheless akin to pain inasmuch as it is a felt deficiency that can make us as miserable as pain itself.[3]

Yet, given these qualifications, it remains true that there is a strong connection between pain and illness which we use to mark those areas of our lives where it is appropriate to call upon the skills of a physician. When we are in pain we want to be helped. But it is exactly at this point that one of the strangest aspects of our being in pain occurs – namely, it is impossible for us to experience one another's pain. That does not mean we cannot communicate to another our pain. That we can do, but what we cannot do is for you to understand and/or experience my pain as mine.

This puts us under a double burden because we have enough problem learning to know one another in the normal aspects of our lives, but when we are in pain our alienation from one another only increases. For no matter how sympathetic we may be to the other in pain, that very pain creates a history and experience that makes the other just that much more foreign to me. Our pains isolate us from one another as they create worlds that cut us off from one another. Consider, for example, the immense gulf between the world of the sick and the healthy. No matter how much we may experience the former when we are healthy, when we are not in pain, we have trouble imagining and understanding the world of the ill.

Indeed, the terms we are using are still far too crude. For we do not suffer illness in and of itself, but we suffer this particular kind of illness and have this particular kind of pain. Thus even within the world of illness there are subworlds that are not easily crossed. Think, for example, of how important it is for those suffering from the same illness to share their stories with one another, as they do not believe others, who may be just as ill only they have a different illness, can understand their particular kind of pain. Thus people with heart disease often find their particular kind of pain gives them little basis of communion with those suffering from cancer. Pain itself does not create a shared experience, but only pain of a particular kind and sort. Moreover, the very commonality thus created but separates the ill from the healthy in a decisive way.

Pain not only isolates us from one another, but even from ourselves. Think how quickly people with a terribly diseased limb or organ are anxious for surgery in the hope that if it is just removed they will not be burdened by the pain that makes us not know ourselves. This gangrenous leg is not mine. I would prefer to lose the leg rather than face the reality of its connection to me.

The difficulties pain creates in terms of our relation with ourselves is compounded by the peculiar difficulties it creates for those close to us who do not share our pain. For no matter how sympathetic we may be, no matter how much we may try to be with and comfort those in pain, we know we do not want to experience their pain. I not only cannot, but I do not, want to know the pain you are feeling. No matter how good willed we may be we cannot take another's pain as our pain. Our pains divide us and there is little we can do to restore our unity.

I suspect this is one of the reasons that chronic illness is such a burden. For often we are willing to be present and sympathetic with someone with an intense but temporary pain – that is, we are willing to be present as long as he works at being a 'good' sick person who tries to get well quickly and does not make too much of his discomfort. We may initially be quite sympathetic with someone with a chronic disease, but it seems to be asking too much of us to be compassionate year in and year out. Thus the universal testimony of people with chronic illness that their illness often results in the alienation of their former friends. This is a problem not only for the person who is ill but also for those closely connected with that person. Thus the family of a person who is chronically ill often discovers that the very skills and habits they must learn to be present to the one in pain creates a gulf between themselves and their friends. Perhaps no case illustrates this more poignantly than a family that has a retarded child. Often they discover it is not long that they have a whole new set of friends who also happen to have retarded children [10].

Exactly because pain is so alienating we are hesitant to admit that we are in pain. To be in pain means we need help, that we are vulnerable to the interests of others, that we are not in control of our destiny. Thus we seek to deny our pain in the hopes that we will be able to handle it by ourselves. But the attempt to deal with our pain by ourselves or to deny its existence has the odd effect of only increasing our loneliness. For exactly to the extent I am successful I create a story about myself that I cannot easily share.

No doubt more can be and needs to be said that would nuance this account of pain and the way it tends to isolate us from one another. Yet I think I have said enough that our attention has been called to this quite common but all the more extraordinary aspect of our existence. Moreover, in the light of this analysis I hope we can now appreciate the remarkable behavior of Job's friends. For in spite of the bad press Job's comforters usually received, they at least sat on the ground with him for seven days. Moreover, they did not speak to him, "for they saw that his suffering was very great." That they did so is truly an act of magnanimity, for most of us are willing to be with a sufferer, especially one that is in such pain that we can hardly recognize them, only if we can 'do something' to relieve their suffering or at least distract their attention. Not so with Job's comforters. They sat on the ground with Job doing nothing more than being willing to be present in the face of his suffering.

Now if any of this is close to being right, it puts the task of physicians and others who are pledged to be with the ill in an interesting perspective. For I take it that their activity as physicians is characterized by the fundamental commitment to be, like Job's comforters, in the presence of those who are in pain.[4] At this moment I am not concerned to explore the moral reason for that commitment, but rather only to note that, in fact, physicians, nurses, chaplains, and many others are present to the ill as none of the rest of us are. They are the bridge between the world of the ill and healthy.

Certainly physicians are there because they have been trained with skills that enable them to alleviate the pain of the ill. They have learned from some sick people how to help other sick people. Yet every physician soon learns of the terrible limit of his craft, for the sheer particularity of the patient's illness often defies the best knowledge and skill he has. Even more dramatically, we learn that using the best knowledge and skill we have on some patients often has terrible results. You do not have to go as far as Illich to believe medicine can be bad for your health [12].

Yet the fact that medicine, through the agency of physicians, does not and cannot always 'cure' in no way qualifies the commitment of the physician. At least it does not do so if we remember that the physician's basic pledge is not to cure, but to share through being present to the one in pain. Yet it is not easy to carry out that commitment on a day-to-day, year-to-year basis. For none of us has the resources to see too much pain without that pain hardening us. Without such a hardening, something we sometimes call by the name of professional distance, we fear we will lose the ability to feel at all.

Yet the physicians cannot help but be touched and, thus, tainted by the world of the sick. Through their willingness to be present to us in our most vulnerable moments they are forever scarred with our pain – a pain that we, the healthy, want to deny or at least keep at arm's length. They have seen the world we do not want to see until it is forced on us, and we will accept them into polite community only to the extent they keep that world hidden from us. But when we are driven into that world we want to be able to count on their skill and their presence, even though we have been unwilling to face that reality while we were healthy. No doubt there are many reasons both sociological and psychological why physicians and their families tend only to know other physicians and their families, but we cannot discount the fact that they are drawn together by their common exposure to the ill. The extent and intensity of the pain they have seen creates a wedge between them and us we do not want removed.

But what do these somewhat random and controversial observations have to do with helping us understand better the relation between medicine and the church and/or the story of my boyhood friendship with Bob? To begin with the latter, I think in some ways the mechanism that was working during that trying time with Bob is quite similar to the mechanism that works on a day-to-day basis in medicine. For the physician and others concerned with our illness are called to be present during times of great pain and tragedy. Indeed physicians, because of their moral commitments, have the privilege and the burden to be with us when we are most vulnerable. The physician earns our deepest fears and our profoundest hopes. As patients, that is also why so often we fear the physician because he may know us better than we know ourselves. Surely that is one of reasons that confidentiality is so crucial to the patient-physician relation since it is a situation of such intimacy.

But just to the extent that the physician has been granted the privilege of being with us while we are in pain, that very experience creates the seeds of distrust and fear. We are afraid of one another's use of the knowledge gained, but even more deeply we fear remembering the pain as part of our history. Thus every crisis that joins us in a common fight for health also has the potential for separating us more profoundly after the crisis. Yet the physician is pledged to come to our aid again and again no matter how we may try to protect ourselves from their presence.

The physician, on the other hand, has yet another problem, for how can anyone be present to the ill day in and day out without learning to dislike them, if not positively detest their smallness in the face of pain. People in pain are omnivorous in their appetite for help and they will use us up if we let them. Fortunately, the physician has other patients that can give him distance from any one patient who requires too much. But the problem still remains how morally those who are pledged to be with the ill never lose their ability to see the humanity that our very suffering often comes close to obliterating. For the physician cannot, as Bob and I did, drift apart and away from those whom he or she is pledged to serve. At least they cannot if I am right that medicine is first of all pledged to be nothing more than a human presence in the face of suffering.

But how can we account for such a commitment – the commitment to be present to those in pain? No doubt basic human sympathy is not to be discounted, but it does not seem to be sufficient to account for sustaining a group of people dedicated to being present to the ill as their vocation in life. Nor does it seem sufficient to account for the acquiring of the skills necessary to sustain that presence in a manner that is not alienating and the source of distrust in a community.

To learn how to be present in that way we need examples – that is, a people who have learned to embody such a presence in their lives that it has become the marrow of their habits. The church at least claims to be such a community, as it is a group of people called out by a God who we believe is always present to us both in our sin and our faithfulness. Because of God's faithfulness we are supposed to be a people who have learned how to be faithful to one another by our willingness to be present, with all our vulnerabilities, to one another. For what does our God require of us other than our unfailing presence in the midst of the world's sin and pain. Thus our willingness to be ill and to ask for help as well as our willingness to be present with the ill is no special or extraordinary activity, but rather a form of the Christian obligation to be present to one another in and out of pain.

Moreover, it is such a people who should have learned how to be present with those in pain without that pain driving them further apart. For the very bond that pain forms between us becomes the basis for alienation, as we have no means to know how to make it part of our common history. Just as it is painful to remember our sins, so we seek not to remember our pain, since we desire to live as if our world and existence were pain-free. Only a people trained in remembering, and remembering as a communal act, their sins and pains can offer a paradigm for sustaining across time a painful memory so that it acts to heal rather than to divide.

Thus medicine needs the church not to supply a foundation for its moral commitments, but rather as a resource of the habits and practices necessary to sustain the care of those in pain over the long haul. For it is no easy matter to be with the ill, especially when we cannot do much for them other than simply be present. Our very helplessness too often turns to hate, both toward the one in pain and toward ourselves, as we despise them for reminding us of our helplessness. Only when we remember that our presence is our doing, when sitting on the ground seven days saying nothing is what we can do, can we be saved from our fevered and hopeless attempt to control others and our own existence. Of course, to believe that such presence is what we can and should do from this perspective entails a belief in a presence in and beyond this world. And it is certainly true many today no longer believe in or experience such a presence. If that is the case, then I do wonder if medicine as an activity of presence is possible in a world without God.

Another way of raising this issue is to inquire into the relation between prayer and medical care. Nothing I have said about the basic pledge of physicians to be present to the ill entails that they should not try to develop the skills necessary to help those in pain and illness. Certainly they should, as theirs is an art that is one of our most valuable resources for the care of one another. But no matter how powerful that craft becomes, it cannot in principle rule out the necessity of prayer. For prayer is not a supplement to the insufficiency of our medical knowledge and practice;

nor is it some divine insurance policy that our medical skill will work; rather, our prayer is the means that we have to make God present, whether our medical skill is successful or not. So understood, the issue is not whether medical care and prayer are antithetical, but rather how medical care can ever be sustained without the necessity of continued prayer.

Finally, those involved in medicine need the church as otherwise they cannot help but be alienated from the rest of us. For unless there is a body of people who has learned the skills of presence, the world of the ill cannot help but become a separate world both for the ill and those who care for them. Only a community that is pledged not to fear the stranger (and illness always makes us a stranger to ourselves and others) can welcome the continued presence of the ill in our midst. The hospital is, after all, first and foremost a house of hospitality along the way of our journey with finitude. It is our sign that we will not abandon those who have become ill simply because they currently are suffering the sign of that finitude. If the hospital, as too often is the case today, but becomes a means of isolating the ill from the rest of us, then we have betrayed its central purpose and distorted our community and ourselves.

If the church can be the kind of people who show clearly that they have learned to be with the sick and the dying, it may just be that through that process we will better understand the relation of salvation to health, religion, and medicine. Or perhaps even more, we will better understand what kind of medicine we ought to practice, since too often we try to substitute doing for presence. It is surely the case, as Paul Ramsey reminds us, "that not since Socrates posed the question have we learned how to teach virtue. The quandaries of medical ethics are not unlike that question. Still, we can no longer rely upon the ethical assumptions in our culture to be powerful enough or clear enough to instruct the profession in virtue; therefore, the medical profession should no longer believe that the personal integrity of physicians alone is enough; neither can anyone count on values being transmitted without thought" ([14], p. xviii). All I have tried to do is remind us that neither can we count on such values being transmitted without a group of people who believe in and live trusting in God's unfailing presence.

*Divinity School,*
*Duke University,*
*Durham, North Carolina, U.S.A.*

## NOTES

[1] For a defense of the concentration of medicine on the body see [11].

[2] Ramsey's position is complex and I can certainly not do it justice here. His emphasis on 'love transforming natural law' would tend to qualify the point made above. Yet it is also true that Ramsey's increasing use of covenant language went hand in hand with his readiness to identify certain 'covenants' that needed no 'transformation'. Of course, he could object that the covenant between doctor and patient was the result of Christian love operating in history.

[3] For a fuller account of the complex relation between pain and suffering see [9].

[4] I am indebted to a conversation with Dr. Earl Shelp for helping me understand better the significance of this point.

## BIBLIOGRAPHY

[1] Amundsen, D. and Ferngren, G.: 1982, 'Medicine and Religion: Pre-Christian Antiquity', in M. Marty and K. Vaux (eds.), *Health/Medicine and the Faith Traditions*, Fortress Press, Philadelphia, pp. 53–92.

[2] Amundsen, D. and Ferngren, G.: 1982, 'Medicine and Religion: Early Christianity Through the Middle Ages', in M. Marty and K. Vaux (eds.), *Health/Medicine and the Faith Traditions*, Fortress Press, Philadelphia, pp. 93–131.

[3] Childress, J.: 1981, *Priorities in Biomedical Ethics*, Westminster Press, Philadelphia.

[4] Curran, C.: 1978, *Issues in Sexual and Medical Ethics*, University of Notre Dame Press, Notre Dame, Indiana.

[5] Fletcher, J.: 1954, *Morals and Medicine*, Beacon Press, Boston.

[6] Gustafson, J.: 1975, *The Contributions of Theology to Medical Ethics*, Marquette University Press, Milwaukee.

[7] Haring, B.: 1973, *Medical Ethics*, Fides Publishers, South Bend, Indiana.

[8] Hauerwas, S.: 1983, 'On Keeping Theological Ethics Theological', in A. MacIntyre and S. Hauerwas (eds.), *Revisions: Changing Perspectives in Moral Philosophy*, University of Notre Dame Press, Indiana (in press).

[9] Hauerwas, S.: 1979, 'Reflections on Suffering, Death and Medicine', *Ethics in Science and Medicine* **6**, 229–237.

[10] Hauerwas, S.: 1982, 'The Retarded, Society and the Family: The Dilemma of Care', in S. Hauerwas (ed.), *Responsibility for Devalued Persons*, Charles C. Thomas Publisher, Springfield, Illinois, pp. 42–65.

[11] Hauerwas, S.: 1982, 'Authority and the Profession of Medicine', in G. Agich (ed.), *Responsibility in Health Care*, D. Reidel Publishing Co., Dordrecht, Holland, pp. 83–104.

[12] Illich, I.: 1976, *Medical Nemesis: Expropriation of Health*, Pantheon Press, New York.

[13] MacIntyre, A.: 1981, 'Can Medicine Dispense with a Theological Perspective on Human Nature?' In D. Callahan and H. Engelhardt (eds.), *The Roots of Ethics*, Plenum Press, New York, pp. 119–138.

[14] Ramsey, P.: 1970, *The Patient as Person*, Yale University Press, New Haven.

[15] Ramsey, P.: 1973, 'The Nature of Medical Ethics', in R. Veatch, *et al.* (eds.), *The Teaching of Medical Ethics*, A Hastings Center Publication, Hastings-on-Hudson, New York, pp. 14-28.

JAMES F. CHILDRESS

# LOVE AND JUSTICE IN CHRISTIAN BIOMEDICAL ETHICS

My task is to identify and analyze some major themes in biomedical ethics from a Christian standpoint. In particular, I want to indicate the importance of the distinctions and relations between the principles of love and justice in several different areas of biomedical ethics, to suggest that more careful analytical and constructive attention to these norms – their content, their distinctions, and their relations – would clarify and perhaps resolve some of the debates in biomedical ethics, and to underline the impossibility of either clarifying or resolving these disputes without attention to their broader theological, metaphysical, and anthropological contexts. My analysis of both problems and theologians will be selective. Although I will mention several problems and identify several major issues in schematic form, I will concentrate on debates about the relation of agape (neighbor-love) to justice in the distribution of the burdens of research involving human subjects and the benefits of medical care. I will concentrate on the writings of Paul Ramsey, Richard McCormick, S.J., and Gene Outka, with some attention to discussions by Arthur Dyck and Joseph Fletcher, among others.

## MORAL PERPLEXITIES NOT CONFRONTED BY THE GOOD SAMARITAN

It is impossible to take the parable of the Good Samaritan (Luke 10:25–37) as a definitive statement of the principle of agape, which Jesus offers as a summary and perhaps even as the constitutive principle of the whole law and the prophets (Matthew 22:34–40 and Mark 12:28–34).[1] The parable directs the agent and his/her actions toward the neighbor's welfare, but it does not raise or resolve several moral perplexities: (1) The conflict between the welfare of the neighbor and the welfare of the agent, e.g., how much risk should the Good Samaritan have assumed for the neighbor? (2) The conflict between different neighbors, e.g., if the Good Samaritan had encountered the robbers while they were attacking their victim, could he have preferred the victim's life over theirs if lethal force



*E. E. Shelp (ed.), Theology and Bioethics,* 225–243.

had been necessary to stop them? Or if the Good Samaritan had encountered several victims of the robbers but could not have taken care of all of them, how should he have selected the recipients of his care? (3) The conflict between the neighbor's needs and the neighbor's wishes, e.g., if the victim had requested that he be left to die (perhaps even flashing his 'Living Will'), should the Good Samaritan have respected his wishes?[2]

Since versions of the first and second conflicts will be addressed throughout this essay, I want to make a few points about the third conflict: whether the agapistic agent or the recipient should have the authority to determine which interests (needs or wishes) are relevant to human well-being and how much weight they should have in the action. This conflict has received less attention in Christian ethics than the other two conflicts. Consequently, the constraints upon paternalism (nonacquiescence in a person's wishes, choices, and actions for that person's own benefit) within a love-ethic are not as secure and as strong as they should be.[3] It is easier for a love-ethic to appreciate the role of *justice* in conflicts among the interests of the agapist and others or among the interests of different people whose welfare the agapist seeks. It is widely conceded that agapistic actions should not fall below justice toward neighbors, and that the agapist should sometimes sacrifice his or her own claims of justice in order to meet the needs of others. When, however, the agapist is not concerned with competing needs (his/her own welfare in relation to the neighbors' or the welfare of several different neighbors) but is rather concerned about the determination of the neighbor's needs, he or she may not perceive the persistent claims of justice in the form of various rights that the recipient may assert (such as a right of self-determination). For example, early in the 20th century, many Progressives and advocates of the Social Gospel (the latter often being a religious version of the former) urged various reforms in the name of welfare. Some of these reforms were ambiguous because they sacrificed 'rights' to 'needs', which the state, conceived as a benevolent parent, could meet. This benevolence was considered more important than procedural protections. For example, the reform of the juvenile justice system emphasized 'needs' rather than 'rights' and removed some procedural protections of the rights of youthful offenders and their families for their own best interests ([27], [35], [36]).

The relations and potential conflicts between love and justice, particularly respect for persons, raised dramatically by debates about paternalism, also appear in debates about euthanasia, where other moral considerations are also prominent. Philippa Foot writes,

> Charity is the virtue that gives attachment to the good of others, and because life is normally a good, charity normally demands that it should be saved or prolonged. But as we so define an act of euthanasia that it seeks a man's death for his own sake – for his good – charity will normally speak in favour of it. This is not, of course, to say that charity can require an act of euthanasia which justice forbids, but if an act of euthanasia is not contrary to justice – that is, it does not infringe rights – charity will rather be in its favour than against ([12], p. 54).

By contrast, Arthur Dyck, a Christian ethicist, argues that love of neighbor, mercy, or kindness requires both nonmaleficence and beneficence, but that nonmaleficence takes precedence over and controls beneficience. The prior, more stringent duty is to do no harm, which includes killing; for example, the Decalogue, which specifies requirements of love for communal life, prohibits killing. Over against an ethic of beneficent euthanasia, Dyck proposes an ethic of 'benemortasia', which would extend the following kinds of care to patients whose death is imminent: relief of pain and suffering, respect for the right to refuse treatment, and provision of health care without regard for ability to pay ([7], pp. 72–91; see also Dyck's article in [16] and contrast Kohl's argument in [15] and [16]).

Against Dyck and in favor of Foot, it is possible to argue that while death is always a harm, it is not always a *net harm*. Thus, killing may not always violate the duty of nonmaleficence, and it may be an expression of love, mercy, kindness, and care. A different version of the Good Samaritan parable appears in Jack Cady's short story, 'The Burning' [2] in which a trucker was trapped in the cab of a burning tanker after an accident. Unable to release the victim who screamed as the flames roared about him, another trucker finally took a pistol and shot him. Even if such direct killings are illegal, they are frequently excused, and they may even be morally praiseworthy as well as morally right in some circumstances. They may manifest both care and respect for personal wishes and choices when death is requested.

But even if we admit, in contrast to Dyck, that some *acts* of 'mercy killing' do indeed express mercy, are beneficent, are not nonmaleficent (at least in the sense of net harm), are not unjust and disrespectful (because they are in accord with a competent person's wishes), we may still be suspicious of efforts to set up a *rule* or *practice* of euthanasia. Whatever is said about particular acts of euthanasia, there are strong reasons to oppose a rule or practice of euthanasia because it would probably lead to abuses of the relevant moral principles, including both love and justice [1]. But it is hard to see how the principle of neighbor-love leads to a moral condemnation of each act of euthanasia.

## RELATIONS BETWEEN LOVE AND JUSTICE IN RECENT CHRISTIAN ETHICS

As Gene Outka [24] contends, there is no single relationship between agape and justice in Christian ethics because of variations in the conceptions of both agape and justice. For example, four major conceptions of agape can be identified; they compete for dominance, often in the work of a single author, and their own relations are matters of complex debate: (1) agape as seeking the neighbor's welfare, (2) agape as self-sacrifice, (3) agape as mutuality, and (4) agape as equal regard. It can be expected, for example, that a theologian defending (2) will have a different conception of the relation of love and justice than a theologian defending (4).

Likewise, there are many different conceptions of justice. On the one hand, it is viewed as a virtual synonym for all of morality, except perhaps for agapistic actions; for this broad conception, justice is almost equivalent to righteousness. On the other hand, justice is viewed more narrowly as one standard among others for regulating institutions, practices, policies, and acts. For the narrow conception it is possible to view an act or policy as just but unjustified, or as unjust but justified, because justice may conflict with other standards.

Within the narrow conception of justice as part of morality, some conflict about goods, etc., is presupposed; otherwise the claims of justice, i.e., claims about what is due individuals or groups, would be irrelevant. Standards of justice as part of morality are particularly applicable to (1) the distribution of social benefits, such as health care, and social burdens, such as military service and taxation, (2) voluntary and involuntary transactions, such as contracts and torts, and (3) assignment of penalties for violation of legal norms. I will concentrate on the distribution of the benefit of health care and the burden of participation in research.

There is widespread agreement about the formal definition of justice as (1) giving each person his/her due (*suum cuique*) and (2) treating similar cases in a similar way. While there are important questions about the relation of agape to formal justice (such as whether agape must treat similar cases in a similar way), most major debates focus on the material criteria that identify what is due individuals and groups and identify relevant similarities and dissimilarities: for example, needs (egalitarian theories), merit (meritarian theories), free actions, including contracts (libertarian theories), and societal contribution (utilitarian theories).

Several different metaphors for relating love and justice emerge from the literature of Christian ethics in the last fifty years. These patterns of relationship include: love and justice as united – justice as the form of reuniting love (Paul Tillich); love and justice as identical (Joseph Fletcher); love and justice in dualism (Emil Brunner); love and justice in dialectical relation (Reinhold Niebuhr); love transforming justice (Paul Ramsey); love completing but not contradicting justice (much traditional Roman Catholic theology); love requiring or translated into justice (much liberation theology); and love and justice as distinct but overlapping principles (Gene Outka).

Debates about love and justice involve several theological, metaphysical, and anthropological presuppositions. For example, convictions about God's relation to the world, creation and fall, nature and grace, reason and revelation all play a significant role in the formulation of the content, distinctions and relations of the principles of love and justice. As these categories suggest, of particular importance is the question of *knowledge* of the norms and *capacity* to act in accordance with them. Does the principle of agape presuppose revelation for its recognition and grace for its fulfillment whereas principles of justice are knowable by reason and are within human capacity? Space does not permit a full discussion of various theological, metaphysical, and anthropological *presuppositions* of different positions regarding love and justice, but some of them will emerge as I explore some *applications* of love and justice to debates about the distribution of the burdens of research involving human subjects and the distribution of the benefits of medical care.

## LOVE AND JUSTICE IN RESEARCH INVOLVING HUMAN SUBJECTS

In their longstanding debate about the use of children in nontherapeutic research, Paul Ramsey and Richard McCormick appeal to the distinction between love (charity) and justice in order to determine which research, if any, may be performed on human beings without their consent, i.e., without their expressed will or against their expressed will.[4] Ramsey and McCormick agree that acts of charity cannot be demanded or enforced. Such acts can be performed only by individuals who can act voluntarily and the acts themselves must be voluntary. Ramsey and McCormick also agree that some standards of justice may be enforced even if individuals do not consent to them. Their debate in part focuses on where the line between charity and justice should be drawn, and their respective positions depend on several theological, metaphysical, and anthropological convictions.

Ramsey's and McCormick's different views about which research activities fall under charity and which under justice appear on two levels: (1)

the society's pursuit of scientific and medical research, e.g., through the allocation of funds, and (2) the individual's responsibility to participate in such research. For Ramsey, scientific and medical research is optional from the society's standpoint, whereas, for McCormick, such research is so important that it is imperative for the society to pursue it. Back of this difference is McCormick's conviction that through God's creation both individuals and communities are inclined toward certain objective, non-moral values that define human flourishing. By examining pre-rational inclinations or tendencies (such as to preserve life and health and to mate and raise children), it is possible to identify these objective, nonmoral values and then to use practical reason to develop 'mediating principles' between these values and concrete choices [23]. Through some of these 'mediating principles', it is possible to establish duties and obligations. This teleological orientation is, of course, largely absent from Ramsey's discussion, where deontological considerations dominate, and where societies, by and large, have more moral discretion to pursue various goals. Deontological considerations set limits on *how* society may pursue those goals. Indeed, whether the topic is society's reasons for going to war or its priorities, Ramsey holds that moral assessments of goals cannot be as definite or as conclusive as moral limits on the pursuit of those goals.

In response to Ramsey's charge that he accepts a research imperative, McCormick distinguishes two meanings of 'research imperative': (1) excessive zeal to achieve the goals of research at any cost, and (2) "an urgent concern to prevent, or alleviate disease, especially crippling disease" ([2], p. 46, n. 4). McCormick rejects the former but accepts the latter. Using the parable of the Good Samaritan, he argues that if it is right and charitable to heal wounds, it is as right and charitable to prevent them; and if research is necessary to prevent wounds, then "research (experimental procedures) is an imperative and indeed a Christian one" ([21], p. 46, n. 4).[5] Thus, it appears that the society has moral reasons to pursue research and may be even blamed for its failure to do so.

McCormick's use of the parable of the Good Samaritan to build a moral and even Christian imperative for research is similar to Ramsey's use of the parable to justify war and Christian participation in war ([29], pp. 142f.). But Ramsey denies the relevance of this elaboration of the parable to research. For Ramsey, "if war is ever justified, that has to be under the head of the prevention of evil. For this reason, a draft or a lottery may be appropriate to fulfill a national duty of 'perfect obligation' " ([34], p. 41). So far he would appear to concur with McCormick, but he insists that research provides positive benefits rather than preventing evils.

McCormick makes no distinction between positive and negative duties. The just defense of a nation means the prevention of a great evil to the common good, even to the international common good, e.g., by repulsion of an invader or the containment of armed and aggressive tyranny. The conquest of diseases is an achievement, goal or ideal. Participation in the first may be mandatory; volunteers for the latter are praiseworthy ([33], p. 230).

Because of his evaluative description of the research enterprise (sometimes viewed as the pursuit of 'progress'), Ramsey contends that it is optional for both societies and individuals.

It is hard, however, to follow Ramsey's distinction, which is probably, in the final analysis, a distinction between positive and negative *goals* rather than positive and negative *duties*. Surely the duty to *prevent* evil is a *positive* rather than a *negative* duty because it requires positive actions. A common example of a negative duty is the duty of nonmaleficence: the duty not to injure or harm others. The duty to prevent such harm or injury involves positive actions and is more plausibly located under some other principle such as justice or beneficence. Nevertheless, within Ramsey's schema, the duty to refrain from injuring others and its associated duty to prevent such injury are more stringent than the duty to help others (beneficence) ([34], p. 41). And Ramsey believes that war can sometimes be construed as preventing evil but that research can only be construed as producing benefits.

McCormick and Ramsey also apply their general perspectives on charity and justice directly to judgments about the obligations of individuals to participate in nontherapeutic research. Just as the society has an 'imperative' based on beneficence or justice to pursue research in order to prevent diseases, according to McCormick, individuals have an 'obligation' based on 'social justice' to participate in some research. 'Social justice' requires "one's personal bearing of his share of the burden that all may flourish and prosper" ([18], p. 16). Choices of social justice are 'choices that all *ought* to make' and that all can be *expected* to make. The 'baseline' for the distinction between social justice and individual generosity and charity is clear: "works that involve no notable disadvantages to individuals and yet offer genuine hope for general benefit" ([18], p. 16). If research involves "discernible risk, undue discomfort, or inconvenience", participation is not a matter of social justice, but rather of charity. From this general perspective on love and justice, McCormick concludes that we may enforce social justice or reasonable sociality.

In such cases he contends that it is not unjust to draft adults to participate if there are not enough volunteers and that it is not unjust to construct or presume consent on the part of children who cannot consent.

Regarding children, proxy consent is appropriate for them where they *would* consent if they could consent because they *ought* to consent – the 'ought' being a duty of social justice [18]. Clearly this is an enforcement of morality, as Ramsey argues [32], but McCormick holds that it is a morality of justice rather than charity, a morality of duty rather than aspiration or inspiration.

It is important to identify McCormick's 'ought' of social justice more precisely. It is possible to argue that many people ought to participate in nontherapeutic research because of a debt of gratitude. This debt would rest on their prior or current voluntary acceptance of the benefits of research (e.g., drugs). But this debt would be limited to those who had voluntarily *accepted* (and not merely received) such benefits, and it is generally not morally appropriate to enforce debts of gratitude in contrast to implicit or explicit contracts. A stronger obligation might be based on fair play: if individuals have accepted the benefits of a practice to which others have contributed, they now have a duty of fair play to take their turn in shouldering the burdens of the practice. But, like the obligation of gratitude, the obligation of fair play presupposes voluntary acceptance of benefits and thus the capacity to consent. By contrast, McCormick's conception of social justice, which grounds the obligation to participate in research, does not depend on the voluntary acceptance of the benefits of research; it does not even presuppose the prior reception of benefits. It rather depends on a non-voluntaristic, organic conception of the relation of the individual and the community. McCormick writes,

> to share in the general effort and burden of health maintenance and disease control is part of our flourishing and growth as humans. To the extent that it is good for all of us to share this burden, we all *ought* to do so. And to the extent that we *ought* to do so, it is a reasonable construction or presumption of our wishes to say that we would do so....sharing in the common burden of progress in medicine constitutes an individual good for all of us *up to a point* ([18], pp. 12–13).

According to McCormick, where the research involves minimal risk and "is of overriding importance to the public health, it would not be unjust of the government to recruit experimental subjects, for example, by lottery, just as it is not unjust for government to draft soldiers for national self-defense" ([22], p. 2197). Nevertheless, factors other than justice are also relevant to a decision about a policy of conscription for research. First, since "consensual community is something to be promoted wherever possible", volunteers are preferred [22]. Second, *feasibility* is also a standard of public policy; it is "that quality whereby a proposed course of action is not merely possible but practicable, adaptable, depending on the circumstances, cultural way, attitudes, traditions of a people..." ([23], pp. 71–73, 190, quoted from Paul Micallef).

It is not surprising then that Ramsey rejects McCormick's proposed analogy between conscription for research and conscription for military service. But one part of Ramsey's argument against the analogy appears to break down; that is his effort to distinguish conscripted (adult) research subjects from conscripted soldiers. He contends,

> conscripted soldiers are citizens who, though young, have lived in a political society and shared in its safety and other benefits. They have tacitly accepted the benefits of that common good that flow back upon them as individuals, and so also arguably have entered into a community of shared expectations concerning the common defense ([33], p. 230).

Similar points could plausibly be made about the conscription of adults for participation in nontherapeutic research involving minimal risk, especially if, as Ramsey himself notes, our society tends to view death always as a disaster and tries to avoid it through research, medical care, etc. 'Tacit acceptance' of the benefits of research could be construed as entrance into a community of 'shared expectations' regarding the war against diseases. (But, as I emphasized, McCormick does not build his position on such voluntaristic and individualistic actions as explicit, implicit, or tacit acceptance of benefits.)

It would appear that Ramsey's effort to distinguish conscripted (adult) research subjects from conscripted soldiers ultimately depends on his interpretation of legitimate 'shared expectations', which in turn depends on his evaluative description of research as optional for the society because it promotes benefits rather than preventing evil. He does grant one 'exception' (though he is not sure that it counts as an 'exception') that has a "remote analogy" with military service. His example involves children:

> the supposable case must be that these *particular* children are going to be at risk of an illness of calamitous proportions in the near future and if the cure or management of that illness beyond doubt requires their use in medical research. Then they *have* a precisely identifiable share in the common defense against illness, as soldiers do ([34], p. 230).

But if this example is plausible in Ramsey's framework, it is not because of the analogy with war, but because this research involving these particular children approximates therapeutic research. The situation is one of "epidemic conditions that bring upon the individual child proportionately the same or likely greater danger" ([30], p. 25).

Hans Jonas also recognizes the possibility of conscription for experimentation when the society faces a "clear and present danger", a health emergency, and can avert the disaster only through conscription. This is a situation of saving, not improving the society [14]. It is not clear whether Ramsey would recognize this sort of exception, since he tends to favor qualifications of the meaning of moral rules rather than explicit exceptions, and since he tends to view medical research almost exclusively in terms of optional improvement for the future, i.e., what Jonas calls a "melioristic goal" [14]. At any rate, if a situation such as Jonas depicts were to exist, it could conceivably justify conscription for research of more than minimal risk. Ramsey worries that McCormick's statement of the analogy between military conscription and research conscription would in fact allow the society to impose more than minimal risks in research when the goal seems to be of "overriding importance to the public health" ([33], p. 230). For these reasons, Ramsey appears to be unwilling to concede an exception of emergency or necessity regarding the public health.

However, a society may morally choose to undertake "mankind's war against diseases" ([32], p. 29). Holding that "the larger questions of medical and social priorities are almost, if not altogether, incorrigible to moral reasoning" ([30], p. 240), Ramsey allows the society great latitude in its priorities, e.g., research vs. medical care, prevention vs. rescue, or health care vs. other goods. But the society's *distribution* of its *benefits* and its *burdens* is morally limited. (I will return to the question of the distribution of medical benefits later.) The burdens of research, at least nontherapeutic research, can morally be distributed only to those who voluntarily agree to bear them. *Whether* the society undertakes a war against diseases is morally within its discretion, but *how* the society pursues and conducts that war is not morally optional. It is not permitted to use research subjects against or without their will. (This discussion is analogous to Ramsey's treatment of *just ad bellum* – the right to wage war – and *jus in bello* – right conduct within war; indeed, the criteria of just research are analogous to the criteria of just war [4], [40].)

Since participation in research cannot be expected or demanded of adults, it cannot be enforced, and it cannot be presumed of an uncomprehending subject. Such " 'construed' altruism" ([32], p. 29) would be violent even if the risks were low or minimal. Morally, the issue is battery, not negligence. The 'altruism' in question concerns the *integrity* of one's body rather than *risks*. Hence, the "sacrifice" is "any exaction without one's will", even if it would not be considered supererogatory or heroic if performed voluntarily [34], p. 42, n. 8). The issue is not merely charity or altruism, but voluntariness ([32], p. 30, n. 7).

Furthermore, if altruistic or charitable actions could be construed or constructed for others, there would be no limit, no principle to constrain them, not even the standard of low or minimal risk. Once the consequentialist calculation is accepted, it tends to allow the imposition of greater and greater risks. Ramsey writes, "I do not see where one could rationally stop in construing all sorts of works of mercy or self-sacrifice on the part of persons, not themselves capable by nature or grace yet of being the subjects of charity" ([31], discussion, quoted in [18], p. 16).

Because research is optional for the society and for the individual, the researcher and the subject should be voluntary "joint adventurers or partners in the enterprise of medical advancement" ([30], p. 25). This model of joint adventurership or partnership excludes the use of subjects against their will (draft of adults) or without their will (the use of uncomprehending persons such as children). This model of research also coheres with Ramsey's theological theme of covenantal relations and a conception of research and participation in research as morally optional. It is buttressed at points by Ramsey's theological anthropology, which emphasizes human sinfulness. Appropriating Lincoln's view that "no man is good enough to govern another without his consent", Ramsey applies this to medical and research relations ([30], p. 7). Finally, while Ramsey's emphasis on consent has affinities with the Kantian tradition of respect for persons, as he admits, his explicit grounds are Christian. Specifically, in relation to the use of children in nontherapeutic research, Ramsey appeals not to the Kantian conception of humanity, but instead to God's agape: "the wholeness of God's care for the least and the littlest ones and their preciousness to Him" ([32], pp. 26, 30, n. 19).

The themes of love and justice resound or echo through the Ramsey-McCormick debate about nontherapeutic research, which focuses on children who cannot consent (though I have also emphasized subjects who refuse to consent). In the context of this debate, McCormick tends to stress the distinction between charity and justice in determining what we can minimally expect and demand of others, even to the extent of exacting it against or without their will. Ramsey accepts but tends to downplay that distinction, emphasizing instead what both love (as based on covenant-faithfulness) and justice (or love-transformed-justice) require of the society, professionals, and family members when they confront potential research subjects who cannot consent or refuse to participate. He thus emphasizes God's care for the weak and the vulnerable and what this implies for our care for them. Human agape shaped by divine agape requires expression in rules of consent in nontherapeutic research, itself an optional endeavor for both the society and its members.

LOVE AND JUSTICE IN THE DISTRIBUTION OF MEDICAL CARE

Gene Outka, the author of one of the most important analyses of agape in this century, has extended his analysis to societal policies for the distribution of critical care ([24], [25], [26]). When we consider both agape and the nature of health crises, Outka argues, we will reject such material criteria of justice as (1) "to each according to his merit or desert", (2) "to each according to his societal contribution", and (3) "to each according to his contribution in satisfying whatever is freely desired by others in the open marketplace of supply and demand" in favor of (4) "to each according to his needs".

Before considering Outka's arguments in more detail, it is important to note an ambiguity in his interpretation and application of the fourth criterion, which he construes as an egalitarian standard. That criterion is susceptible to two different interpretations, which Outka does not adequately distinguish: He sometimes defends a societal goal of "the assurance of comprehensive health services for every person irrespective of income or geographic location", while at other times he contends "that every person in the entire resident population should have equal access to health care delivery" ([25], pp. 11–12). The satisfaction of needs, such as the need for medical care, is distinct from equal access to what the society provides to satisfy the needs. These goals are not necessarily identical. As Albert Weale argues, "the satisfaction of needs is itself a benefit to be distributed in accordance with some principles, and this means that the principle of satisfying needs is on a distinct logical level from the principle of equality" ([41], p. 70). The principle of needs picks out which benefits are to be distributed; the principle of equality or equal access determines their distribution.

It is easy to see how these two principles could be conflated and confused. In part Outka's conflation of needs and equal access is shaped by his conception of agape, which combines both welfare and distributional concerns. According to Outka, agape is "equal regard", i.e., regard ("active concern for the neighbor's well-being") that is independent of the neighbor's action and is unalterable because it is equal ([24], p. 260). It focuses on generic characteristics, such as need. When Outka approaches questions of social justice under the 'influence' or 'pressure' of this conception of agape, it is natural that he should conflate welfare and distributional concerns, when they should be distinguished, even though not separated. A conception of agape as seeking the neighbor's welfare, which may not imply a conception of equality (or equal access to one's agapistic actions), would probably not lead to this conflation.

There are also other reasons for the conflation. Defenders of equal access to medical care rule out any of the first three material criteria of justice that Outka identifies, and they tend to make need a *necessary* condition for the distribution of medical care. If need is a necessary condition and if we exclude other standards such as merit, societal contribution, and the marketplace, it might appear that the principle of needs and the principle of equal access are identical. But such an identification would hold only if needs offered a *sufficient* condition for the distribution and receipt of medical care. In fact, Outka permits various macroallocation decisions that would severely limit the goal of 'comprehensive health services' under conditions of scarcity; it is not morally required that health care consume most of the societal budget that also provides resources for other needs and desires. Equal access finally means (a) that only needs will be considered when a person seeks medical care, (b) subject to the limitations of the society's allocation decisions, and (c) that scarce medical care will then be distributed in accord with the formal principle of justice – treat similar cases in a similar way (which might permit queuing, a lottery, or exclusion of some types of disease or illness from treatment). The principle of equal access may thus permit unequal outcomes.

There are two major sets of critics of the principle of equal access. Some critics contend that it implies too much – that if anyone in the society has any form of medical care, it must be extended to everyone [13]. According to this interpretation, the principle of equal access is almost identical to the principle of 'to each the same thing', which is not what Outka defends. Other critics contend that it implies too little – equal access only to whatever is available in the society, even if it is not a decent minimum. As a principle of distribution, equal access does not necessarily imply any level of societal allocation. Yet some of Outka's reasons for adopting a principle of equal access may imply priority for medical care in the society's budget.

Before considering in more detail how Outka's view of agape as equal regard affects his conception of justice in medical care, I want to examine his conception of health crises, which leads him to reject some material criteria as irrelevant to the distribution of medical care. At the heart of Outka's argument is his belief that health crises are, for the most part, randomly distributed, unpredictable, and undeserved. His interpretation of "undeserved" presupposes the other two features: health needs are undeserved largely because they are randomly distributed and unpredictable. Because health needs are undeserved, Outka concludes that standards of justice other than needs are "unfair as well as unkind" in the distribution of health care ([25], p. 23).

As Tristram Engelhardt argues, one of the central issues in debates about the distribution of health care is the interpretation of the 'natural lottery' [8]. This metaphor of a lottery suggests that health needs, various diseases and illnesses, largely result from an impersonal lottery and are thus undeserved. But even if these needs are largely undeserved insofar as they result from chance, this point does not determine *why* the society ought to respond to them in certain ways. For undeserved needs may be viewed as either unfortunate or unfair. For example, Lance Stell distinguishes strong and weak senses of 'undeserved'. In its *strong* sense, 'undeserved' implies that "some sort of injustice (is) present, that someone has received more or less than his rightful share and that redress is morally appropriate...." [38]. In its *weak* sense, 'undeserved' implies that "the moral category of desert does not apply to states of affairs of that type." Winning or losing a lottery is an example of the second, weak sense of 'undeserved'. Needs that are undeserved in the strong sense are unfair and require redress; needs that are undeserved in the weak sense may be unfortunate but they are not unfair. Unfortunate needs may be the object of compassion and kindness, and individuals, voluntary associations and even the society may try to meet those needs out of compassion or kindness. There are special obligations of fairness to meet needs that result from the actions of others (e.g., assaults), or failures of institutions (e.g., environmental pollution), or actions on behalf of the society (e.g., military service). There is, however, less agreement about whether the principle of fairness *requires* the society to blunt the effects of the natural lottery. It is much easier to establish a societal 'ought' on grounds of compassion or kindness to the unfortunate, but that sort of 'ought' does not arm the recipient of health care with the weapon that a moral right, based on fairness, would provide.

Outka also contends that the 'influence' and 'normative pressure' from agapeic considerations are 'in line' with an egalitarian conception of justice that appeals "to the generic characteristics all persons share rather than the idiosyncratic attainments which distinguish persons from one another" and downplays "desert considerations". ([25], p. 13, [24]). At work here is Outka's general conception of relations between love and justice. First, he emphasizes that there is no single relation between love and justice, because there are different material criteria of justice. Love, however, tends to affect "priorities", for example, emphasizing needs before merits ([24], p. 92). Second, Outka's most common metaphors for relating love and justice are spatial: overlap and conjunction. Rejecting positions that either identify (e.g., Joseph Fletcher) or oppose (e.g., Anders Nygren) love and justice, Outka argues that love and justice are distinguished, but not separated. Rather than being separated or identical, they *overlap*: "the closer one approaches egalitarian notions of justice, the greater the material overlap with agape" ([24], p. 309). They are not, however, interchangeable, since agape is "more inclusive" and applies in some contexts, such as intimate friendship, where justice is not directly relevant ([24], pp. 309f.). This overlap or conjunction implies that agape may require more, but never less than justice ([25], p. 23). This overlap or conjunction is greater for Outka's conception of agape as equal regard than it would be for some other conception of agape, such as mutuality, self-sacrifice or even seeking the neighbor's welfare.

Another implication of Outka's conception of agape as equal regard can be seen in his argument against the utilitarian criterion of justice: to each according to his/her societal contribution ([25], especially pp. 17–19). Utilitarianism counts each person as one and only one and sums up interests in determining the greatest good for the greatest number. In the distribution of medical care, some utilarians propose selecting recipients according to their potential societal contribution, concentrating on the future rather than on the past (unless neglect of the past would undermine future productivity). Such a proposal appears, for example, in Joseph Fletcher's writings, where love is identified with beneficence and utility [10], [11].

Outka's conception of agape as equal regard has a very different implication. It cannot, however, be viewed as emphasizing intention rather than results, whereas Fletcher's conception tends to highlight beneficience and then utility because of the need to produce and balance benefits (and harms). Rather, Outka's conception of equal regard is closely related to the principle of respect for persons.[6] Indeed, Outka's affinities with philosophers in the Kantian tradition who almost identify agape with respect for persons is evident ([24], pp. 16–17, fn. 17). Outka avoids identification, again using the language of 'overlap' in 'content' between agape and respect for persons. When he considers the distribution of medical care according to potential societal contribution, he contends that such a standard must be rejected, not only because of the nature of health needs (i.e., their lack of connection with freedom), but also because of agape: "If one agrees, for whatever reason, with the agapeic judgment that each person should be regarded as irreducibly valuable, then one cannot succumb to a social productiveness criterion of human worth" ([25], p. 19). Since agape "enjoins one to identify the neighbor's point of view, to try imaginatively to see what it is for him to live the life he does, to occupy the position he holds", it requires 'minimal consideration' that cannot be set aside even for long-range social benefits. If we set aside that 'minimal consideration', we treat the other person merely as a means or an instrument ([24], p. 311). Furthermore, Outka continues, the "immediate and reasonably foreseen" effects on the neighbor should shape one's actions and policies more than remote social goals ([24], p. 311, see also Dyck's important argument in [6]).

Finally, Paul Ramsey also invokes agape in his argument against the use of utilitarian criteria in microallocation decisions in health care. Some of his arguments against the selection of recipients of scarce lifesaving medical resources by criteria of social worth are independent of religious beliefs: for example, the pluralistic nature of our 'unfocused' society makes it difficult if not immpossible to develop and apply criteria of social worth. Furthermore, his argument that criteria of social worth deny the equal worth of persons may be accepted without religious beliefs. But Ramsey also introduces a specifically Christian reason for using a lottery or queuing rather than criteria of social worth: God's agape or indiscriminate care. Because God makes the rain fall and the sun shine on the just and the unjust alike, there is an appropriate way to 'play God' – the provision of indiscriminate care without regard to social worth ([30], chap. 7). Although Ramsey did not emphasize his earlier conception of love-transforming-justice in this connection ([28], [33], p. 191), his interpretation of agape is clearly more consistent with some conceptions of justice than others. In this case, agape requires equality of opportunity when scarcity means that medical care cannot be extended to everyone. Along with Outka, and in contrast to Fletcher, Ramsey rules out appeals to social worth in this setting.

These explorations of two areas of controversy in biomedical ethics – distribution of the burdens of nontherapeutic research and distribution of the benefits of medical care – indicate the importance of conceptions of love and justice – their content, their distinctions, and their relations – in the context of broader theological, metaphysical, and anthropological convictions. These explorations also suggest that analyses and constructive proposals in Christian biomedical ethics need more careful attention to these norms in their broader contexts. But such an endeavor will require explicit and systematic theological discourse and not simply application of norms – the main focus of Christian biomedical ethics to the present.

*University of Virginia, Charlottesville, Virginia, U.S.A.*

## NOTES

[1] Although the great commandment has two parts, I am concentrating on the commandment to 'love your neighbor as yourself' (which I will refer to as the principle of agape or the principle of neighbor-love or charity) rather than the commandment to 'love the Lord your God...'. For an important discussion of the ethical relevance of the latter, see Dyck [6]. I am also focusing on both agape and justice as principles for judging institutions, practices, policies, and acts. Thus, I do not consider them as virtues, i.e., settled dispositions to act in certain ways.

[2] Some other difficult questions include determination of who counts as one's neighbor (e.g., the fetus and future generations) and determination of the neighbor's needs and priorities among those needs.

[3] For a fuller discussion of paternalism, including its basis in love, care, beneficence, and patient-benefit, see Childress [5], from which much of this paragraph has been drawn.

[4] The debate between Ramsey and McCormick about research has focused on the use of children, but I will concentrate on their arguments about love and justice as these have emerged in the course of the debate. Regarding the language of 'burdens of research', it is important to note that McCormick generally denies that low risk or minimal risk research counts as a 'burden', and Ramsey views 'sacrifice' as "any exaction without one's will" ([34], p. 42). For another discussion of justice in research, see Lebacqz [17].

[5] Maurice Visscher has argued this point even more strenuously with reference to the duty of nonmaleficence (*primum non nocere*), which is at the heart of medical care. This duty requires efforts to reduce *unintentional* harm as well as intentional harm. But the only way physicians can avoid unintentional harm is through research that can establish the safety and efficacy of various diagnostic and therapeutic procedures [39].

[6] Fletcher's conception of love was more personalistic and less utilitarian in his earlier writings, such as *Morals and Medicine* [9].

## BIBLIOGRAPHY


[1] Beauchamp, T. L. and Childress, J. F.: 1983, *Principles of Biomedical Ethics*, 2nd ed., Oxford University Press, New York.

[2] Cady, J.: 1965, 'The Burning', *Atlantic Monthly* **216**, 53–57.

[3] Childress, J. F.: 1979, 'A Right to Health Care?' *Journal of Medicine and Philosophy* **4**, 132–147.

[4] Childress, J. F.: 1981, *Priorities in Biomedical Ethics,* The Westminster Press, Philadelphia.

[5] Childress, J. F.: 1982, *Who Should Decide? Paternalism in Health Care*, Oxford University Press, New York.

[6] Dyck, A. J.: 1968, 'Referent-Models of Loving: A Philosophical and Theological Analysis of Love in Ethical Theory and Moral Practice', *Harvard Theological Review* **61**, 525–545.

[7] Dyck, A. J.: 1977, *On Human Care: An Introduction to Ethics*, Abingdon Press, New York.

[8] Engelhardt, H. T., Jr.: 1981, 'Health Care Allocations: Responses to the Unjust, the Unfortunate, and the Undesirable', in E. Shelp (ed.), *Justice and Health Care*, Vol. 8 of *Philosophy and Medicine*, D. Reidel, Dordrecht, Holland, pp. 121–137.

[9] Fletcher, J.: 1954, *Morals and Medicine*, Princeton University Press, Princeton, New Jersey.
[10] Fletcher, J.: 1967, 'Love and Justice are the Same Thing', in *Moral Responsibility*, The Westminster Press, Philadelphia, pp. 42–57.
[11] Fletcher, J.: 1979, *Humanhood: Essays in Biomedical Ethics*, Prometheus Books, Buffalo, New York.
[12] Foot, P.: 1978, 'Euthanasia', *Virtues and Vices*, Basil Blackwell, Oxford.
[13] Fried, C.: 1976, 'Equality and Rights in Medical Care', *Hastings Center Report* **6**, 29–34.
[14] Jonas, H.: 1969, 'Philosophical Reflections on Experimenting with Human Subjects', in P. A. Freund (ed.), *Experimentation with Human Subjects*, George Braziller, New York, pp. 1–31.
[15] Kohl, M.: 1974, *The Morality of Killing*, Humanities Press, New York.
[16] Kohl, M. (ed.): 1975, *Beneficent Euthanasia*, Prometheus Books, Buffalo, New York.
[17] Lebacqz, K.: 1981, 'Justice and Human Research', in E. Shelp (ed.), *Justice and Health Care*, Vol. 8 of *Philosophy and Medicine*, D. Reidel, Dordrecht, Holland, pp. 179–191.
[18] McCormick, R. A., S.J.: 1974, 'Proxy Consent in the Experimental Situation', *Perspectives in Biology and Medicine* **18,** 2–20.
[19] McCormick, R. A.: 1975, 'Transplantation of Organs: A Comment on Paul Ramsey', *Theological Studies* **36**, 503–509.
[20] McCormick, R. A., S.J.: 1975, 'Fetal Research, Morality, and Public Policy', *Hastings Center Report* **5**, 26–31.
[21] McCormick, R. A., S.J.: 1976, 'Experimentation in Children: Sharing in Sociality', *Hastings Center Report* **6**, 41–46.
[22] McCormick, R. A., S.J.: 1976, 'Experimental Subjects: Who Should They Be'? *Journal of the American Medical Association* **235**, 2197.
[23] McCormick, R. A., S.J.: 1981, *How Brave a New World? Dilemmas in Bioethics*, Doubleday, Garden City, New York.
[24] Outka, G.: 1972, *Agape: An Ethical Analysis*, Yale University Press, New Haven, Connecticut.
[25] Outka, G.: 1974, 'Social Justice and Equal Access to Health Care', *Journal of Religious Ethics* **2**, 11–32.
[26] Outka, G.: 1976, 'Letter to Editor', *Perspectives in Biology and Medicine*, Spring, 449–452.
[27] Platt, A.: 1974, 'The Triumph of Benevolence: The Origins of the Juvenile Justice System in the United States', in R. Quinney (ed.), *Criminal Justice in America*, Little, Brown, Boston, pp. 356–388.
[28] Ramsey, P.: 1962, *Nine Modern Moralists*, Prentice-Hall, Inc., Englewood Cliffs, New Jersey.
[29] Ramsey, P.: 1968, *The Just War*, Charles Scribner's Sons, New York.
[30] Ramsey, P.: 1970, *The Patient as Person*, Yale University Press, New Haven, Connecticut.
[31] Ramsey, P.: 1973, 'Medical Progress and Canons of Loyalty to Experimental Subjects', Proceedings of Conference on Biological Revolution/Theological Impact, Institute for Theological Encounter with Science and Technology, St. Louis, Missouri.
[32] Ramsey, P.: 1976, 'The Enforcement of Morals: Nontherapeutic Research on Children', *Hastings Center Report* **6**, 21–39.

[33] Ramsey, P.: 1976, 'Some Rejoinders', *Journal of Religious Ethics* **4**, 185–237.
[34] Ramsey, P.: 1977, 'Children as Research Subjects: A Reply', *Hastings Center Report* **7**, 40–42.
[35] Rothmann, D. J.: 1978, 'The State as Parent: Social Policy in the Progressive Era', in Willard Gaylin *et al., Doing Good: The Limits of Benevolence*, Pantheon Books, New York, pp. 67–96.
[36] Schlossman, S. L.: 1977, *Love and the American Delinquent: The Theory and Practice of 'Progressive' Juvenile Justice, 1825–1920*, University of Chicago Press, Chicago.
[37] Shelp, E. (ed.), 1981, *Justice and Health Care*, Vol. 8 of *Philosophy and Medicine*, D. Reidel, Dordrecht, Holland.
[38] Stell, L. K.: 1978, 'Rawls on the Moral Importance of Natural Inequalities', *The Personalist* **59**, 206–215.
[39] Visscher, M.: 1975, *Ethical Constraints and Imperatives in Medical Research*, Charles C. Thomas, Springfield, Illinois.
[40] Walters, L.: 1977, 'Some Ethical Issues in Research Involving Human Subjects', *Perspectives in Biology and Medicine,* Winter, 193–211.
[41] Weale, A.: 1978, *Equality and Social Policy,* Routledge and Kegan Paul, London.

RONALD M. GREEN

# CONTEMPORARY JEWISH BOETHICS: A CRITICAL ASSESSMENT

In his novel *The Agunah*, The Yiddish author Chaim Grade portrays the course of a bitter controversy in Vilna, Lithuania, during the late 1920s and early 1930s between an impoverished younger Rabbi and the city's authoritative Rabbinic Council. The immediate cause of the controversy is a young widow whose husband has disappeared in combat (along with his entire regiment) during the opening days of the First World War. Because no one had actually seen the husband fall in battle or found his body, the woman remained an *agunah*, abandoned wife, who, according to standard Jewish legal thinking, was forbidden to remarry. Consequently, when the woman sought permission for remarriage after fifteen years of waiting, she was coldly rebuffed by the Council's expert on such matters.

As a last resort, the woman turned to the younger Rabbi. Moved by compassion, he searched the legal precedents, and basing his view upon that of a lone dissenting authority in the earlier tradition, the Rabbi gave the woman permission. Unfortunately, this proves to have disastrous results for everyone. Hounded into still deeper poverty by the Rabbinic Council, the young Rabbi sees his infant child die. The *agunah* eventually takes her own life. And the Rabbinic Council becomes an object of scorn, with the expert who had denied permission driven, on the edge of madness, into premature retirement [7].

Grade's novel is not a simple morality tale. There are no 'good' guys or 'bad' guys. Although the younger Rabbi has many of the marks of sainthood, his motives are not altogether clear. The older Rabbis may be corrupted by power and prone to arrogance, but they nevertheless have a genuine zeal for Torah. What does emerge from this novel is a complex picture of a religious community in crisis. Currents of free thought, of secularism and of Bolshevism are sweeping through Vilna. New voices are being heard. Faced with these challenges, the established religious authorities can find no more creative response than to reassert some of the tradition's most problematical and seemingly inhumane teachings. For them, World War I is not a novel historical event that must first be assimilated and only then compared with previous events upon which the tradition's normative rulings are based. And the *agunah* is not a person



*E. E. Shelp (ed.), Theology and Bioethics*, 245–266.

so much as a principle, one that must be reaffirmed precisely because it is under attack. If the younger Rabbi emerges as the people's – and perhaps the novel's – hero, it is because he is able to perceive that an ancient legal tradition can continue to command respect only so long as it remains attractive to the changing circumstances of human need.

Why do I begin an article on Judaism and bioethics by recounting the plot of this novel? In part because I believe that an important body of thinking and writing in Jewish bioethics today is being influenced by some of the same forces depicted by Grade and because I believe that a number of writers and authorities in this area have taken a course similar to that of Grade's fictional Rabbinic Council. Threatened from without and within by secular attitudes, perceiving themselves as the sole custodians and defenders of an embattled tradition, and confronted with difficult new problems to which the older precedents do not always clearly apply, a number of Jewish 'bioethicists' and Rabbinic decisors have often reacted by reasserting some of the most rigorous and anti-modern elements of the Jewish tradition. The result is a body of writing on biomedical topics – much of it being aggressively published in English – which represents only one side, one voice, in a complex tradition but which often presents itself as the normative Jewish view.

Foremost among the writings I have in mind in this respect are the works of Rabbi J. David Bleich and the physician-*halakhic* scholar Dr. Fred Rosner. Their several volumes on biomedical topics, because of their concision and coverage of issues, have become for many persons initial references for Jewish thinking on problems of medical ethics [1, 2, 3, 4, 21, 22]. With important qualifications, the work of Rabbi Immanuel Jakobovits, dean of contemporary Jewish bioethicists, could also be placed on this list [10, 12]. On select issues – notably that of abortion – the writings of several other leading Jewish scholars of medical ethics might be characterized in similar terms [14, 18, 23, 24, 26].

In defense of these writers, it might be said that, on many of the issues they discuss, their role is that of scholarly reporters, not primary decisors. Jewish ethical thinking is profoundly rooted in *halakhah*, the normative body of Jewish law. This is founded on the Pentateuch, on the oral law believed given to Moses at Sinai and on the associated commentary tradition – all of which form the body of authoritative Talmudic teaching. It further comprises the major codes of Jewish law based upon the Talmud and subsequent Rabbinic *responsa* to specific problems and questions. Since the work of the writers I have mentioned largely draws upon this vast corpus of legal teaching, it could be said that they have limited room for independent interpretation. Whatever character their work has, therefore, whether it is conservative or antimodern, lies within the Talmudic tradition itself.

In response to these claims, I might make two observations. First, the tradition itself is a complex one. Opinions are divided on many issues and, in my view, there is a major division between classical and more contemporary approaches. The early tradition seems particularly marked by a flexibility and openness of spirit that seems to have waned in more recent writings. Indeed, much of the narrowness and defensiveness to which I allude can in fact be located within more contemporary Rabbinic treatments of biomedical or sexual issues. In this respect it is correct to say that Jewish writers on bioethics are merely reporters and are reflecting a conservative turn already present within the tradition itself. This strikes me as particularly true of the work of Jakobovits, who often feels compelled to report rulings or a consensus on an issue with which he does not entirely agree.

But it is also true – and this is my second observation – that all reportage is interpretive. Even to identify precedents and rulings that have a bearing on novel issues is a matter of decision. This largely secondary literature on Jewish bioethics, like any moral literature, necessarily represents a process of moral judgment and expresses to some extent the preferences of its authors. The conservative caste of much current writing on Jewish bioethics, therefore, results both from the contemporary *halakhic* process itself and from the views of writers who have moved to the fore as mediators and interpreters of the Jewish tradition in this area. At both levels, that of Rabbinic decision and of compilation and publication of these decisions, a deeply conservative and defensive tendency is at work which I believe threatens to obscure the more progressive traditions of classical Jewish law. For this reason, what follows is something of a warning. My aim is to prompt the reader untutored in *halakhah* to treat contemporary Jewish biomedical writing with caution. I also hope to suggest why this conservative reaction has moved to the fore and what it means for our broader understanding of the relationship between religion, theology and bioethics.

I have made some strong claims about the nature of contemporary writing in Jewish bioethics. The best way to substantiate these claims, I think, is to look at the ways in which some of the broadest principles of Jewish medical-ethical-legal thinking have been handled by contemporary Jewish 'bioethicists'. These principles are established in the very earliest teachings of Jewish law on medical and sexual matters. They constitute some of the distinctiveness of Jewish teaching in this area and are what makes Jewish law a resource for those interested in understanding how a religious tradition might approach questions posed by medicine and health care. In my own view, each of these principles merits our consideration, and each has applicability to some of the new questions we confront. My cavil with the writers I have mentioned is not that they are unaware of these principles or that they fail to present them. On the contrary, these principles often form the organizing basis for these writers' discussions. Rather, it is that in interpreting and applying these principles to novel biomedical problems, these writers strike me as losing sight of the humanistic intentionality of many of these fundamental teachings.

## THE LEGITIMACY OF HEALTH AND THE DUTY TO PRESERVE LIFE AND HEALTH

This is one of the most firmly established principles of Jewish law: human beings have the right, and indeed the obligation, to preserve health, to combat disease and to employ medical means to do so. At first sight, this principle may seem unexceptional, but we must recall that Judaism, like many of the theistic traditions to which it is related, emphasizes the divine providence and insists on faith and trust in God. This means that there is a strong impulse in Judaism – as there has been in Christianity and Islam – to regard disease or physical suffering as divinely inflicted, whether as punishment or as a form of edification, and as not subject to human avoidance or control. We know that both Christianity and Islam sometimes succumbed to this impulse (witness the debates in the last century over the use of anaesthetics in childbirth). This impulse also found expression in Judaism among the Karaites, early sectarians who based their rejection of human medical intervention on a literal reading of Exodus 15:26: "I will put none of the diseases upon you which I put upon the Egyptians, for I am the Lord your physician" ([10], pp. 2f.).

This text in Exodus provides important support for an anti-medical position. In view of the seriousness with which scholars studied every word of Torah, therefore, it is remarkable that the Karaite position was never adopted by the main tradition of Jewish legal thinking. Instead, from the earliest date, authority for healing was found in other texts, even when these were admittedly of limited or ambiguous significance. For example, Exodus 21:19–20, where a wrongdoer is mandated to have his victim healed, became a major text supporting medical intervention

([5], *Baba Kamma* 85a), even though, as some scholars were forced to note, it applies at best only to humanly inflicted wounds ([10], pp. 4f.). In the interests of healing and preserving human life, therefore, the early decisors were fully willing to bend scripture to their needs or – more fairly – to their deeper understanding of the divine will. In any event, authority for healing and health care in general was found and was reinforced. Just as man is obligated to tend the earth, some scholars argued, so is he required to tend to his body and to ward off afflictions whatever their source ([10], pp. 303f., n.7). God may be the ultimate cause of healing or recovery; but the physician can be his proximate agent ([5], *Berakhoth* 60a).

None of the bioethical writers we are considering fails to insist upon this principle. Nor do these writers, in the variety of their discussions of biomedical problems, refuse when necessary to carry it into new areas of biomedical advance. For example, despite their reservations on other grounds about aspects of genetic medicine (especially, as we shall see, where selective abortion is involved), none of these writers opposes the concept of genetic therapy involving basic manipulation or alternation of human DNA. None voices objections, sometimes heard in other traditions, that such manipulation represents an intrusion on divine prerogative, an illicit tampering with human life or a violation of some divinely constituted natural order. Instead, therapeutic genetic engineering, whether at the cellular or gametic level, seems to be regarded as on a par with any other chemical or surgical intervention permitted by Jewish law ([20], [22], pp. 409–420).

To a substantial degree, therefore, the work of these writers evidences the same commitment to healing and to health manifest in the earliest discussions in Jewish law. Nevertheless, it can be asked whether the same degree of boldness and initiative shown by the early decisors characterizes contemporary discussion. The problem I have in mind here will become clearer when we turn to technologies or procedures that may run counter to stern *halakhic* prohibitions, as may some of the sexual therapies now available. In such cases, it seems to me, these writers appear more committed than were their earliest predecessors to respecting the letter of specific commandments than to promoting the broader (but also *halakhicly* mandated) goal of health. But the very same tendency toward unimaginative conservativism of reasoning can also be seen with respect to the relatively minor matter of cosmetic plastic surgery.

Discussing this matter in his book *Judaism and Healing* [4], J. David Bleich notes that *halakhah* ordinarily prohibits the mutilation or wounding of one's body ([5], *Baba Kamma* 91b), but he also points out that the Biblical injunction to healing (Exodus 21:19) has been interpreted so as to exclude the surgical correction of deformed or malfunctioning organs from the sphere of this prohibition. Such procedures were regarded as of a different character from those prohibited by the ban on wounding ([4], pp. 126–128). What, then, is to be said of surgical procedures undertaken for purely cosmetic purposes – rhinoplasty or face-lifts, for example? Along with many other contemporary Rabbinic decisors, Bleich classes such procedures under the heading of 'wounding' rather than healing. This means that they are ordinarily prohibited unless needed to relieve serious pain and suffering. For Bleich, in particular, such procedures become legitimate if, in their absence, an individual would shun ordinary social intercourse, would be unable (or unwilling) to contract marriage or to secure employment ([4], pp. 127f.).

I do not want to put myself here in the position of morally defending procedures of this sort. It may be that on psychological or economic grounds we do not wish to see costly medical resources employed in this way, and it may even be that Bleich's specific strictures are ones to which reasonable persons might agree. My discomfort with his discussion has to do more with the way in which he reaches his conclusions than with the conclusions themselves. There is no doubt that many individuals who resort to cosmetic plastic surgery do so because their appearance causes them distress, even though their psychic suffering might not interfere with their normal functioning. The question then becomes why, even in a *halakhic* context, such procedures should be prohibited. The reply that they constitute a form of 'wounding' begs the question of whether modern plastic surgery really belongs in this category at all. Surely procedures in this area are not on a par with forms of cosmetic disfigurement (e.g., bodily incision or tattooing) practiced by some cultures and long prohibited by Judaism. If anything, they aim not at mutilation by at the restoration of a physical 'normalcy' of which the patient feels deprived.

Nor do some of the simpler procedures undertaken today represent any significant physical risk, as almost any form of surgical intervention did in the past. The point here is that cosmetic plastic surgery is something almost entirely novel – and it perhaps merits a novel approach. It might, for example, be regarded not as wounding but as a purely cosmetic procedure. Or it might be regarded as a form of healing – on a par with valid medical efforts to relieve a digestive disorder that troubles the patient without threatening his life. But Bleich himself does not venture out onto this uncertain ground. Neglecting the novelty of the issue before him, he applies reasoning drawn from another era. In doing so, moreover, he seems less solicitous of the human suffering the issue involves than were his predecessors in approaching what for them were equally novel matters.

On a far more serious plane, the problem could also be illustrated by considering the restrictions some of these writers would place on modern autopsy procedures. The bases of these restrictions are various Talmudic prohibitions against the desecration of the dead or utilization of the body for purposes of gain or personal advantage ([5], *Baba Bathra* 15b; *Hullin* 11b; *Avodah Zarah* 29b). Two centuries ago, an important *responsum* by Rabbi Ezekiel Landau applied these teachings to the matter of medical autopsy and concluded that such desecration would only be permitted to save an identifiable life at risk (for example, the life of a child threatened by a genetic disease similar to that which killed the parent or the life of a fellow participant in an experimental therapy) ([10], pp. 144f.). Following on this teaching, many authorities have prohibited autopsies – even those authorized by the decedent before his death – where only general scientific or medical benefit might accrue. Bleich, once again, appears to agree with this position and to present it as the normative Jewish view. Only the saving of the life of a patient at hand (*Holeh lefaneinu*) justifies autopsy on his interpretation of Jewish law.

But are the classical prohibitions really applicable to modern autopsy procedures? To his credit, this is the question asked by Jakobovits, whose treatment of autopsy ([10], pp. 282–283) represents an exception to the generally uncreative approach taken by these writers (including Jakobovits himself on other issues). He observes, for example, that under modern conditions of communication a patient 'at hand' may not necessarily be a known individual in the same community. And he suggests that information obtained today from even routine autopsy procedures may have lifesaving value for some patient at some time. On this basis, he elaborates a far more lenient view on this entire matter than Bleich or the conservative decisors on whom Bleich depends. If Jakobovits' reasoning here stands out from the norm it is not because his conclusions strike me as reasonable. Rather, it is because he has taken the trouble, within a rigorously *halakhic* framework, to examine the issues in terms of their novelty and uniqueness. Just this approach is lacking, however, in many discussions by these writers. In the end, one is prompted to ask not only whether they retain the creativity of their predecessors, but more importantly, whether they carry on the classical concern with saving human life and promoting health.

## THE SANCTITY OF HUMAN LIFE

The principle of the sanctity of human life contributes to Judaism's basic commitment to medical care, just as it underlies the rule of *pikku'ah nefesh* which mandates violation of the commandments when such violation might preserve a human life.[1] But in addition to this, the sanctity of life has come to have a very specific connotation in Jewish biomedical ethics today. It is taken to mean that every life and every moment of life, regardless of its condition or quality, has absolute and infinite value. This, in turn, underlies a series of norms prohibiting qualitative evaluation or selection among human lives because of the individual's physical or mental condition or likelihood of surviving.

The bases for this teaching are found in a number of Biblical and Talmudic texts. The unusual plural in the Biblical phrase "the bloods of thy brother cry out" (Genesis 4:10), for example, was interpreted in the Talmud as an admonition to witnesses in cases involving the prospect of capital punishment. Such witnesses were cautioned that upon their decision rested not only the life of the accused but of the whole line of persons who might descend from him ([5], *Mishnah Sanhedrin* 4:5). A related text holds that the human race was created from the single individual Adam so that we might learn that "[i]f any person has caused a single soul... to perish, Scripture imputes it to him as if he caused the whole world to perish" ([5], *Sanhedrin* 37a). Nor was the duration or condition of a life regarded by the classical writers as grounds for qualifying its sanctity. For example, it was ruled that efforts on a sabbath to free a person buried under a collapsed building must be continued even if the victim was so injured that he could not live more than a short time ([13], *Orah Hayyim* 329:4). It was also held that anyone who killed a child while it was falling from a high roof would, in principle, be regarded as a murderer, even if the child would otherwise have died immediately ([10], p. 794 and interpretations of [5], *Baba Kamma* 26b). The very same teaching is expressed in relation to medical care by Maimonides: "He who kills, whether [the victim be] a healthy person or a sick person approaching death, or even a patient in his death throes, is treated as a capital criminal" ([12], p. 794).

This broad principle of the sanctity of life whatever its quality or likely duration was also applied by classical writers to the difficult matter of the treatment of terminally ill or dying patients. In general, teaching in this area absolutely ruled out any form of active killing in such circumstances, however compassionate its motives ([10], pp. 123f.). It was also held that normal forms of care and medical efforts must be continued for a patient, no matter how ill he might be. A very limited exception to this rule in the classical sources concerns the case of the *goses*, one already in the dying process. Such a patient is on the very threshold of death: his breathing is labored, his chest "narrows" and he brings up "a secretion" in his throat ([10], p. 121). While such patients must be treated with extreme solicitousness – they must not be jostled, placed on the ground or otherwise disturbed ([16], Volume 14, p. 174) – and while no active efforts may be made to hasten their death, it is permissible to remove certain impediments to their dying. One source includes among these 'impediments' any trace of salt on the tongue which rivets the soul's attention or the noise of a woodchopper in the vicinity that prevents release ([10], p. 123).

With this as their classical background, the Jewish biomedical writers I have been discussing draw a number of conclusions regarding the difficult contemporary issue of what care should be given dying patients whose suffering might be accentuated or prolonged through the aggressive use of therapies or of artificial means of life support. Generally, they are of the opinion that Judaism regards the maintenance of life as a higher obligation than the prevention of suffering ([4], pp. 22, 134–156; [22], p. 18). Non-treatment or withdrawal of treatment to allow the patient to be freed by death from pain is regarded (for all but the *goses* or imminently dying patient) as impermissible, even if the patient urgently requests that efforts cease. Rejecting the Roman Catholic distinction between ordinary and extraordinary means, for example, Bleich contends that any medication or procedure needed to sustain life must be employed, and he categorically rejects as permissible for the devout Jew any sort of 'living will' in which the patient stipulates a limit to medical care *in extremis* ([4], p. 139). On the other hand, since relief of suffering is a valid objective according to *halakhah*, Bleich and Jakobovits would both permit the extensive use of analgesics like opium or heroin to relieve pain for a dying patient, even if these medications might have the side-effect of hastening death ([4], p. 138; [10], p. 276).

It is difficult to fault these judgments. The matter of weighing continued life against a deteriorating quality of life is one of the most uncertain matters of rational and moral decision. It is the kind of issue on which reasonable and well-intentioned persons can disagree. In my own view, religious-ethical positions often have their greatest value in matters like this: by applying a heritage of experience and reflection, they provide a course through such difficult and tormenting problems. Certainly Judaism's teaching about the value of life constitutes one of its most distinctive contributions to bioethics. Furthermore, as we have just seen, contemporary Jewish thinkers are not inattentive to the problem of suffering for the terminal patient and would permit efforts to relieve pain even if these shortened the patient's life.

Why, then, am I uneasy with contemporary Jewish bioethics at this point? In part because I believe that once again these writers tend to be insufficiently sensitive to the radically new situation medical technology has created for dying patients. By means of repeated surgical interventions, the use of antibiotics, intravenous feeding, or ventilators, patients today can be kept alive far longer than was ever conceivable in the past, but in some cases at a terrible price in suffering for the patient and for those around him. I question how well contemporary scholars or the Rabbinic decisors on whom they draw have understood this fact. Is it really correct, for example, to liken a terminal cancer patient who has repeatedly undergone chemotherapy and surgery to an individual buried under the rubble of a collapsed building? We can understand even desperate efforts to save the latter individual, no matter how faint the prospects of his survival. Such efforts fully express what it means to value and care for a person. But is the same thing true when we fight the infection that would otherwise grant the cancer patient release?

My difficulty here can be expressed in another way. We have noted that classical Jewish teaching is more lenient with respect to the duty to maintain life when the patient is a *goses*, one already in the dying process. In an effort to lend more precision to this concept, Bleich draws on the classical sources and concludes that such a patient is one who cannot survive for more than 72 hours ([4], p. 141). Any patient, no matter what his condition, who can survive longer than this is not a *goses* on this view and hence is subject to unceasing medical efforts. Can it not be said, however, in view of the capabilities of modern medicine, that the category of the *goses*, as it was formerly understood, no longer exists? When the patient today begins choking on secretions, a tracheotomy is performed or the breathing passage is cleared out. When the 'last breath' is drawn, the ventilator is turned on. One conclusion, of course, is that very few patients any longer are *goses* so that whatever permission existed in such cases for the cessation of efforts no longer applies: everything must be done to extend life. This seems to be Bleich's conclusion. But it is equally open to the *halakhic* scholar to conclude that medical advance forces a radical reconsideration of the classical sources in order to discern the intent of rulings that mandated life-saving efforts or that created the special category of the *goses*. It may be that temporal limits no longer suffice to identify the imminently dying patient, for example, and that some consideration of the hopelessness of the patient's condition or continued quality of life are more relevant to the determination of this status. Once again, my discomfort with some of these contemporary writers has less to do with their conclusions than with their unresponsiveness to the changed conditions confronting the tradition they have inherited.

## THE OBLIGATION TO PROCREATE AND THE VALUE OF SEXUALITY

One of Judaism's relatively distinguishing features is its strong insistence on the positive religious value of sexuality and family life. Rabbinic Judaism has always rejected celibacy as a way of life, has stressed the essential goodness of sexual expression within marriage, and has insisted on the obligation to found a family [6].

Within the *halakhic* tradition, these basic values and principles have been given expression in a series of positive and negative commandments. Sexual expression, for example, was confined to the setting of monogamous marriage, where, for reasons of procreation and for the sexual satisfaction of the wife, it was held to be a duty incumbent upon the male. Masturbation, fornication, extramarital or homosexual relations were all condemned. Within marriage, the commandment to procreate (*p' ru ur' vu*) was interpreted to require having at least one son and one daughter (provided each was able to found a family), but a tradition of Rabbinic commentary also insisted upon a life-long obligation incumbent upon the male to try to have as many children as possible ([6], pp. 48f.). Unless needed to save a life, surgical sterilization and contraception were both prohibited, although an important tradition exists which allows the woman to use contraceptive measures or a sterilizing potion when pregnancy would endanger her health or cause grave discomfort (such as great pain in childbirth) ([5], *Yebamoth* 65b). Male efforts at contraception (for example, coitus interruptus) were explicitly forbidden as a form of wastage of seed (*hash-h.atat zera*).

While there is a great deal in classical Jewish teaching about sexuality that merits consideration and respect, the area of family life and sexuality is also one that has undergone enormous change during the modern period. Changing mores, social realities, and medical technologies have all combined to force reexamination of many religious teachings in this area and have also yielded concrete change in people's conduct. The widespread use of contraception and the heated debates about its permissibility in many religious communities are evidence of these profound transformations.

The need for a reconsideration of Jewish law in these circumstances is clear. A pioneering study of birth control and abortion in Jewish law published over a decade ago by Rabbi David Feldman [6] sought to address this need. Drawing on neglected or overlooked elements in the tradition, especially the important classical stress on maternal health and well-being and the woman's right to sexual satisfaction, Feldman's study pointed the way toward a full legitimation of contraception by the woman in situations of serious need. Going beyond even the traditional emphasis on maternal health, Feldman suggested at the very close of his study that the spacing of children and limitation of family size might also find support in Jewish law. For example, those teachings which permit the postponement of marriage, Feldman suggested, might have applicability as well within marriage when a limitation on family size or the frequency of births is desired ([6], p. 304). Unfortunately, even this very cautious adaption of Jewish teaching to modern circumstances is largely lacking in the work of contemporary Rabbinic decisors or the bioethicists who mediate it. Throughout, the normative ideal remains the large family and a lifelong commitment to procreation ([4], p. 61–64; [10], pp. 156–169; [22], Chapters 3, 5, 6). In this respect, *halakhicly*-oriented Jewish discussions in this area resemble those found in some of the sterner expressions of Roman Catholic teaching on birth control. True, there are some differences. The Jewish perspective always permits contraception where the mother's life is in peril, and there are some very new motives behind Jewish teaching, among them concern with the perilous demographic situation of modern Jewry. But in both cases, the tendency is to carry an ideal elaborated in quite different circumstances into the conditions of modern life, and to respond to the hardships this might cause by enjoining reliance on God's providential care. What very few Jewish thinkers in this area have tried to do is follow Feldman's lead in going to the core of Jewish teaching about marriage and family in order to determine how traditional values and norms might best be served today.

Before leaving the matter of sexuality and procreation, a brief word is in order about the issue of homosexuality. In fact, this is a very complex problem whose discussion by Jewish thinkers merits independent treatment, but the difficulties involved here can be briefly suggested. On the one side, we have a body of classical Biblical and Rabbinic teaching which regards homosexual conduct as an 'abomination' (*to 'evah*) and which staunchly prohibits any behavior which blurs the distinction between the sexes. On the other side, we have a confused and constantly changing series of scientific perspectives on homoerotic preference or homosexual conduct, ranging from views which regard this orientation as pathological to views which see it as part of a normal repertoire of human sexual possibility.

As we might suppose, the thinkers I have mentioned generally feel compelled to reassert the traditional prohibitions in this area ([4], Chapters 12, 13; [22], Chapter 11; [26]). Some *halakhicly*-oriented writers, it is true, imaginatively try to bring elements of Jewish law into congruity with modern perspectives in order to alleviate the harsher elements of the classical prohibitions. Norman Lamm, for example, has drawn on scientific discussions of the compulsory nature of much homosexual conduct and has sought to relate this to the facet of Jewish law which mitigates punishment and condemnation where involuntary behavior is involved [15]. But there is little evidence in this writing of a willingness to reconsider in the light of some modern views what the Biblical prohibitions really involve or whether homoerotic preference was even an object of these prohibitions. On the contrary, Jewish writers in this area tend to see in the Biblical and Talmudic prohibitions a subtle confirmation of the more pejorative views of homosexuality found in current psychological literature ([26], p. 55).

I mention this here for two reasons. First, because it provides further evidence for the generally conservative caste of contemporary Jewish ethical literature. As a rule, it is on sexually related matters that contemporary decisors and writers are least innovative. Nowhere is this truer than on the matter of homosexuality, perhaps because it represents an orientation so alien to most persons.

Second, it seems to me that on this issue, the unimaginativeness of some *halakhicly*-oriented biomedical writing reaches astonishing proportions. It is fully understandable that writers and thinkers oriented to the tradition would be reluctant to override or bend traditional teaching in this area. But at some moments, this reluctance leads to an abandonment of even traditional norms encouraging a concern for human health. One illustration will suffice. In an article considering homosexuality in the light of *halakhah,* one writers appears to find the view that homoerotic preference is a form of illness to be persuasive ([26], p. 55). He then proceeds to the question of whether it is *halakhicly* permitted to try to remedy this condition by means of behavioristic therapies, which use homosexual and heterosexual pornography in conjunction with a system of rewards and punishments to alter a patient's sexual preferences. The writer's conclusion is that such therapy, because of its incitement to illicit sexual fantasy, would probably not be permissible ([26], p. 65). Although it is ironic that this conclusion would please some homosexuals who regard such therapy as unwarranted, it is an extraordinarily hide-bound conclusion in any case. Reading a discussion of this sort, one is led to wonder whether one is really in the same tradition that centuries ago permitted desecration of the sabbath or eating on the holiest fast day when human life is in peril. Even if homosexual oriention, in view of the current epidemic of AIDS, is not as life-threatening as some conditions (and this is by no means clear), surely the 'medical' use of erotic materials falls into a different category than other forms of sexual stimulation. That this matter receives only passing treatment by this writer supports my claim about the uncreativeness and insensitivity of much of this literature.

## THE PRIORITY OF MATERNAL OVER FETAL LIFE

This constitutes the final major principle of classical Jewish bioethics whose handling by contemporary decisors and Jewish biomedical writers I want to examine. Merely to state and explain the principle, however, poses a problem, since its meaning is not really clear even in the earliest Rabbinic sources. We will see that my major objection to contemporary discussions of abortion is not that these discussions usually adopt only the most conservative of the possible interpretations of this principle, but that they do so without adequately suggesting the diversity of interpretations that have reigned in this area.

We can see the problem if we regard one of the classical loci for Jewish teaching about feticide. This is a Talmudic passage in the tractate *Oholoth* (7:6):

> If a woman is in hard travail, we cut up the child in her womb and bring it forth member by member, because her life comes before that of the child. But if the greater part had proceeded forth, one may not touch it, for one may not set aside one person's life for that of another.

Now while this passage clearly indicates Judaism's priority on maternal as against fetal life (in contrast, for example, to the Roman Catholic tradition), what is not indicated by this text is the status of the fetus in lesser instances of conflict. Is abortion permitted when the mother's health, but not her life, is in jeopardy? Is it permitted when a pregnancy threatens a family's well-being? Or is it allowed simply when the mother finds the pregnancy embarrassing or inconvenient? *Oholoth* permits – indeed, mandates – abortion when the mother's life is at risk. It is silent on other indications for feticide.

Proceeding from this silence, a major tradition of *halakhic* scholarship finds the text's specification of abortion in only this case as a sign that abortion is not permitted for lesser reasons than risk to the mother's life. This reading of *Oholoth* is reinforced by Maimonides' commentary on the passage. Discussing this ruling, he explains that in cases of this sort, the fetus may be likened to a pursuer (*rodef*) who places another's life in jeopardy and who, according to Jewish law, might be killed without the defender's incurring capital punishment. Once the child leaves the womb and emerges at birth, however, it loses the status of pursuer according to Maimonides and its life is on an equal footing of sanctity ([16], Volume 11, p. 196).

Maimonides' discussion is the point of departure for a restrictive tradition of Jewish teaching on abortion. Since the fetus is like an adult pursuer, it follows that its life may be taken only when another's life is in peril, but not for lesser reasons. This restrictive tradition can be traced in Rabbinic discussions up to the modern period. Indeed, it appears to have moved to the fore in contemporary Jewish writing and it figures substantially in the views of all of the writers on Jewish bioethics I have mentioned. The untutored reader who surveyed the discussions of these writers on abortion and abortion-related topics (such as *in vitro* fertilization or research involving the human fetus) might easily conclude that in this area *halakhic* Judaism broadly parallels Roman Catholic thinking. Apart from Judaism's non-agreement with the Catholic prohibition of any deliberate killing of the fetus even to save the life of the mother, the two traditions seem to agree on the inadmissibility of abortion in all other cases ([4], Chapter 17; [10], Chapter 4; [22], Chapters 8, 9).

The fact that contemporary discussions in Jewish bioethics are profoundly shaped by this restrictive tradition on abortion can be illustrated with reference to the matter of prenatal diagnosis and abortion for genetic reasons. Clearly some of the most difficult decisions in medical ethics lie in this area. Is a life ever worth not living? May we ever – as in the case of some sex-linked genetic diseases – terminate one potential life merely on the *probability* that it may fall victim to a genetic disease? What is the extent of our responsibility to future persons who may inherit the genetic diseases we propagate today?

To these and other questions, the writings of Bleich, Rosner and (with some qualifications) Jakobovits offer fairly straightforward answers. In and of themselves, genetic considerations do not furnish a reason for deferring procreation or for permitting abortion. Since Jewish law teaches that life, whatever its quality, is sacred and that a pregnancy may be terminated only to save the mother's life, abortion in these cases may be allowed only if the birth of a genetically deformed child would likely drive the mother to suicide. But where her mental instability is not at issue, the parents must welcome the child into their midst, however serious its disease condition ([4], Chapter 18: [22], Chapter 8, pp. 123f., Chapter 9, pp. 160f., Chapter 10).[2]

In general, these writers are also not well-disposed to programs of screening for genetic disease. The relatively high incidence of Tay-Sachs disease among Jews of Ashkenazic background has made this an important issue in the Jewish community, but these writers tend to lay down fairly restrictive norms for the conduct of screening programs. For example, they would discourage the screening of married couples, since know ledge of carrier status in this case would lead either to a decision not to have more children or to the use of selective abortion – neither option being viewed by these writers as *halakhicly* viable ([4],Chapter 18; [22], Chapter 10; [10], p. 263).

These views with respect to genetic medicine illustrate the powerful impact on these writers' thinking of their understanding of the permissibility of abortion. If we consider the enormous tragedy associated with a genetic condition like Tay-Sachs disease, it would be reasonable for the reader unfamiliar with Jewish law in this area to conclude that the classical norms regarding abortion are unambigious and immutable. But this is not at all the case. In fact, if anything, the bulk of the early tradition supports a view contrary to that of contemporary writers, a view which accords only a very minimal moral status to prenatal life.

The foundation of *halakhic* reasoning in this area is Exodus 21:22, where killing of the fetus is regarded as a tort (against the father primarily) and not an act of homicide. This text is explained by later commentators in terms of the fact that the fetus is not a *nefesh*, a living person in the juridical or moral sense ([19], *Sanhedrin* 72b). This *halakhic* evaluation of fetal life, incidentally, has nothing to do with Rabbinic ideas of ensoulment, which comprise a spectrum of quite different views ([6], pp. 271–275). In this case, it is not speculative material on the origin of personhood nor *aggadic*, imaginative narrative discussions of uterine life that are normative, but the clear text of the Pentateuch.

This classical Talmudic attitude toward prenatal life is also dramatically evidenced by a text which is either neglected or touched on only in passing by the contemporary bioethicists I have been discussing. This text in the tractate *'Arakin* (7a) establishes norms for the execution of a woman who has been convicted of a capital crime but who is found to be pregnant after her trial has concluded. The question before the sages here was whether the execution might be delayed – even a matter of hours – to allow the child to be born. Their answer is an unequivocal 'no', and the reason for the answer is significant: since anyone about to be executed is in a state of great anxiety, it is not permissible to prolong the anxiety or suffering. As the *gemara* or commentary on this teaching proceeds, the question is asked why this ruling is even made, since the conclusion is an obvious one, given the teaching that the fetus is not a *nefesh*. The answer is that otherwise some might have argued that the father's property rights in the child take precedence over the mother's suffering. In other words, the moral and the material value of the fetus are here both subordinated to the suffering of a person, a *nefesh*. The text concludes with the ruling that, in such cases, it is appropriate to strike the condemned woman on the abdomen shortly before the execution in order to kill the child. This is done to prevent her disgrace (*nivvul*) should the still living child emerge from her body after the execution.

This is a ghastly discussion. Small wonder that it is neglected even by liberal Jewish proponents of the right of abortion. But we must keep in mind the real intent of the teaching here: in a context where capital punishment is presumed, the sages are clearly demonstrating their concern for the mother rather than for the child. Saving the child's life does not merit inflicting even a few hours more suffering on her. Likewise, the prospect of her disgrace also bulks larger in the sages' view than any claims of prenatal life.

This text was not simply abandoned by the *halakhic* tradition. Among other things, it plays a part in a ruling by the 18th century Rabbi Jacob Emden which permitted an adulteress to conceal her crime by having an abortion. Emden reasoned that if the woman's crime became known, she would be convicted of the capital offense and the child would be dead in any case. Hence the woman might seek to save her life by having an abortion ([6], pp. 288f.; [25], pp. 124 f.). Beyond this, Emden drew upon the concepts underlying the classical ruling to conclude that a woman might have an abortion in any case of 'great need'. Recently, Emden's ruling in this case was appealed to by Rabbi Eliezer Waldenberg in a decision which permitted selective abortion where the fetus is likely to be genetically diseased ([25], pp. 126f.). Following Emden, a tradition of commentators has also legitimated abortion in instances of even 'slight need' on the mother's part. By and large, however, this minority tradition of recent scholarship (rooted in important Talmudic judgments) is largely neglected by the contemporary Jewish bioethicists we are discussing. Bleich and Rosner assert that the normative tradition does not sanction abortion for genetic reasons. Although Jakobovits' more recent views on this issue have shifted, perhaps under the impact of Waldenberg's ruling, he still retains a very conservative perspective on the abortion issue ([10], pp. 273–275).

These discussions of abortion help illustrate what I have been trying to say about some of the more salient writings in the area of Jewish bioethics today: despite their authority and their erudition, these writings are not genuinely representative of the Jewish tradition as a whole. Not only the available secondary discussions of Jewish bioethics, but many of the contemporary Rabbinic decisions upon which they rest, display a markedly conservative tendency that is out of keeping with much of the spirit of the earlier tradition as well as with many of the explicit rulings of the tradition.

The existence of this conservative tendency raises at least two important questions. First, how can we account for the fact that important thinkers and writers in this area have moved the tradition in this direction? Second, what does this movement imply for our broader theme in this volume: the question of the relationship between religion, theology, and bioethics? Let me hazard some very tentative answers to these questions.

The conservative drift of contemporary Jewish bioethics is surely a result of some major sociological factors today. Jakobovits himself is not unaware of this conservative tendency and he suggests that it may have something to do with the shift in Rabbinic scholarship from the practical rabbinate to the academy. It seems to be his view that purely academic scholars may be less in touch with the urgent realities of medical practice and family life than those who maintain a more pastoral role ([10], p. 259). Somewhat differently, David Sinclair has suggested that this conservative tendency is traceable to the desire on the part of many orthodox scholars to demonstrate that on matters of sexual and moral concern, Judaism is no less exigent and no less rigorous in its views than are the most conservative Catholic or Protestant teachings. Indeed, Sinclair finds such sentiments embedded within some contemporary discussions of abortion as justifying reasons for elaborating a very conservative Jewish view ([25], pp. 125, 130). To these explanations, I would add the suggestion that the existence of this tendency has something to do with the divided state of contemporary Judaism. With progressive or liberal opinions regarded as characteristic of less *halakhicly* oriented Reform or Conservative scholars, those who deal with *halakhic* materials may feel compelled to distinguish themselves by elaborating a deliberately antimodern perspective. No less than their most conservative Catholic counterparts, in other words, some Jewish scholars may find it institutionally and personally important that Jewish teaching represent a 'sign of contradiction' to secular and liberal attitudes.

If this conservative tendency exists and is partly attributable to these social forces, what does this tell us about the relationship between religion, theology, and bioethics? Most obviously, that no tradition of profound moral and religious reflection proceeds in a vacuum, unaffected by the social realities which surround it. More importantly, this whole matter may give us some insight into the very specific perils which a religious-ethical tradition faces in the course of its development. At its best, the elaboration of a religious-ethical view is a difficult enterprise. Delicate oppositions between theological and moral elements must be reconciled to create a rich and viable tradition. But this effort is likely to be least successful when difficult intellectual matters become entangled in institutional conflicts and social disputes.

Let me develop this point a bit. In another context, I have argued that religious and moral conceptions play an important part in the development of a full moral view [8]. Moral commitment requires metaphysical beliefs to sustain and motivate it, while specific religious beliefs only make rational sense wedded to moral commitment and to the discipline of moral reason. Obviously, conjoining these two vitally important aspects of a full moral view is no easy task. There is always a tendency for the specific religious elements of a position – for example, the affirmed certainty of ultimate moral recompense – to vitiate or undermine aspects of moral obligation. The specific problem we saw in Jewish medical-ethical teaching of this nature concerned how a commitment to medical care and healing could be harmonized with the tradition's absolute insistence on God's providence and loving care of the righteous.

That classical Jewish teaching steered the difficult course required of a vital religious ethic strikes me as evident. Despite substantial textual problems, the earliest sages founded a live, compassionate tradition in which healing and the dedicated protection of human life were given an important place. Moral understanding was enriched, as well, by painstaking analysis of possible conflict situations. With few exceptions, the revealed sources were consistently interpreted so as to enjoin minimization of human suffering and the protection of independent persons over merely potential human life. And in those identifiable instances where the tradition took an adamant stand, it was very often to protect individuals from the subtle coercions that might lead them to renounce their deepest interests. Hence classical teaching is fairly restrictive on matters of voluntary self-sacrifice, on questions of subordinating the one to the many, and on the degree of latitude possessed by an individual in neglecting his or her own life and health. It is just in connection with such matters that we encounter the tradition's insistence on the absolute sanctity of life and one's duty to preserve oneself. Undeniably, the tradition is rigorous at these points. But it is a rigor always in the service of understandable human values [9].

In the course of this discussion I have tried to suggest that contemporary writings in Jewish bioethics have lost some of the balance and wisdom that marked the earlier tradition. An attentive discernment of human need has often been replaced by rigor for the sake of rigor. Norms elaborated initially to protect human beings from neglect have been construed so as to mandate cruelty. The claims of uterine life have been overvalued at the expense of the pressing needs of women and families. And, in the sexual-moral realm generally, a concern with propriety and strictness has sometimes overridden careful attention to the bearing of older norms on contemporary lives.

I have also tried to suggest that this loss of balance is partly attributable to the difficult social and intellectual context in which contemporary Jewish thinkers must work. As John Noonan showed so decisively in his pioneering study of teaching about contraception in the Roman Catholic tradition [17], religious-ethical thinking tends to fare badly when it is made hostage to deeper social, political, or religious conflicts. Religious traditions can make an important contribution to our thinking about new ethical questions in the area of medicine or elsewhere. They can sharpen our sensitivities to the crucial values involved and they can help guide us through difficult and ambiguous choices by drawing on a rich heritage of experience and reflection. Judaism's own teachings regarding the sanctity of life form a body of moral guidance of just this sort. But, as the characters in Grade's novel learned, if a tradition is to be heeded and respected, it must remain open to the new issues it faces and to their impact on human beings.

*Dartmouth College,*
*Hanover, New Hampshire, U.S.A.*

## NOTES

[1] The only exceptions to this rule occur in the cases of orders to commit homicide, incest (or equally serious sexual sins), or to engage in idolatry ([5], *Sanhedrin* 73a).
[2] Although Jakobovits' earliest writings on abortion suggest a position substantially in agreement with Bleich's view, he appears to have moved toward a more liberal view in the more recent edition of his *Jewish Medical Ethics* ([10], pp. 273–275) and elsewhere [11].

## BIBLIOGRAPHY

[1] Bleich, J.D.: 1977, *Contemporary Halakhic Problems,* Ktav Publishing House, New York.
[2] Bleich, J. D.: 1979, 'Abortion in Halakhic Literature', in F. Rosner and J. D. Bleich (eds.), *Jewish Bioethics*, Sanhedrin Press, New York, pp. 134–177.
[3] Bleich, J. D.: 1980, 'Sexuality and Procreation', in I. Frank (ed.), *Biomedical Ethics in Perspective of Jewish Teaching and Tradition*, College of Jewish Studies, Washington, D.C., pp. 1–11.
[4] Bleich, J. D.: 1981, *Judaism and Healing*, Ktav Publishing House, New York.
[5] Epstein, I. (ed.): 1935–1961, *Babylonian Talmud*, Soncino Press, London.
[6] Feldman, D. M.: 1974, *Marital Relations, Birth Control and Abortion in Jewish Law*, Schocken Books, New York.
[7] Grade, C.: 1974, *The Agunah*, Menorah Publishing Company, New York.
[8] Green, R.: 1978, *Religious Reason*, Oxford University Press, New York.
[9] Green, R.: 1982, 'Jewish Ethics and Beneficence', in E. E. Shelp (ed.), *Beneficence and Health Care*, D. Reidel, Dordrecht, Holland, pp. 109–125.
[10] Jakobovits, I.:1975, *Jewish Medical Ethics,* revised edition, Bloch Publishing Company, New York.
[11] Jakobovits, I.: 1977, 'Tay-Sachs Disease and the Jewish Community', *Proceedings of the Association of Orthodox Jewish Scientists* **5**, 11–17.
[12] Jakobovits, I.: 1978, 'Judaism', *Encyclopedia of Bioethics*, Vol. 2, pp. 791–802.
[13] Karo, J. (1488–1575): 1911, *Shulhan 'Arukh*, Romm (ed.), Vilna.
[14] Kirschner, R.: 1981, 'The Halakhic Status of the Fetus with Respect to Abortion', *Conservative Judaism* **34**, 3–16.
[15] Lamm, N.: 1974, 'Judaism and the Modern Attitude toward Homosexuality', *Encyclopedia Judaica Yearbook,* pp. 194–205.
[16] Maimonides, M. (Moses ben Maimon, 1135–1204): 1949, *The Code of Maimonides* (*Mishneh Torah*), 19 vols., Yale University Press, New Haven.
[17] Noonan, J. T.: 1966, *Contraception – A History of Its Treatment by the Catholic Theologians and Canonists*, Harvard University Press, Cambridge, Massachusetts.
[18] Novak, D.: 1979, 'Judaism and Contemporary Bioethics', *The Journal of Medicine and Philosophy* **4**, 347–366.
[19] Rashi (Solomon Yitzhaki, 1040–1105): 1928, *Pentateuch, with Commentary*, Horeb (ed.), Berlin.
[20] Rosenfeld, A.: 1979, 'Judaism and Gene Design', in F. Rosner and J. D. Bleich (eds.), *Jewish Bioethics,* Sanhedrin Press, New York, pp. 401–408.
[21] Rosner, F.: 1972, *Modern Medicine and Jewish Law*, Yeshiva University Press, New York.

[22] Rosner, F. and J. D. Bleich (eds.): 1979, *Jewish Bioethics,* Sanhedrin Press, New York.

[23] Siegel, S.: 1975, 'Experimentation on Fetuses which are Judged to be Viable', *Report of the National Commission for the Protection of Human Subjects*, Appendix: 'Research on the Fetus', DHEW Publication No. (OS) 76–128, Article I–7.

[24] Siegel, S.: 1980, 'Healing and the Definition of Death', in I. Frank (ed.), *Biomedical Ethics in Perspective of Jewish Teaching and Tradition*, College of Jewish Studies, Washington, D.C., pp. 28–33.

[25] Sinclair, D.: 1980, 'The Legal Basis for the Prohibition on Abortion in Jewish Law', *Israel Law Review* **15**, 109–130.

[26] Spero, M. H.: 1979, 'Homosexuality: Clinical and Ethical Challenges', *Tradition* **17**, 53–73.

DAVID H. SMITH

# MEDICAL LOYALTY: DIMENSIONS AND PROBLEMS OF A RICH IDEA

> [A] self is a life in so far as it is unified by a single purpose. Our loyalties furnish such purposes, and hence make of us conscious and unified moral persons. Where loyalty has not yet come to any sort of definiteness, there is so far present only a kind of inarticulate striving to be an individual self. This very search for one's true self is already a sort of life-purpose, which, as far as it goes, individuates the life of the person in question, and gives him a task. But loyalty brings the individual to full moral self-consciousness. It is devoting the self to a cause that, after all, first makes it a rational and unified self, instead of what the life of too many a man remains, namely, a cauldron of seething and bubbling efforts to be somebody, a cauldron which boils dry when life ends ([17], pp. 922 f).

Loyalty or fidelity has been a central concept in much twentieth century literature in theological and medical ethics. I hope to show something of the richness of the concept by exploring the different emphases and interpretations to be found among some of its users and to suggest some proposals of my own. The picture that should emerge is one of a core concept developed in several alternative directions. I do not mean to imply that the writers discussed are always and everywhere in agreement, that they constitute a 'party' or even a 'school of thought'. Apparent, however, are an identifiable tradition, patterns of family resemblance, and agreement against some alternative viewpoints in ethical theory.

Although *loyalty* and *fidelity* are sometimes distinguished ([17], p. 955), I shall use the terms interchangeably. Etymologically, loyalty relates to the French word *loi* or *law*; it may be said to have its natural context in political community. *Fidelity* stems from *fiducia*; it can denote truth, and its common English connotation in conjugal. I will treat these differences as secondary, however, and focus on the conceptual content shared by the two words. Further, I shall not refer as much as I might to Gabriel Marcel's discussion of fidelity but will confine myself to twentieth century American materials.

## JOSIAH ROYCE ON LOYALTY

The work of Josiah Royce, especially his *Philosophy of Loyalty*, is the natural starting point for our considerations. His preliminary definition of loyalty is "the willing and practical and thoroughgoing devotion of a person to a cause" ([17], p. 861), but his book is really an attempt to move



*E. E. Shelp (ed.), Theology and Bioethics*, 267–282.

from this rough root conception to a more precise and nuanced understanding of loyalty. He begins with some aspects of our moral lives; recurrent issues cluster around our search for a will of our own and our tendency to mimic and conform. "By nature," Royce says, "I am a sort of meeting place of countless streams of ancestral tendency" ([17], p. 865). Plans of life are modeled for us by those around us. "We in so far learn what our own will is by first imitating the wills of others" ([17], p. 868). A loyalty enables persons to discover themselves and to relate to others.

Individual persons need to 'centralize' their lives around their own purposes. What purposes will bring fulfillment? Neither happiness nor power nor conformity to the existing social order is an adequate conception of the chief good, for happiness and power are unattainable and our inevitable tendency to idealize exposes the inadequacies of existing social institutions and relations. On the other hand, freedom and independence are also inadequate goals, for they ignore the social character of selfhood. Selves gain definiteness from the specificity of the cause to which they are loyal, through social involvements. The only way to be an ethical individual is to choose a cause and serve it; "the only way to be practically autonomous is to be freely loyal" ([17], p. 892).

Royce realized that *loyalty* connoted material acts and rote obedience. He was eager to dissociate his proposal from those things. The problems, as he saw them, concerned, first, the choice of cause to be loyal to and, second, conflicting loyalties. In order to harmonize self and world, a loyalty must involve my natural interest and must be freely chosen. People must begin by limiting their 'personal range' to a special and personal cause. Persons will find themselves loyal to more than one cause, but any cause to which they are loyal must be attractive to their temperaments, likes, and interests. Loyalties should be related in a kind of 'plan', and they will evolve and grow through observation and example. However, in order to be worthy of any loyalty at all, a cause cannot live by overthrowing the loyalties of other persons. A good or worthy cause manifests "loyalty to loyalty, that is, is an aid and a furtherance of loyalty in my fellows" ([7], p. 900).

Loyalty is the basis for duties to self, and it leads to concern with rights. My rights and the self matter because of the cause served. Self-cultivation and assertion are pointless except as related to my loyalty; "rights which are not determined by my loyalty are vain pretence" ([17], p. 912). Loyalty is the basis for some more specific duties: I should be *truthful* because of loyalties – to my auditors and to the whole human community. *Justice* is the formal side of loyalty: keeping promises, respecting the commitments of others, engaging in conflict only for the sake of one's own loyalties or those of others. Without loyalty this justice is "vicious formalism" ([17], p. 912). *Benevolence* refers to a loyal person's concern for the inner lives of those affected by his actions. Our major concern should be with the other person's loyalty. Benevolence not so focused degenerates into sentimentalism. Loyalty to loyalty provides the supreme principle for choosing a cause, for resolving conflicts among causes, and for discerning our moral duties. Conscience is the ideal of life determined by the self's cause.

The stress on idealization was very important for Royce. "Too great literalness in the interpretation of human relations is...a foe to the development of loyalty" ([17], p. 958). His specific referent for this claim is children, but adults too need to learn to idealize a cause. Indeed, loyalty is never raised to the highest levels without grief. Idealization comes through suffering and the inspiration of leaders. So transfigured, loyalty is most precisely understood and "the will to manifest, so far as is possible, the Eternal, that is, the conscious and superhuman unity of life in the form of the acts of an individual self" ([17], p. 996). Less metaphysically, using James' terminology, it is "the Will to Believe in something eternal, and to express that belief in the practical life of a human being" ([17], p. 997). Serving universal loyalty thus involves viewing the interests of all conscious life as a kind of unity.

The crucial fact that makes this view possible is the certain defeat of all our causes. Our commitment to a lost cause brings not only grief and suffering but realization of the supra-human nature of that to which we were really loyal. Our loyalty is directed to something beyond the particularities of our cause. Religious imagination symbolizes for us our involvement in the 'world-life'.

This crude summary of Royce's analysis must suffice for now. It is an imaginative proposal rooted in his idealistic metaphysics but open in form and spirit to ordinary moral experience and counter – argument. What does this viewpoint suggest about problems in bioethics, problems which, so far as I know, Royce never addressed? Two general implications are clear. They concern the point of departure for medical ethics and the problem of suffering.

It must be clear that Royce means to reject a utilitarian analysis of medical ethics. For him the main moral problem with utilitarianism is its failure to root ethics in the particularities, affections, needs, and abilities of the individual. Mill was proud of this aspect of his thought, but Royce found this impersonal objectivity has been purchased at too great a price.

Evidently he thinks that the individual self is transcended in the moral life of loyalty, but the form of this loss of self, for Royce, must be such as to fulfill the *particular* individual. Utilitarianism is not wrong in stressing a good beyond that of the self, but its conception of that good, and of the relation between individual fulfillment and higher goods, is mistaken.

On the other hand, Royce would also reject an analysis of medical ethics strictly in terms of individual rights. For Royce rights exist *for the sake of something*, namely, one's cause or loyalty. They are important, but secondary, moral conceptions. The problem with granting primacy to rights is the implication that individuals are complete in themselves, whereas Royce's whole point is that the self only comes to identity and autonomy in relation to something beyond itself.

Perhaps an illustration will clarify the point. Consider a treatment refusal by a Jehovah's Witness. Royce could never discuss this issue strictly in utilitarian terms, for his primary concerns are with the patient, health professionals, and families – their causes and their relationships. At the same time, the question for him is more than a matter of rights. The Witness must think through the full nature of his loyalty. Is it a cause that supports the loyalties of others? Is its symbolism destructive? Similarly, health professionals are committed to the self, conscience, or loyalty of the patient. They may not 'sentimentally' choose physical life in the face of that loyalty. On the other hand, they are not forbidden discussion or argument about the truth of that loyalty. Furthermore, health professionals and patients share a loyalty. It begins with their complementary commitments to cause or self and patient, but ultimately it is rooted in the loyalty 'to loyalty' or 'the conscious and superhuman unity of life' that they should share if they are sufficiently reflective. For Royce the ultimate fact of metaphysical community means one can dare to speak about morally and mortally serious matters to the end. It would be disloyal not to discourage an act of disloyalty at the end of life.

I do not think that Royce would support hectoring of patients or deathbed evangelism. It is against just such abuses that his principle of *loyalty to the loyalties* of others is meant to guard. At the same time, both the relative truth of the patient's loyalty and the real community that exists among the loyal must be relevant facts for analysis. A patient's choices are to be respected – but no abandonment of care, concern, or conceptual engagement is tolerable. One may have to respect the choice of the true believer, but only after taking the loyalty seriously enough to talk it over.

A second issue illuminated by Royce's analysis of loyalty is the question of suffering. This issue is in some ways the most fundamental in medical ethics and an issue that is more than moral. Speaking very roughly, moral theories divide over the question of the importance of our commitment to the eradication of suffering. Do other goals or constraints bear on medicine? Certainly Royce would argue for at least one such constraint – loyalty. If I have to choose between being loyal to you and eliminating your suffering, I ought to take the former.

His more particular reflections on suffering take him in two directions. To start with, as I have suggested, suffering for Royce must be associated with the loss of a cause: death of a loved one, betrayal by a friend, corruption in one's government. These sufferings have the effect of purifying or idealizing one's loyalty, for they reveal the possibility of remaining loyal after the tangible material with which one began has been lost from view. Thus Royce can say that suffering is a good in the sense that without it there could be no moral growth and the heights of loyalty could never be achieved. Although Royce clearly thinks these sufferings are inevitable, he could pity someone whose cause had never faltered, for that rich man would find it very hard to squeeze through the needle's eye of moral growth. Indeed, any cause adequate to a life plan, Royce thinks, will be lost. To be adequately human is, and always will be, to suffer.

These reflections do not lead Royce into a moralistic theodicy. He rather insists that the religious response to suffering should stress the identification of God with the sufferer. The so-called free-will defense of God's justice ignores the obvious innocence of some sufferers: and if I, innocent, am said to suffer for someone else, then this attempt to vindicate God is really abandoned. Rather, we should begin by noting the complexity of human moral experience. The best life is not innocence but "states of inner tension, where our conscious lives are full of a warfare of the self with itself... [O]ur highest states of activity are the ones which are fullest of this crossing, conflict, and complex interrelation of loves and hates, of attractions and repugnances" ([16], pp. 848–849). In the best life evil is not abolished but subordinated. Thus for God (the world consciousness on Royce's terms) as for us, it is "impossible...to know a higher good than comes from the subordination of evil to good in a total experience" ([16], p. 851). God's love is "fulfilled in the endurance of physical ill, in the subordination of moral ill...and in the discovery that the endless tension of the finite world is included in the contemplative consciousness of the repose and harmony of eternity" ([16], p. 851). God is not the One beyond the many but "the being whose unity determines the very constitution, the lack, the tension, and relative disharmony of the finite world" ([16], p. 851). Thus God does not sympathize with the sufferer from without; instead he identifies with him. "God here sorrows, not *with* but *in* your sorrow. Your grief is identically his grief, and what you know as your loss, God knows as his loss, just in and through the very moment when you grieve" ([16], pp. 852–853). Relief for the sufferer comes from a fragmentary grasp of this truth.

Royce's idealism is as central to this discussion of suffering as to his ethics in general. It gives him a way of explaining the unity of self and God, which allows him to assert that the correct core of a religious response to sinful or dying persons is neither judgment nor exhortation but identification. In Royce's terms there is no way of taking an event or person more seriously than by identifying God (the Absolute) with it. The analysis seems to me to be profoundly correct in its avoidance of moralism and as an application of religion to the loneliness and desolation that are central to suffering. Of course, if someone rejected the idealistic metaphysics, for logical or historical reasons, and shared Royce's concerns with self, loyalty, and suffering, a new synthesis would have to be worked out. To such an attempt I now turn.

## H. R. NIEBUHR: LOYALTY TO THE ONE

H. R. Niebuhr held that faith was the center of a person's being. It is made up of trust in and loyalty to an object. Selves depend on one or more centers of value in order to give meaning to their lives, and Niebuhr agrees with Royce, "the essence of the moral life is loyalty to a cause" ([10], p. 21).

> ...selfhood and loyalty go together...[and] however confused the loyalties of selves may be, however manifold their causes and however frequent their betrayals, yet it is by fidelity that they live no less than by confidence in centers of value which bestow worth on their existence ([10], p. 22).

For H. R. Niebuhr, everyone has a center of value or cause; most people probably have more than one. There may be men without countries but no selves without loyalties. In this sense of the word, atheism is impossible.

Upon the other hand, the God confessed by Western theism differs significantly from the faiths of persons today. The most visible faith arises when one's cause is a single specific and identifiable social group: a race, family, religious body, or – especially – a country. This social group then serves to define morals and artistic value. This *henotheism*, as Niebuhr calls it, amounts to having a single unidealized (or never lost) cause in Royce's terms. In the form of nationalism it is, Niebuhr thought, the most powerful force in the modern world. When clearly articulated it inevitably has a kind of monomaniacal or fanatical character. Of course, monotheism is a prominent attitude within medicine which, in reality as well as in fiction, may become a magnificent obsession. The physician who lives only for her profession, or her patients, may come to find guild or broken body unworthy of total devotion.

Although the life of the individual with such a cause is *unified*, inspection of the god with open eyes usually reveals its feet of clay. Thus people "take recourse to multiple centers of value and scatter their loyalties among many causes. When the half gods go the minimal gods arrive" ([10], p. 28). This pluralism or polytheism "in which an unintegrated, diffuse self-system depends for its meanings on many centers and gives its partial loyalties to many interests" ([10], p. 29) is the best we have when an all-sufficient community breaks down. We then value ourselves only for the interests we have or the functions we perform – not in our wholeness. This plurality of loyalties is tolerant and nonfanatical, but the self thus created knows little of what we might call integrity: it is hollow.

Beyond these possibilities Niebuhr recommended what he called *radical monotheism* in which one is loyal to "One beyond all the many, whence all the many derive their being, and by participation in which they exist" ([10], p. 32). The true loyalty is *not* loyalty to loyalty but "loyalty to all existents as bound together by a loyalty that is not only resident in them but transcends them" ([10], p. 34). Morally this means loyalty to whatever is, rather than to the loyal. It is inclusive. Henotheism tends to see only duties to members of one's group, polytheism only limited duties insofar as particular values are shared. In contrast: "in radical monotheism my neighbor is my companion in being; though he is my enemy in some less than universal context the requirement is to love him" ([10], p. 34).

Niebuhr drew three moral principles from this analysis: the goodness of all things as related to the One; the limited value of all things as other than the One; and, third, the need to understand all human relations on analogy with the relation to the One: as covenant relations, as matters "of promise-keeping or of keeping faith" ([10], p. 41). This last principle takes its great force from a final stage in Niebuhr's argument. For him the One to whom we should be loyal, the principle of being, is a person. The world, he thought, cannot be understood in nonpersonal categories. The opposite of impersonal truth is error; that of personal truth is a lie.[1]

Because there is a unity in the world, a One beyond the many, a unity of the self is possible. We can respond to the One in more than one way – in distrust and fear or trust and loyalty ([11], p. 118 ff.). We can ignore or fight or attempt to appease the power in our world that leads to death and suffering ([11], p. 141). Basically these are kinds of defensiveness. Salvation occurs if and when we learn to trust in the One, thus loyally interpreting all actions as His action ([11], pp. 142–145).

If we apply this analysis to issues of medical ethics, we can take several different directions. We can proceed, for example, from Niebuhr's analysis of human nature, its pathologies and virtues, to exploration of character and institutions in the medical world. William F. May's work [4, 5, 6, 7] can fruitfully be seen as an imaginative outworking of these lines of thought. On the other hand, the preoccupation with fidelity between physician and patient is nowhere more central than in Paul Ramsey's medical ethics, to which I shall soon turn. Finally, James Gustafson's thought reflects the Niebuhrian preoccupation with the action of God everywhere, the theocentrism bordering on pessimism, and the concern with particularity so characteristic of Niebuhr.[2] This is not the context in which to trace out all the family resemblances; it is important to note the fruits of Niebuhrian theology in medical ethics.

Some of Niebuhr's ideas about suffering are of particular relevance. The fact of suffering is for him the decisive argument for an ethic of loyalty (or responsibility), rather than a teleological or deontological ethics. Suffering is "that which cuts athwart our purposive movements...the intrusion into our self-legislating existence of that which is not under our control" ([11], p. 60). Thus it cannot be explained in teleological or deontological terms. The fact of suffering shows us that we are less powerful, less in control, than teleological or even deontological ethics wants to allow. Therefore, all we can do is remain loyal to the One, interpreting and responding to His actions as best we can. The deathbed or cancer ward is for Niebuhr a prism of the human condition. It is not a context in which duty or goals are most appropriately invoked but a context in which it is best to speak the language of coping. Suffering and death should not be thoughtlessly combatted, or denied, or perversely glorified. Rather, they are to be interpreted in a context of ultimate trust.

How much does Niebuhr differ from Royce? At first glance the difference seems great. For Royce, self and absolute are one, and the self finds fulfillment in loyalty to a lost cause. Suffering is made bearable by the identification of self and God. Niebuhr takes a more pessimistic tone throughout. His preoccupation is with the limits on humankind revealed by suffering. Certainly he meant to reject the metaphysics of idealism.

Nevertheless, some congruities are apparent. Both writers are concerned with salvation, and, for both, suffering plays an important epistemological role. Both agree that a true loyalty is only made possible – or at least only revealed – in suffering, and that an attempt to rid the world of all suffering is Promethean. Medicine should not be seen as a saving but as a palliating art, at best a try to gain a little time. For both thinkers, the ultimate focus of loyalty is beyond those other persons whom we know. Only when moral relations are set in this complex perspective – of trust or loyalty – can they be ordered aright. And neither is sanguine about working out this ordering.

## PAUL RAMSEY AND COVENANT LOYALTY

Paul Ramsey's ethic begins with the idea that God has saved persons from 'anxious self-centeredness'. The appropriate response is faith in God and love or 'covenant fidelity' to other persons. Fidelity for Ramsey is primarily a matter of proper relationships between individuals: Christian ethics, love or fidelity "may claim to be relevant in criticism of every situation precisely because its standard...is not accommodated to man's continuing life in normal historical relationships..." ([12], p. 44). Our lives involve a plurality of relationships and conflicting loyalties; analysis of our problems is aided by a standard that can be applied to every one of them individually. The task of theological morality is formulating "canons of loyalty or canons of faithfulness" ([13], p. 70). "[I]n Christian ethics we are mainly concerned about the requirement of loyalty to covenants among men..." ([13], p. 125).

How are the requirements of loyalty to be discerned? It is not clear that Ramsey has a systematic method for answering this question. He has tended instead to reflect on certain types of situations, which he once called prismatic cases, and to generalize about the requirements of loyalty in those situations. Two examples from medical ethics illustrate his method of work and suggest likely directions for constructive uses of fidelity as a normative principle.

Experimentation on human subjects is, for Ramsey, a peculiar kind of medical activity; its essence is doing something to someone that is not to that person's benefit. The crucial moral fact for Ramsey is that the medical relationship is not therapeutic, and he concludes that only consent can justify it: No nontherapeutic medical touching without consent can be an act of fidelity. Ramsey then proceeds to apply this principle of fidelity to any medical situation that can plausibly be described as nontherapeutic. (He has remarkably little to say about therapeutic experimentation, probably because it must seem to him to raise issues that are consequential but not theoretically interesting.) Fidelity involves some rules discerned in paradigmatic situations; it relates to another person's basic needs or interests; it is 'singleminded' in its concern for the other. The general social benefits that might follow from a well-conceived nontherapeutic experiment on a child are, to Ramsey, morally irrelevant ([14], Chapter I).

Care for the dying shows another side of Ramsey's use of the fidelity principle. He does not concern himself, as do Royce and Niebuhr, with the moral agent's own suffering. Rather, he writes about our responsibilities to other persons who are suffering in a medical context. Care for them is always required, but there comes a time when they cannot be cured. Then loneliness is their greatest suffering, and we should stop trying to cure and concentrate on companionship and comfort. Loyalty is incompatible with desertion, or getting it over with, or quality of life judgments; it requires acknowledgement that the situation or human needs change. In his more recent writings on care for the dying, Ramsey has been particularly concerned to insist that this univocal focus on the individual patient be maintained.

## THE FIDELITY PRINCIPLE, SUFFERING, AND THE OBJECT OF LOYALTY

I hope the foregoing makes clear that a fidelity principle with some consistency can be developed in several different directions. Issues remain for its users. One concerns fanaticism. All are aware of the problem, but it is unclear that loyalty to loyalty has sufficient content to evaluate specific social movements or causes; a genuinely radical monotheist might well take his children or country or patient a little less seriously than we would like; and Ramsey's concern for this needy individual can leave other persons and social institutions untouched by moral reflection. If Royce makes a god of the cause and Niebuhr removes god from the world, Ramsey makes a god of the patient. None of these strategies seems altogether adequate.

For all these writers, the fidelity principle has a kind of ambivalent relation to particular institutional contexts. Ramsey offered a scriptural justification early in his work, but his main interest has been in using the principle, not its derivation. Niebuhr really introduced the sociology of knowledge into American Christian theology, and he was preoccupied with the historically conditioned character of selfhood, but his premature death deprived us of a full statement of his view on the relationship between Scripture, tradition, and his philosophical analysis of selfhood. Before Niebuhr, Royce, in *The Problem of Christianity*, suggested and began a tradition of theological reconstruction, and in his moral analyses he consistently reinterpreted traditional concepts. But that was seventy-five years ago. *Loyalty* naturally suggests institutions (countries, companies, churches, or marriages) that have traditions. In part, at least, disloyalty is a betrayal of these historic and specific rules, customs, and perceptions ([3], Chapter 11). The relation of loyalty to tradition – as a justification for it or content of it – calls for further work.

As a start in this direction I wish to indicate two lines of thought that may reveal both the ongoing relevance of a fidelity ethic and a form it might take. It is convenient to begin as the fidelity tradition has begun, with the issue of suffering.

Suffering does not so much raise the question of purpose or guilt, but the question of identity. Genuine suffering is something that appears purposeless, a violation of the principles, perceptions, or rules that had constituted myself and my world. If my prior interpretive framework were adequate to handle it, it would not really be suffering. At such a time the subject is not in charge. He is trying to cope, to respond. He has to reassess who he is, what he stands for. The great moral lesson to be learned from medicine is this simple fact: the weakness of human persons and their need for something to trust.

Fundamentally, two responses are possible to the crisis of selfhood that suffering provokes. One kind of response involves an attempt to secure the self against invasion, assault, and attack. We look for territory in which we can be secure and powerful. This search leads to a preoccupation with our own independence, as commitments or alliances always limit or bind a self and make it vulnerable. For the independent self, the only real powers beyond the self that are acknowledged are powers that are seen as a threat, and the only kinds of relationships with other persons that are comfortable are those that are distant and superficial or those in which the self is in charge – as parent, teacher, or physician. (We do not like to be children, pupils, or patients.) Medicine suggests this is a mistaken model of selfhood, as does Christian theological tradition.

In an original essay, *Suffering: A Test of Theological Method,* Arthur McGill pointed out the relationship between this defensive response to suffering and the Arian view of God. For Arius, God's identity is a mirror of the independent self just described. God is 'unbegotten and unbegetting'. His power is unshared and His individuality uncompromised. While He does relate himself beneficently to the world, this relationship ultimately has no effect upon Him. God is the great king par excellence. Jesus, therefore, cannot be His equal but must be construed to be His close confidant. God could never suffer; His junior colleague or child certainly can.

McGill notes that a major segment of the Christian movement (led by Athanasius) rejected this understanding of God. For them, God's identity and power were associated with His giving Himself, His involvement with the world, rather than with His independence. Their idea was that in the Incarnation God surrendered independence, bound Himself to the world and, in doing so, made Himself vulnerable. Thus the incredible debates about the status of Jesus in the fourth century were not 'a furor over a diphthong', as Gibbon would have it, but concerned the very nature of God. In insisting that the Son was of the same identity as the Father, the Athanasians were claiming that God Himself suffers, that true selfhood involves vulnerability to suffering.

The Christological symbolism, in other words, coheres with medical experience suggesting that persons are insufficient in themselves. They need a cause. This conclusion is not itself a moral directive, but a descriptive statement. However, it is morally relevant. Furthermore, the theological tradition goes on to suggest that God has identified with this fact of human insufficiency. Human selves, therefore, can accept it. The need for a cause is not something unfortunate about persons, but something owned even by God. As examplar of an ideal the incarnate God demonstrates what persons should do. God's cause becomes that of humankind; persons formed by acceptance of this cause find it essential to identify with the sufferings of others. That is the cause of God.

Human selves modeled on this normative image would be very different from those preoccupied with independence and power over their own turf. They would see suffering as inseparable from their being(s), for they would seek identity in loyalties to others. The commitments evoked by those loyalties inevitably involve vulnerability, which begins with identification of themselves with their own imperfect bodies and includes their friendships, marriages, children, and jobs. Living *in* those identities means living with suffering. Life is actually a quest for loyalty; the good life is only possible when this fact is accepted; selves who accept it accept suffering, but not alone: "When you suffer, *your sufferings are God's sufferings*, not his external work, not his external penalty, not the fruit of his neglect, but identically his own personal woe. In you God himself suffers, precisely as you do..." ([16], p. 843).

Suppose then that there are traditional and experiential grounds for saying that to be a self is to live for a cause, that to be moral is to be loyal. This does not solve many of our problems: it only gives us a principle, as yet vague and diffuse, from which to proceed. Can we do so? The task is complicated.

Earlier I mentioned that fanaticism is an inevitable problem for a theory making loyalty central, as selfishness is for an egoistic ethic. The resolution of the problem may surface if we see that the writers I surveyed share a kind of studied ambiguity about the object of loyalty. It is both an interesting cause – and the principle that cause embodies; it is the One beyond the many – but human relations are made covenantal; it is the needy neighbor – but one's true faith lies in God. The cause of God and other causes are never identical; for the more recent thinkers they are necessarily a little at odds. Thus, for these users of the fidelity principle the solution to the problem of fanaticism is not deemphasis of loyalty. Rather, it is ordering or thinking about loyalties. The great moral mistake occurs when we say to the fanatic that his loyalties do not matter. He needs to take them more seriously, not ignore them.

Perhaps it is possible to do more, to suggest some general moral principles that are related to the core loyalty principle. Can one articulate some dimensions of loyalty in advance? Let me suggest some lines of thought that might be pursued, as we think about medical loyalty. Two are especially pertinent. First, fidelity to another means fidelity to his or her body. This body has a complex relation to our friend. It matters as the friend's body, but as body it has a kind of logic of its own. In loyalty to another we commit ourselves not just to the uses he or she makes of the body, but to the body that he or she, strictly speaking, *is*. Thus loyalty to our friends includes a commitment to the health or good functioning of their bodies.

To a point the requirements for bodily health are the same for all human beings, and I assume it is plausible to speak of a healthy (or ill) human being in some cross-cultural ways. Healthy bodies function within certain broad biochemical, anatomical, or physiological parameters. Medicine is the art of understanding, and healing, these processes. Thus, medical loyalty is initially a commitment to the health of another's body. If educational loyalty means commitment to another's mind and economic loyalty involves special concern for another's financial affairs, medical loyalty – what a physician primarily owes a patient – concerns care for another's body.

Second, fidelity means fidelity to the *temporal* self of the other. Others with whom we are entrusted have both a past and a future. Their past has given them a history of loves, disappointments, successes and failures, projects and hopes, children, parents, and lovers. These loyalties of theirs have made them who they are as we know them. They constitute our friends' particularity. It cannot be loyal to disregard this particularity by ignoring the causes and commitments for which our friends have lived. Their commitments inform us about our friends, and our loyalty to our friends colors our attitude toward their commitments.

At the same time our friends have a future, and that future always, to a point, is open. Of course most tomorrows are controlled by yesterdays, but lives in fact change. In the normal situation we do not know what our friends will do tomorrow, or what will happen to them. Sinners convert, saints fall away, Prince Hal becomes Henry IV. A traitor to one cause is a hero to another. Who can chart or predict, much less evaluate, the various involvements and commitments of a life? Loyalty requires sticking with one's friends through these modulations; more precisely it means expecting such changes – growth or decline – in a friend.

This aspect of loyalty to others has considerable relevance to medical decision-making. It is not, however, as definitive of *medical* loyalty per se as is concern for the body.

In sum, loyalty to others requires concern for their bodily health, their personal particularities and their future prospects. The point could be made negatively: I betray one to whom I should be loyal, if I do not act for the sake of her health, if I am not peculiarly responsive to her unique complex of needs and cares, if I do not leave her room to maneuver, change, and blossom. These derivative principles of nature, justice, and liberty may well conflict with each other. My friend's commitment to music may be ruinous to her health, and it may foreclose other options if she has become a slave to that cause. It is not immediately clear what loyalty would require in that kind of dilemma. In general, all I hope to suggest is that loyal reflection must bring at least this range of considerations into focus. Moreover, there may be a kind of moral vocationalism that we should acknowledge, for it is incumbent on doctors to be particular spokesmen for the *health* of their friends.

So far we have been thinking strictly in terms of my loyalty to one other human being. Helpful as this paradigm is, it is oversimple. We often find ourselves in situations where our loyalties conflict, when fidelity to one person leads to infidelity to another. This issue regularly confronts parents, or indeed all of us insofar as we exercise power and responsibility. One way to deal with it is by denying responsibility in some relations;

another is by denying responsibility in *all* relations and reducing the problem to a distributional puzzle, perhaps to be resolved on utilitarian grounds.

While I do not have a full resolution of this issue to propose, I do have two suggestions. The first is that insofar as the conflicting loyalties (e.g., to different patients) are *sui generis, equality* is an appropriate proximate principle of conflict resolution. Insofar as I stand in the same kind of relation to persons with the same stake in my action, equal cases should be treated equally.

Second, it is important for us to distinguish this kind of thinking about balancing from the kind we do when we attempt to specify our fiduciary responsibilities. It is one thing to be a referee, another to be a friend. Often we play these roles on the stage of our own mind, but as a practical matter this leads to obvious problems of self-deception, betrayal, and favoritism. At times, when the stakes are particularly high, it is fitting that this play between modes of reasoning be embodied in actual social institutions and persons – when the conflict should be transposed to a more public forum in which double casting will not occur. We may not be able conceptually to adjudicate between conflicting loyalties; the best we can do may be a good social system of conflict resolution. These considerations suggest two supplements to the principles of nature, integrity, and freedom mentioned before. One is a principle of equality that serves to force consideration of unacknowledged loyalties and to limit idolatrous ones. The other is a kind of acknowledgement of finitude suggesting the need for institutional processes and supports. These institutional processes (advisers, committees, courts) are not necessarily right or corrective, but at the very least they help the individual gain some perspective. And when they are functioning well, they provide an invaluable forum for the discussion of some conflicts.

## A SUMMARY STATEMENT

At least since the time of Royce, loyalty or fidelity has been a moral idea that has captured the imagination of some major American moralists. Heavy use of the concept forces a writer to make some claims about human nature, and it is associated with serious concern with suffering. Medical work suggests the plausibility of the concept, and it coheres with some theological reconstruction. The danger of fanaticism may be countered by formulation of some derivative principles of fidelity, e.g., concern for body, integrity, liberty, equality, and due process.[3]

*Indiana University*
*Bloomington, Indiana, U.S.A.*

## NOTES

[1] For an early statement of Niebuhr's argument for the personal aspect of reality, see [9].
[2] See, for instance, [1] and [2].
[3] Portions of this essay have appeared in Carol Levine (ed.), *Essays on Death, Suffering and Wellbeing* (forthcoming). I am grateful to the Hastings Center for permission to reorganize and reuse this material.

## BIBLIOGRAPHY


[1] Gustafson, J. M.: 1975, *The Contribution of Theology to Medical Ethics*, Marquette University Press, Milwaukee.
[2] Gustafson, J. M.: 1981, *Ethics from a Theocentric Perspective, Vol. 1: Theology and Ethics,* University of Chicago Press, Chicago.
[3] Kirk, K.: 1927, *Conscience and Its Problems*, Longmans, Green and Company, London.
[4] May, W. F.: 1973, 'Attitudes Toward the Newly Dead', *Hastings Center Studies* **1**, 3–13.
[5] May, W. F.: 1974, 'The Metaphysical Plight of the Family', *Hastings Center Studies* **2**, 19–30.
[6] May, W. F.: 1975, 'Code, Covenant, Contract or Philanthropy', *The Hastings Center Report* **5**, 29–38.
[7] May, W. F.: 1976, 'Institutions as Symbols of Death', *Journal of the American Academy of Religion* **2**, 211–223.
[8] McGill, A.: 1967, *Suffering: A Test of Theological Method*, Geneva Press, Philadelphia.
[9] Niebuhr, H. R.: 1941, *The Meaning of Revelation*, MacMillan, New York.
[10] Niebuhr, H. R.: 1960, *Radical Monotheism and Western Culture*, Harper, New York.
[11] Niebuhr, H. R.: 1963, *The Responsible Self: An Essay in Christian Moral Philosophy,* Harper & Row, New York.
[12] Ramsey, P.: 1950, *Basic Christian Ethics*, Charles Scribner's Sons, New York.
[13] Ramsey, P.: 1968, 'The Case of the Curious Exception', in G. Outka and P. Ramsey (eds.), *Norm and Context in Christian Ethics*, Charles Scribner's Sons, New York, pp. 67–135.
[14] Ramsey, P.: 1970, *The Patient as Person: Explorations in Medical Ethics*, Yale University Press, New Haven.
[15] Ramsey, P.: 1978, *Ethics at the Edges of Life: Medical and Legal Intersections*, Yale University Press, New Haven.
[16] Royce, J.: 1897, 'The Problem of Job', reprinted in J. McDermott, (ed.): 1969, *The Basic Writings of Josiah Royce*, University of Chicago Press, Chicago, pp. 833–854.
[17] Royce, J.: 1908, *The Philosophy of Loyalty*, reprinted in J. McDermott (ed.): 1969, *The Basic Writings of Josiah Royce*, University of Chicago Press, Chicago, pp. 855–1013.

PAUL LEHMANN

# RESPONSIBILITY FOR LIFE: BIOETHICS IN THEOLOGICAL PERSPECTIVE

The present endeavor to give some account of certain issues and dilemmas of bioethics in theological perspective finds this perspective given in Christian faith, thought, and experience. In this context, the perspective is a single vision which illuminates and informs four inter-related claims. The claims are: (1) that providence, eschatology, and destiny are singularly significant perspectives upon bioethics; (2) that when these perspectives are brought to bear upon critical issues and dilemmas of bioethics, the central ethical foci are freedom, responsibility, and justice; (3) that when these ethical foci are understood and practiced in the context of providence, eschatology, and destiny, they are the criteria for taking responsibility for life; (4) and that when bioethical research, discovery, and practice are pursued with careful attention to the context and criteria thus identified, certain particularly awesome and disquieting ethical perplexities surrounding bioethical endeavor – e.g., the questions of limits, alternatives, possibilities, and uncertainties – receive a liberating light upon the responsible direction to take.

## OF PROVIDENCE, ESCHATOLOGY AND DESTINY

*Providence* is a theological 'root-word' denoting a universe purposed for human fulfillment and directed towards a universe at once fulfilling and fulfilled. In such a universe, mystery and meaning, happening and hope, promise and possibility, expectation and remembrance, trust and time, cohere. This coherence surrounds, sustains, and shapes continuity and change, identity and diversity, uncertainty and ambiguity, achievements and failures, frustrations and faults, dilemmas and deceits. The purpose is that what is coming to be shall have room to displace and replace what has been, in and for the consummation of what is. In short, providence is the *here and now* of that *then* referred to in the concluding chapter of the last book of the Bible (cf. Rev. 21:1–4 RSV).

*Eschatology* is the inclusive theological 'root-word' which denotes both the distance and the immanence of the future coming our way. Both vision and conceptualization, discernment and meaning are included in this denotation. Eschatology is intrinsically correlative with providence,



*E. E. Shelp (ed.), Theology and Bioethics*, 283–302.

as a doctrine of Last Things is correlative with a doctrine of beginnings, understood as Creation. Creation is the experience of the world as *there*, in consequence of an original and originating coherence purposed for fulfilling consummation. In such a world, God and humanity do not, and need not, play against each other. They are together on a covenantal journey. On that journey, the absence of alternatives is never a reason for holding to what *is* against what is coming to be. On that journey, limits are *not* constraints upon ingenuity, resourcefulness, experimentation; nor, above all, on freedom. Limits are the gateways of responsibility for the world as purposed and for the purposed human fulfillment being fulfilled. Limits are the pauses that refresh, not the points to be breached.

The prevailing contemporary view of limits, however, is that limits are overextensions of what is and, thus, are barriers to the ingress of what is coming to be upon what has been. Understood in this way, limits are perforce to be breached. But the breach is not perceived as what it is: a transgression of the providential order of time, and things and people by confusing the freedom *for* possibilities with the freedom *of* possibilities. Freedom *for* possibilities finds the perils of security and the pressures for immediate gains in knowledge and power safer than the perils of patience and of trust in the promise of a future only beginning to be disclosed. At least since Lot's wife, it has been a matter of record that looking back was thought to be the prudential way ahead (Gen. 9:26), and that Esau, the bearer of the covenantal destiny, calculating that 'a bird in the hand was worth two in the bush,' operationally "despised his birthright" and settled for "bread and a pottage of lentils" (Gen. 25:34). Freedom *of* possibilities, on the other hand, is the pursuit of the open road towards that liberating future already on the way for the sake of the only past worthy of a present. Humanity, on this march to a liberating future, from a perspective of faith, has discovered that the promise of the unlikeliest of uncertainties signals a forward and sustaining purpose. Evidence of this belief is found in Abraham who risked the slaying of Isaac (Gen. 22), in Sarah who harnessed disbelief with humor (Gen. 8:11–16), and in Mary who surrendered incredulity to humility and burst into a *Te Deum* in celebration of the radical reversal of power, place, and opportunity that marks the providential displacement of possibilities that have played themselves out by fresh and healing possibilities on the way (Lk. 1:46–55). "Things that are not" are chosen "to bring to nothing things that are" (cf. I Cor. 1:29).

The operative word for such a covenantal journey is *Destiny*. Destiny is experienced as the liberating point of intersection between Eschatology and Providence. The accent is always upon what is coming towards one as the way forward from where one has been towards where one is being invited to be. Since nature crossed the frontier into life and life crossed the frontier into human life, the future bears the secret of meaning and responsibility which transfigures the past into a prelude of the present. In a world purposed for human fulfillment, the experience of Destiny identifies Eschatology as the warrant for Providence and Providence as the foretaste of Eschatology. So Calvin notes that:

> to represent God as a Creator only for a moment, who entirely finished all his work at once, were frigid and jejeune; and in this it behooves us especially to differ from the heathen, that the presence of divine power may appear to us no less in the perpetual state of the world than in its first origin.... What is called providence describes God, not as idly beholding from heaven the transactions which happen in the world, but as holding the helm of the universe, and regulating all events.... What to us seems a contingency, faith will acknowledge to have been a secret impulse of God. It is not always, indeed, that there appears a similar reason; but it should be considered as indubitably certain, that all the revolutions visible in the world proceed from the secret exertion of the Divine power.... The providence of God is to be considered as well in regard to futurity as in reference to that which is past; ... it governs all things in such a manner as to operate sometimes by the intervention of means, sometimes without means, and sometimes in opposition to all means; ... it tends to show the care of God for the whole human race... ([6], Volume 1, Book. 1, Chapter 16:1, 4, 9; Chapter 17:1).

Thus, in theological perspective, the apperception and experience of the world and human life, of wholeness and fulfillment in the world, continue to be formed and nurtured by the dynamic interconnections between Creation and Consummation, forged by the coordination of Providence and Destiny, mystery and meaning, commitment and trust.

## OF FREEDOM, RESPONSIBILITY, AND JUSTICE

When Providence, Eschatology, and Destiny provide the context from which theological perspectives upon biomedical findings and problems are to be drawn, a reciprocity emerges between the rationality of belief and responsibility for life. A reverse *Aufklärung* is under way that makes a move from God to Morals to Medicine at once more pertinent to and promising for the relations between theology and bioethics than the reductionist *Aufklärung* in its symmetrical confidence in God, Freedom, and Immortality has turned out to provide.

The original syndrome of Enlightenment presupposed and sought to effect a reduction of the Christian Creed and to convert the *fortissimi* of Christian faith and obedience into whisperings in the wings. Consequently, God, Freedom, and Immortality (the terms by which the rationality of belief was explored in the 18th Century) were indeed denotative of the nub of the matter in a world from which neither God nor humanity could easily be excluded and within which one could not get on happily with horizontals only (i.e., with no reference to the Transcendent). The nub of the matter, as Kant rightly discerned, concerned the well-being or happiness of human beings in a world of natural necessity. The experience of necessity disclosed itself to understanding through the power of rationality to bring sense experience and the categories of thought meaningfully together within an order of law. But, as we all recall, Kant went on to show that the question 'What can I know?' exhausts neither the range of human curiosity nor the height and depth dimensions of human experience. Two even more insistently human questions keep cluttering up the tidy and orderly landscape sketched by Newtonian physics and mechanics. The questions are: 'What ought I to do?' and 'What may I hope for?' ([10], A805/B833, p. 635). They are the questions of responsibility and destiny. They presuppose an order of freedom, distinct from the order of necessity, and an order of fulfillment beyond the boundaries of the world as experienced. They are, indeed, the very stuff of morals and they point to indispensable correlations and convergences of freedom and volition, adjudication and consummation, ends and means, of responsibility for humanity and of the human meaning and direction of responsibility. In the Kantian world, the kingdom of ends was the foretaste of the kingdom of God. God and Immortality were the twin supports of an order of freedom, responsibility, and fulfillment, disjoined from which, the order of necessity could neither take account of nor do justice to the reality and highest good of human life.

The reductionist *Aufklärung* had identified freedom and responsibility as pivotal instances of the experience of God and of the life beyond death. The trouble was that under the reductionist Enlightenment, a deficient sense of Providence, Eschatology, and Destiny allowed freedom and responsibility to effect a too easy linear passage between God and Immortality. Consequently, freedom and responsibility became the provenance of Morals. Morals became a manageable calculus of good and evil, virtues and vices, rights and duties, aims and motives, volition and inhibition, ends and means, permissions and prohibitions, happiness and futility. It was not significantly noticed that the ancient distinction between behavior according to custom and behavior according to reason – i.e., between morality and ethics – had been gradually fused according to an ancient Stoic perception of a law of reason and nature, anchored in the natural law. It did not matter that, in the tradition, neither theology nor moral philosophy had ever been able to state precisely what the natural law could be said to be, or to sort out the precise relations between the natural law, the law of Moses, and the law of love. It seemed clear and sufficient to enjoin the doing of good and the avoidance of evil as the fundamental precept of the natural law, as St. Thomas had done (*Summa Theologiae* IaIIae, Q. 94, Art. 2, Resp.), and to find in the law of love the quintessence of the law of Moses and of the natural law [13]. Accordingly, the way of freedom and responsibility could be operationally charted between God and Immortality, as the course of virtue and obligation, rooted and grounded in the law of reason and nature anchored in the natural law. Along this course, there were moral virtues which informed individual responsibilities; and there were legal limits and duties which informed the responsibilities of duly constituted authorities. All the while, medicine, being in its infancy, could readily adopt the Hippocratic commitment to preserve life against death as a variant of the natural law and congruent both with Moses and Jesus. As it turned out, the syndrome God, Morals, and Medicine provided a focal instance of the ethical reality and power of the syndrome God, Freedom, and Immortality.

This happy domestication of Deity, Duty, and Disease effected also the domestication of limits. Limits were perceived as constraints to be breached, not pauses that refresh. How should it be otherwise when the custodians of the covenantal journey had already converted limits into constraints that restrict? Thus, the breaching of limits in the power of reason and through the onward march of rationality was a providential break for freedom as a rebuke of the inquisitorial fanaticism that had repressed the covenantal story in its magisterial zeal. In the context of the covenantal story, Galileo's recantation turns out to have been exactly one of those pauses that refresh at the gateway of responsibility. Although inquisitorial fanaticism has erupted from time to time (as with Darwin and Freud, and more recently with Marx and Humanism), the sound and the fury, being deprived of the power to take life, have been set upon a path of waning intensity. On this journey towards irrelevance, this sound and fury have been overtaken by the covenantal journey that has been preparing all the while for an identification of limits as the liberating intersection between freedom and responsibility.

Far from being exceptions that upset the rules and block the forward advance of knowledge, power, and responsibility for life, limits are, in the context of the covenantal story, the decisive pointers to the next responsible move in making room for freedom. Their identifying sign is

*Justice*. Of course, there are risks. But when limits are risked at the gateway of responsibility, they indicate the open road as the forward road, whereas the risks of breaching limits confuse the open road with the road back. A memorable instance of the former is that momentous confrontation on the Selma Bridge, on the way to Montgomery, when Martin Luther King, Jr., faced down Sheriff Bull Connor and turned aside in the power of the future already sweeping the past aside. An ominous instance of the confusion of the open road with the road back is the fateful moment when the bitter fruit of 'Operation Trinity' fell on Hiroshima, on the Feast of the Transfiguration. When limits are the gateways of responsibility, the pauses resist the pressures of power to convert risk into policy. The pauses check the temptation to adopt a policy that justifies breaching limits without regard for the potential human consequences of so doing in the name of increased power, knowledge, or privilege. In so doing, the pauses identify the open road as the point-by-point setting right what is not right in the world. At stake is the human point at which freedom and responsibility intersect, and risks are warranted, not by the prospect of the enlargement of knowledge, possibilities, and power, but by the making of room for being human in the world. Conversely, when limits, as the gateways of responsibility, are breached, the risks signal a conversion of possibilities into policy under the pressures of power. The open road, marked by the setting right of what is not right in the world, is effectively consigned to planned obsolescence and deterioration.

The setting right of what is not right in the world is the meaning of *Justice* in a covenantally structured human world. Justice understood in this way is the discovery of the covenantal journey. It is notably different from justice understood as a moral and political virtue derived from laws of reason and nature. Its focus and tonalities have been hewn and shaped for responsible freedom from happenings, memories, and hopes that chart a course from Micah to the Magnificat, from the transfiguration of Jesus to the Parable of the Last Judgment and into the story of the community of the Kingdom of God, at once present and to come. Although Calvin overlooked saying so, this is why the providence of God, as he did say, shows "the care of God for the whole human race, and especially his vigilance in the government of the Church, which he favors with more particular attention" ([6], Volume 1, Book 1, Chapter 17:1).

When Justice, in this sense, is detached from Providence, Eschatology, and Destiny, it becomes a rational and political virtue designed to subordinate passion to fairness (Cicero and Rawls), what is right to what is good, the urgency of what it takes to be and to stay human in the world to the not yet fully achieved rationality of privilege and power. Conversely, when Providence, Eschatology, and Destiny are detached from Justice, limits which are the gateways of responsibility lose their ordination as harbingers of the freedom to be human. They become arbitrary and heteronomous constraints upon the prospects and power of autonomous ingenuity. 'Justice being absent then', the human prospect is darkened by the debilitating rhythm between autonomy and heteronomy and its destructive oscillation between anarchy and tyranny. 'Justice being absent then', freedom and responsibility lose their reciprocity, and Providence, Eschatology, and Destiny are deprived of their cutting edge in sustaining, building, and fulfilling the point and purpose of life as human life.

Then, when 'The Rationality of Belief', once succinctly and even stridently demonstrable in the syndrome, 'God-Freedom-Immortality', suddenly changes to 'God-Morals-Medicine', something akin to a 'paradigm shift', in Thomas Kuhn's phrase, is being adumbrated. Moses and Jesus, so happily and harmlessly domesticated under the natural law and the postulates of the practical reason in 'the Age of Reason' hastening towards 'the Age of Humanity', have been suddenly joined by Hippocrates in a 'witches brew' of troubles for all three. Medicine, it would seem, has put both God and Morals on the spot; while freedom and immortality have been intriguingly congealed in a cryological eschatology of life beyond death, engineeringly projected as life without death. Hippocrates used to provide medicine with sufficient ethical coverage to include freedom and responsibility within medical practice according to the rule. For some time now, however, this ethical coverage has given way before the passionate pursuit of biological and medical knowledge and achievement, brilliantly extending the range and healing of life, of which 'eye hath not seen nor ear heard', but which draws towards itself such an unprecedented array of technological, foundational, and organizational power as to establish the fact of life as the evidence of responsibility for life. Indeed, the pace has been so rapid that the phrase, 'medical ethics', has scarcely entered the currency of language before being replaced by the current neologism, designed to keep medicine and morals in sight of each other.

The neologism is the hyphenated word: *bio-ethics*. The mysteries, complexities, perplexities, and dilemmas which abound in the experience and interpretation of nature, humanity, and God and in the experience and interpretation of ethics are pressing upon us a 'Copernican revolution' in the taking of responsibility for life. An initial step in response to such an urgency could be the exploration of a liberating reciprocity between theological perspectives and bioethical research, discoveries, limits, and possibilities.

## OF RESPONSIBILITY FOR LIFE

The 'taking of responsibility for life' is poised on the frontiers of (1) freedom and justice drawn by providence and destiny (theologically understood), and (2) genetic engineering (ethically understood) in a world more widely and fully 'come of age' than even the Enlightenment had claimed, and even than Dietrich Bonhoeffer had begun to imagine.[1] The question is, who is prepared to take responsibility for life, and under what conditions? Surely Providence is a better clue than mere coincidence, to the fact that towards the end of the Enlightenment an extraordinary essay by the German physician, John Karl Osterhauser, published in 1798, bore the title: 'Über medizinische Aufklärung?', in deliberate imitation of Kant's 'Was ist Aufklärung?' Borrowing Kant's celebrated definition of Enlightenment, the author declared that medical Enlightenment is: "man's emergence from his dependence in matters concerning his physical well-being."[2] As the 'world come of age' was beginning to be everywhere evident, and at the same time more haunted by the prospect of its end, Bonhoeffer wrote: "I've come to be doubtful of talking about any human boundaries. It always seems to me that we are trying anxiously to reserve some space for God; I should like to speak of God not on the boundaries but at the center. ...God is beyond in the midst of our life" ([5], p. 282).

The reverse *Aufklärung*, therefore, is not to be understood as a move towards the *status quo ante.* On the contrary, it is a move beyond the abstractions of reasons into the bone and marrow of life and of living in two worlds (horizontal and vertical), always at the same time. This move is rooted in the discernment that the morality of freedom, compounded of the bond between virtue and happiness, and of the will directed towards the highest good, disclosed by reason at the service of law and operative in the law at the service of the rational good, is unable to take and to nurture responsibility for life. The morality of freedom based upon Enlightenment doctrines is thus caught in a vise between prudence and circumstance, between the frontiers of knowledge and the dilemmas of living, between individual possibilities and public policies. The awesome consequence is that responsibility for life and the human meaning of life are experienced, interpreted, and practiced at cross-purposes.

On the other hand, the broader context for freedom and responsibility drawn from the wider horizons of the Christian creed to include once again a serious appropriation of the perspectives of Providence, Eschatology, and Destiny, may conceivably bring responsibility for life and the human meaning of life upon a *common decision* in furtherance of *the human good as the healing good.* As once, "in the rhetoric of the Enlightenment, the conquest of nature and the conquest of revealed religion were one: a struggle for health" ([9], p. 16), so, once again, in the nurture of responsibility for nature and of responsiveness to the perceptions and perspectives of revealed religion, the struggle for health is one in which theology and bioethics may be allied in the taking of responsibility for life as fulfilling human life.

The prospect is less remote and less fanciful than it may appear to be. The post-Enlightenment world come of age, in which we now live, is being shaped and accelerated by a technological paradigm shift of enormous range and triple magnitude. The nuclear, genetic, and communications revolutions awesomely echo the Cappadocian doctrine of the perichoresis of the triune God in the dynamics of power, wisdom, and purpose in the world. One controlling and controlled energy, in three modes of identity, differentiation, and inter-relation, seems to have brought us across a great divide. We are on the other and nearer side of a form and organization of energy which marks the present and future from the past by a difference of degree of such precision, intensity, complexity, and inescapability as to be a difference in kind.[3] This technological revolution has catapulted responsibility for life into a radical revision of the relations between knowledge and power, nature and society, freedom and limits, possibilities and priorities, the center and the boundaries in "the human use of human beings" [17].

It seems that Calvin and Augustine have not only widened our horizons in discerning what is holding us together and whither we are going. They have also overtaken us en route, as our contemporaries. With a perspicacity savoring clairvoyance, Calvin has prepared us to consider "as indubitably certain, that all the revolutions visible in the world proceed from the secret exertion of the Divine power" and that "the providence of God is to be considered as well in regard to futurity as in reference to that which is past", and that "it tends to show the care of God for the whole human race" ([6], Volume 1, Book 1, Chapter 16:1, 4, 9; Chapter 17:1). With like perspicacity, Augustine puts us before the decisive option which our providential destiny invites us to pursue. "Two cities have been formed by two loves:.... In the one, the princes and the nations it subdues are ruled by the love of ruling; in the other ..." they serve one another in love while taking thought for all. "The wise men of the one city... have sought for profit to their own bodies or souls, or both, ... glorying in their own wisdom, and ... possessed by pride .... In the other ... there is only godliness, which ... looks for its reward in the society of saints ..." [3]. In *that* city, the human good pre-empts the moral good and takes priority over knowledge and power in determining responsibility for life.

Two critical instances of bio-medical research, reflection, and concern may be addressed in the present attempt to suggest how the human good, providentially arrived at and identified, enables the taking of responsibility for life. One case in point is provided by genetic engineering; the other, by the question of abortion. The first raises the complex and perplexing question of the limit of knowledge, power, and possibility set by the human good. The second underlines the complex and perplexing question of whose good is the human good. Both cases severely test the claim that freedom and justice are the primary indicators of the human good in the taking of responsibility for life.

*Bioethics and Genetic Engineering*

It is of no little significance to the thesis here being explored to be able to note that theology and bioethics converge upon and set out from a common point. As Matthew Meselson has put it, in identifying his purpose as a biologist and as a citizen, he seeks "to build an ethos for the future, one that says a deep knowledge of life processes must be used only to reinforce what is essentially human in us."[4] Meselson dedicated his eminence as a biologist to two major achievements in furtherance of his commitment and purpose as a scientist and as a citizen. The first has to do with his wisdom, learning, and counsel in effecting "the conversion of our biological warfare laboratories to open programs of medical research" ([8], p. 175). The second achievement is more directly related to genetic engineering. It concerns the question whether ongoing research in recombinant DNA should be furthered or prohibited. Crucial to this question is the exploration of the structure of genes, which involves a variety of techniques, including the technique of cloning. The possibility and the pursuit of this exploration became something of a 'cause celèbre' of bioethics in theological perspective when two universities requested of the public authorities of the communities of which they were a part, permission to engage in research in recombinant DNA. The issues of knowledge and its limits, of human values and consequences, of possibilities, policies, and power, forcefully converged upon the question of responsibility for life.

In Cambridge, Massachusetts, a protracted controversey led the Mayor to appoint a Citizens' Committee to make recommendations concerning the continuance of research in recombinant DNA. Pending the report of the Committee, experiments were banned for seven months. Owing more than a little to Meselson's technical competence and humane sensibility, the Committee members came to trust "his quiet uncertainty more than they trusted the loud certainty of his opponents. In the end they voted unanimously to recommend to the City of Cambridge the continuation of recombinant DNA experiments, subject to reasonable restrictions and supervision by local public health authorities" ([8], p. 177). Research had not been proscribed. It had been coordinated instead with a tenuous limit set by public authorities.

While the Cambridge Committee was still at work, in the fall of 1976 Princeton University made a similar request of Princeton municipal authorities. Again a Citizens' Committee was appointed which, after four months, reported a split decision to the municipal authorities who had wanted clear guidance. The vote was 8 to 3. In the Municipal Council, the vote was 5 to 1 to accept the recommendations of the majority of the Citizens' Committee. Professor Dyson, who was a member of the Princeton Committee, explains that although he felt closer personally and philosophically to the minority, he voted with the majority. He did so on legal grounds. "From a legal point of view," Dyson writes, "the municipality of Princeton has a right and a duty to restrict any research at Princeton University that may cause a hazard to the health of the citizens. But no public authority should have a legal right to restrict research merely because the people in positions of authority are philosophically opposed to it. ... As Thomas More says in Robert Bolt's play, *A Man For All Seasons*, 'I know what's legal, not what's right. And I'll stick to what's legal' " ([8], p. 181).

These debates and agonies over recombinant DNA are scarcely confined to the communities of Cambridge and Princeton. They have been mentioned because they exhibit vividly and concretely a major perplexity of bioethics that is pivotal to the question of responsibility for life. They also identify a critical instance of the pertinence of theological perspectives to bioethics in the clarification and furtherance of responsibility for life. This major perplexity and critical instance converge upon the question of limit.

'I know what's legal, not what's right. And I'll stick to what's legal' is one resolution of the question of limit. Philosophy and/or theology are to have no standing in court where the future of scientific knowledge and research are concerned. But in theological perspective and in bioethical pursuance of genetic engineering, what is legal and what is right cannot be disjoined, if 'a deep knowledge of life processes must be used only to reinforce what is essentially human in us.'

An international meeting of biologists in 1975 voluntarily adopted a set of guidelines that prohibits experiments that seem to biologists to be irresponsible and establishes containment procedures for permissible experiments. These guidelines have greatly reduced the immediate public health hazards of DNA experimentation. Given these safeguards, why is the public still scared? Dyson's answer is that "the public sees farther into the future and is concerned with larger issues than immediate health hazards. ...The public is rightly afraid of the abuse of this knowledge. ...The public sees, behind the honest faces of Matthew Meselson and Maxine Singer, the (diabolical) figures of Doctor Moreau and Daedalus" ([8], p. 178). So limits can be the pauses that serve the human good of each through the human good of all! But this interpretation is scarcely an option in a world of mutability and order in which promising possibilities are ultimately ascribed to fate. This interpretation requires instead a world of origin and providence, of destiny and fulfillment to provide and nurture a sustaining apperception and practice of the human good of each through the human good of all. In such a world, the human good preempts the moral good and takes priority over knowledge and power in determining responsibility for life. In taking responsibility for life, the human good is identified by the freedom to be whomsoever one has been given life to be and in that gift fulfilled; and by the justice which sets right what is not right, so that the human good of all may become the human good of each.

The awesome potential danger to the human good, occasioned by genetic engineering, urgently presses the question of limits upon knowledge and power. Matthew Meselson's comment, "to build an ethos for the future, one that says a deep knowledge of life processes must be used only to reinforce what is essentially human in us," itself warrants a *suspension* of genetic engineering as a limit at the gateway of responsibility for life, pending the effective presence in society, culture, and the centers of power of a sustaining context of freedom, justice, and mercy joined to enhance the human good. Such a context not only is available to bioethics in theological perspective. It is indispensable to the nurture of public trust in those who bear responsibility for knowledge and power. Such a context is also indispensable to the nurture of public commitment to the freedom and justice essential to the reciprocity between the human good of all and the human good of each.

*Bioethics and Abortion*

A second critical instance of the taking of responsibility for life is abortion. This present essay must conclude with an all too brief consideration of this issue. And, as in the instance of genetic engineering a conversation was entered upon with Professor Freeman Dyson, so in the instance of abortion a conversation will be taken up with Professor Paul Ramsey. The focus of the conversation will be Professor Ramsey's Bampton Lectures published in 1978 under the title *Ethics at the Edges of Life* [15].

Professor Ramsey notes that the decision of the United States Supreme Court in 1976 legalizing abortion was misleadingly hailed as a 'victory' for unlimited abortion. Actually, the subtleties of the decision, together with the concurrences and dissents of the several justices, made the decision a more carefully circumscribed one than would at first appear. Nevertheless, abortions are no longer unexceptionally proscribed. Strictly, of course, they never have been unexceptionally proscribed, either in common law or criminal law. The 'rights' of the mother and the 'rights' of the fetus have always constituted exceptions under certain circumstances.

I am in complete agreement with Ramsey that 'an opportune moment' has come for 'moral dialogue' about this question, about "perils and problems ahead, if we are to avoid further descent into technological barbarism" ([15], p. 46). With Ramsey, I take with full seriousness "the preciousness of unborn life" ([15], p. 46). Nevertheless, the medical, legal, and institutional evidence brought together within the pages of *Ethics at the Edges of Life* is ominous indeed in its ambiguity, its diversity, and, even worse, its capriciousness.

The Judeo-Christian tradition itself never disallowed exceptions to the commandment proscribing the taking of human life. Ramsey reminds us that both Jewish and Roman Catholic 'moral theology' are "operationally very similar" ([15], p. 47). Jewish teaching undertakes to identify "individual human life only after the head or greater portion of the fetus has passed through the birth canal" ([15], p. 46). At the same time, Jewish teaching is that "one should violate the most holy day to save a fetus which is only potentially a human being. One violates for him this sabbath so that he will remain alive to observe many sabbaths" ([15], pp. 46–49). In Roman Catholicism, "abortion is justified only to save the mother's life" ([15], p. 47). Thus we find exceptions to the unexceptionable. And this is where my troubles with Paul Ramsey began to be troublesome. When Ramsey writes, "a general obligation to provide abortion service cannot be made consistent with freedom of conscientious refusal" ([15], p. 53), I can comprehend the logic but cannot commend the reasoning. The question 'whose good is the human good?' seems to have escaped Ramsey's notice for the moment. Either that, or he has been caught in a logical trap. Rights are amenable to logic but responsibilities always break out of logic into relational reality. Consider a further point. Ramsey writes:

> Most people in all ages act *sub specie boni*. In any case, the issue to which past discussions of cooperation and conscientious refusal to cooperate were addressed had to do with what *the cooperator* understood *himself* to be doing and was actually doing or influencing in the moral order – not first of all with what the primary agent thought *he* was doing. That issue still remains with us, despite all attempts to dissolve it ([15], pp. 85–86).

Agreed! But somehow one senses that for Ramsey 'the refuser' is more faithful to conscience as the guardian of the moral order than is the cooperator. Along that road lies the dehumanizing vulnerability to self-righteousness which is in every ethical view that insists that there is always only one right thing to do; or if not only one right thing to do, there is only the less dark counsel that there are degrees of approximation and faithfulness to the moral order. Luther's perception that the sixth commandment is violated not only when a person actually does evil, but also when he fails to do good to his neighbor is, unhappily, altogether overlooked.

It has been noted that the abortion debate may be divided into three major segments [14]. At one extreme, there is the 'no abortion' position. As the other extreme is the 'abortion on demand' position. And in between, there is a view which may be called the position of 'justifiable abortion'. The common point upon which all three notions converge is the notion of *rights*. 'Abortion on demand' seeks to further and defend the 'rights of the woman'. 'No abortion' seeks to further and defend the 'rights of the unborn'. The position of 'justifiable abortion' seeks somehow to adjudicate the rights of both. In any case, a grievous politicization of a critical human situation has taken and is taking place. At the very least the whole discussion goes on unmindful of a companion perception of Luther's that "it will do you no good to plead that you did not contribute to (your neighbor's) death by word or by deed, for you have withheld your love from him and robbed him of the service by which his life might have been saved" ([12], p. 391).

The human fact is – whatever the social, legal, political, and moral circumstances may be – the unborn has *no* rights but only a divine ordination to the responsibility for life on the part of the born. The human fact is that the woman has no rights on demand, but only a divine ordination to the responsibility for life under which she, together with all the born, *male* and *female*, man and woman and child are called to be. And they are called to be – *born* – as surely as "a world of made is not a world of born" ([7], No. XIV).

Accordingly, 'the right to life movement' is in thoroughgoing disregard of the theological perspectives which enable bioethical research and findings to take responsibility for life, since that movement subsumes 'responsibility for life' so tightly under the 'right to life' as to foredoom the unwanted fetus brought to birth to a less than fully human life. This is most evident, of course, from the fact that the denial of abortion most grievously affects the poor and from the fact that the privileged life into which the forcibly unaborted fetus comes is the life which values property over people.

The 'abortion on demand' position is more open to the dehumanizing reality of a pregnancy for which the woman is left to bear the major torment, pain, and bitterness in a society whose principles and patterns for living increasingly deprive the woman – with a child – of a sustaining community of shared concern and drive her into isolation. The nadir of this societal repudiation of the 'human good of all' as the 'human good of each' was described in a press report some years ago concerning a shelter for pregnant teenage girls in Los Angeles. According to the report, of the fifteen pregnant girls, the overwhelming majority voiced their desire to bring the fetus to birth. The reason given was: 'Then there would be some one to love us!' This report has somehow become unforgettable to me because it seemed to point to yet a further depth in the incapacity and/or unwillingness of the society in which these girls were growing up to nurture the human good. I should have expected the reason to be that the birth of the child would give the girl someone to love. Actually, however, the point was the contrary: 'There would be someone to love her.' How deep and desperate can alienation and loneliness get?

Bonhoeffer, in his discussion of "the right to bodily life" ([4], pp. 155–165), is clear "that the question whether the life of the mother or the life of the child is of greater value can hardly be a matter for human decision." He is not less clear that "the simple fact is that God certainly intended to create a human being and that the nascent human being has (in abortion) been deliberately deprived of life. And," Bonhoeffer goes on, "That is nothing but murder." At the same time, however, Bonhoeffer notes that

> a great many different motives may lead to an action of this kind; indeed in cases of where it is an act of despair, performed in circumstances of extreme human or economic destitution and misery, the guilt may often lie rather with the community than with the individual. Precisely in this connection money may conceal many a wanton deed, while the poor man's reluctant lapse may far more easily be disclosed ([4], p. 176).

Exactly so! The rights approach to abortion is out of phase both with 'providence and destiny' and 'freedom and justice', which provide the context for the answer to the question of whose good. The answer is that the human good of all is the human good of each. The rights approach to the question presupposes and perpetuates a view of conscience according to which the conscience has been cut off from its covenantal context and the individual is left to the devices and desires of his or her own heart which both subvert and are nurtured by a legalism in Christian ethics. To the contrary, the mere claim that an act violates rules is not sufficient to call the deed into question. Responsibility for life rescues the individual both from his or her solitariness and from the tyranny of conscience by drawing him and her into the social, as well as private – making room for the freedom to be human. In this context it is clear that abortion is murder and that all in the midst of whom abortions occur are murderers. All depend upon the gift of forgiveness and the grace of life; and all are called to take responsibility for life in the power of the strength that is made perfect in weakness.

So when a friend of mine bluntly asked whether I was 'for abortion, or against it?', my reply was that I am against it and for it – and in that order – trusting, as the Heidelberg catechism so beautifully puts it, "In my only comfort in life and in death," namely, God who means freedom and who is against sin yet in the midst of sin summons precisely the sinners to take responsibility for life. In short, abortion is not justifiable; but it is forgiveable! The proscription of abortion by law, constitutional amendment, or other means exhibits a sterilization of faith. Such a legal proscription threatens an obedience of faith in which one moves in the direction of freedom and justice regardless of the incongruities between law and morality (making human life human) encountered on the way. Faith, not law, makes room for freedom and justice; or to state the same idea differently, I cannot pursue my own righteousness in disregard of my neighbor. As surely as the letter kills, the spirit gives life.

A HUMANIZING PROSPECT

The spirit does indeed give life. Pending the suspension of genetic engineering and of the moralization, legalization, and politicization of the question of abortion, the primary agenda before bioethical research and reflection in theological perspective would be the search for and furtherance of a convergence of perspectives, purposes, apperception, and value in terms of which the taking of responsibility for life as human life would simply happen. It could be that as biologists and theologians, physicians and pastors, historians of science and culture, sociologists, philosophers and moralists, citizens under public authority and public authorities under consent of citizens increasingly find themselves meeting each other on the frontiers of limits and risks, of knowledge and power, of human possibilities and prospects, they may discover a new and ennobling comradeship in the taking of responsibility for life. In the light of such a discovery, bioethical findings and dilemmas may be drawn towards a livelier sensitivity to humanizing theological perspectives, and theological perspectives and perplexities may be drawn towards a livelier sensitivity to the discernment and integrity of the biologists' commitment 'to reinforce what is essentially human in us'. The stakes are enormously high because irrevocably ultimate: for "life and good, death and evil" (Dt. 30:15), for human destruction or fulfillment.

A moving passage in Freeman Dyson's book reflects upon "the anguish of every human being who faces in ... imagination the implications of modern biology" ([8], p. 169). Dyson writes:

> The progress in biology in general, and the mutability of species in particular, threaten to deprive mankind of two psychological anchors: our sense of our own identity, and our sense of brotherhood with one another.... Our understanding is still fragmentary and partial. But it can hardly take us more than a few decades, or at most a century, to decipher and read the DNA language in all its details. ... When we have learned in all detail how life is reproduced, we shall have learned how life is produced. Whoever can read the DNA language can also learn to write it. Whoever learns to write the language will in turn learn to design living creatures according to his whims. God's technology for creating species will then be in our hands. Instead of the crude nineteenth century figure of Dr. Moreau with his scalpels and knives, we shall see his sophisticated twenty-first century counterpart, the young zoologist sitting at the computer console and composing the genetic instructions for a new species of animal. Or for a new species of quasi-human being. ... Can man play God and still stay sane? In our real world, the answer must inevitably be no ([8], p. 169).

Thus, one prospect is that of an unholy alliance between Faust and the Sorcerer's Apprentice in bartering the human condition to Mephistopheles. There are at least two other possibilities which, if allied, could open the way for human freedom and fulfillment. One such prospect is pointed to by a citation of President Michael Sovern of Columbia University, in conferring upon Isaac Asimov, author and biochemist, the honorary degree of Doctor of Science. Said President Sovern: "writing brilliantly about the future, you have shown a profound understanding of the past. Your respect for fact is equalled only by the penetration of your fantasies" [16].

The other prospect is pointed to by a poet's invitation to discern amidst the perplexities and dilemmas of our times, the wisdom of all times, when the human condition finds itself on the boundary between ultimate prospect or default, between a gift and a barter.

O how the devil who controls
The moral assymetric souls
The either-ors, the mongrel halves
Who find truth in a mirror, laughs.
Yet time and memory are still
Limiting factors on his will;
He cannot always fool us thrice,
For he may never tell us lies,
Just half-truths we can synthesize.
So, hidden in his hocus-pocus,
There lies the gift of double focus,
That magic lamp which looks so dull
And utterly impractical
Yet, if Aladdin use it right,
Can be a sesame to light
([2],p. 42).

*Union Theological Seminary,*
*New York, U.S.A.*

## NOTES

[1] See [9], Vol. II, esp. Ch. 1, Section 2: 'Enlightenment, Medicine and Cure'. Also see [5], esp. the letters of 30 April, 30 June, and 8 July, 1944. Of the Enlightenment, Gay writes: "In the rhetoric of the Enlightenment, the conquest of nature and the conquest of revealed religion were one: a struggle for health. If the philosophers were missionaries, they were medical missionaries.... Philosophic literature abounds in solemn claims for the affinity – the near equivalence of modern medicine with modern philosophy. Perhaps the most striking of these claims came toward the end of the Enlightenment, in 1798, in an extraordinary essay by the German physicist, Johan Karl Ostenhauser, 'Über medizinische Aufklärung'. In its title, as in its content, the essay is a deliberate imitation of Kant's 'Was ist Aufklärung?' " ([9], p. 17).

On 30 April 1944, from his cell in Tegel Prison, Berlin, Bonhoeffer wrote: "What is bothering me incessantly is the question of what Christianity really is, indeed who Christ really is for us today, ...We are moving towards a completely religionless time; people as they are now simply cannot be religious any more. ...Our whole nineteen-hundred-year-old Christian preaching and theology rest on the religious *a priori* of mankind. ...But ... that ...foundation is taken away from the whole of what has up to now been our 'Christianity'. ...I've come to be doubtful of talking about any human boundaries, ...It always seems to me that we are trying anxiously to reserve some space for God. I should like to speak of God not on the boundaries but at the center, not in weakness but strength. ... God's 'beyond' is not the beyond of our cognitive faculties. The transcendence of the epistemological theory has nothing to do with the transcendence of God. God is beyond in the midst of our life" ([5], pp. 279–280, 282).

On 30 June 1944, he wrote: "Let me just summarize briefly what I am concerned about. The claim of a world that has come of age by Jesus Christ" ([5], p. 342).

[2] 'Of Medical Enlightenment', quoted in [9], p. 17. Kant had begun his essay with the sentence: "Enlightenment is man's release from his self-incurred tutelage" ([11], p. 286).

[3] So, Robert Oppenheimer at Los Alamos, responding to the explosion of the atom, quotes a passage from the *Dhaja-Vadghita*: "I am the destroyer of worlds." Similarly, Freeman Dyson borrows the title of his book [8] from T. S. Eliot's *Four Quartets*, from 'The Love Song of J. Alfred Prufrock': "Do I dare disturb the universe?' In that book, Dyson comments on the power difference in kind rather than degree since Hiroshima, 1945 ([8], pp. 41–44).

[4] So, [8], p. 178. The understanding and evaluation of bioethics, with particular reference to genetic engineering, is greatly indebted to Professor Dyson's discussion ([8], Chapter 15 and 16). Meselson is Professor of Biology and head of the Laboratory for Biological Research at Harvard University.

## BIBLIOGRAPHY

[1] Aquinas, T.: 1966, *Law and Political Theory*, in T. Gilby (trans.), *Summa Theologiae*, Blackfriars, Cambridge, England, Vol. 28.
[2] Auden, W.: 1941, *The Double Man*, Random House, New York.
[3] Augustine, A.: 1950, *The City of God*, M. Dods (trans.), Random House, New York.
[4] Bonhoeffer, D.: 1965, *Ethics*, MacMillan, New York.
[5] Bonhoeffer, D.: 1978, *Letters and Papers from Prison*, enlarged edition, E. Bethge (ed.), MacMillan, New York.
[6] Calvin, J.: 1960, *Institutes of the Christian Religion*, J. Allen (trans.), Presbyterian Board of Education, Philadelphia.
[7] Cummings, E.: 1944, *1 x 1*, Henry Holt, New York.
[8] Dyson, F.: 1979, *Disturbing the Universe*, Harper and Row, New York.
[9] Gay, P.: 1969, *The Enlightenment: An Interpretation*, W. W. Norton, New York.
[10] Kant, I.: *Critique of Pure Reason*, N. Kemp Smith (trans.), St. Martin's Press, New York.
[11] Kant, I.: 1949, 'What Is Enlightenment?', in L. Beck (ed.), *Critique of Practical Reason and Other Writings in Moral Philosophy*, Univ. of Chicago Press, Chicago, pp. 286–293.
[12] Luther, M.: 1959, 'The Large Catechism', in T. Tappert (ed.), *The Book of Concord: The Confessions of the Evangelical Lutheran Church*, Muhlenberg Press, Philadelphia, pp. 357–462.

[13] McNeill, J.: 1946, 'Natural Law in the Teaching of the Reformers', *The Journal of Religion* **26**: 168–182.
[14] Potter, R.: 1968, 'The Abortion Debate', in D. Cutler (ed.), *The Religious Situation,* Beacon Press, Boston, pp. 112–161.
[15] Ramsey, P.: 1978, *Ethics at the Edges of Life*, Yale Univ. Press, New Haven, Conn.
[16] Sovern, M. 1983, Columbia University Commencement Address, quoted in '7300 Awarded Degrees By Columbia University', *New York Times*, 18 May, p. B4.
[17] Wiener, N.: 1954, *The Human Use of Human Beings: Cybernetics and Society*, Houghton Mifflin, Boston.

EPILOGUE

JOHN B. COBB, JR.

# DOES THEOLOGY MAKE A CONTRIBUTION TO BIOETHICS?

During the past half century specialization triumphed in the faculties of theology in the separation of ethics from systematics. To those who hoped this would enable the church to speak with greater rigor and authority on the issues of the day the results have been, to put it mildly, disappointing. Having become a separate department in the seminary, ethics had to become also a *discipline*. But that was no simple matter! It seemed to be a mixture of social science, moral philosophy, the history of the ethical teaching and practice of the church, and selected aspects of systematic theology. And it was also expected to address public issues! How could it do so with integrity? Much of the creative energy of Christian ethicists has been devoted to the ongoing effort to establish a *discipline* that is at once Christian and illuminating of the issues. The actual illumination has played a much smaller role.

Bio-ethics has been a happy exception. The issues lent themselves to the application of established principles of moral philosophy. Since moral philosophy itself had become complacent in its role of academic *discipline*, using public issues only to illustrate its academic points, it felt little need to enter the arena of public and professional decision making. Hence Christian ethicists, more highly motivated in this respect, seized the opportunity and helped to shape the new *discipline* of bioethics. But to the question: In what way is the contribution of Christian ethics Christian? the answer has been obscure. Perhaps the real answer is that Christian faith seeks expression in the world and that the problems of bioethics have given Christian ethicists an opportunity.

But this is not the kind of answer that academicians want. Shelp quotes MacIntyre's call for Christian ethicists to state what constitutes them as Christians and what difference this makes in their treatment of ethical issues. This is to call for the solution of the unsolved question of the nature of the *discipline*.

MacIntyre suspects that if these questions are answered, the special features of Christian ethics will turn out to be indefensible. Frankena's essay in this volume spells out the problem in detail. My own suspicion is that, indeed, what is being looked for is simply not to be found. Chris-



*E. E. Shelp (ed.), Theology and Bioethics,* 303–307.

tianity does not provide a set of moral principles distinct from those that have been systematized in the rational reflection of Christian and post-Christian thinkers. Deontological ethics expresses the ultimacy of the divine will. An ethics of consequences based on benevolence rationalizes the call to love others as one loves oneself. And much of the secular quest to clarify the meaning of justice reflects the Biblical conviction that God cares especially for the poor and oppressed. Unless Christian ethicists return to pre-rationalized expressions of these features of their faith – appealing, for example, to the absolute authority of sacred texts – there is no reason for them to differ from moral philosophers who work with these rationalized forms. They would differ from secular moralists only if these turned their backs on concerns for obligation, benevolence, and justice for the poor.

Those who want to know the implications of Christian faith for specific ethical issues that differ from those of moral philosophy in general will need to look elsewhere. It is appropriate that in this volume Shelp has turned to theologians. He is understandably disappointed that they have dealt with bioethical issues so little. But that has followed naturally if not inevitably from the division of disciplinary responsibilities. Theology as a *discipline* no longer treats of public issues, since those belong to the separate theological *discipline* of ethics!

Some of the essays here express the struggle, internal to each *discipline*, to determine how to relate to other *disciplines*. The discussions evoked by that struggle remain quite remote from the question of how faith is related to the issues of bioethics. But other essays do deal with that question. The answer in general is that in addition to grounding and motivating the ethical concern it shares with secular humanism, religious faith provides an ethos and a worldview, and it affects the people and communities involved.

Even when the theological contributors approach bio-ethics in distinctive ways, not all the distinctiveness is distinctively Christian. Farley shows that feminists bring sensitivities and concerns to the issues that cannot be assumed on the part of all moral philosophers. She does not show that *Christian* feminists have a distinctive contribution. Sturm shows that a relational model of reality throws light on the issues of bioethics and can provide a certain unity to its fragmented principles. He does not argue that it is a monopoly of Christians. Mitchell appeals to a traditional Christian view of nature as setting parameters of reflection on current issues, but historically that view has Greek as well as Biblical sources, and it can be defended independently of Christian theology.

Hartshorne shows that a rational way of understanding God and the world undercuts the arguments of many 'right-to-life' advocates, but he presents this view as open to persons of diverse religious traditions. Hauerwas shows the need for communities like the Christian church to provide contexts for dealing with the sick, but there is nothing in the argument to demonstrate that such communities must be avowedly Christian. McCormick seems to make a stronger claim for the relation to Christian faith of the personal depths and sensitivities of those who deal with the sick, but he provides no evidence that equally valuable depth and sensitivity may not be produced in other ways. Lehmann proposes that when the Christian perspectives of providence, eschatology, and destiny are brought to bear some painful issues receive a liberating light, but he does not make clear just what distinctive policies would eventuate.

This suggests three conclusions. (1) The relevance of Christian commitments to practical life is not mediated by moral teaching alone. (2) But neither in moral teachings nor elsewhere will this relevance be, in principle, unique to Christians. And (3) the diversity among Christians is such that on almost any issue one group of Christians will ally themselves with some who are not Christian against similarly mixed groups.

To persons of rationalistic inclinations this is a frustrating situation. They want to know what the teachings of Christianity are, how they are justified, and how they apply. Inability to answer these questions is tantamount, in their eyes, to confession that the word 'Christian' has lost all meaning.

Another interpretation of the situation is that 'Christian' has never been well understood in terms of a central core of distinctive doctrines, theological or ethical. It has not been well grasped in terms of a particular ethos or a worldview or personal style of life either. The unity of Christians lies in a shared memory, story, or internal history centering in the Jewish teacher, Jesus, and in the apostolic witness to him. Living from that history always involves theological doctrines, ethical teachings, an ethos, a worldview, and a quality of life as well as religious practices, social institutions, and many other things. No aspect of life or society is unaffected.

There *are* doctrines, moral teachings, types of ethos, worldviews, and styles of life that are simply incompatible with the unifying center of faith. But every effort to specify *the* theological doctrines, *the* moral teachings, *the* ethos, *the* worldview, or *the* style of life that alone is appropriate to all Christians at all times and places expresses a misunderstanding of the nature of the faith – a misunderstanding that has of course been common and widespread. Fortunately, it is always countered by those who find other implications in the shared story.

This does not imply that it is useless to ask theologians to address the issues. It does imply that one should not expect them to lay out Christian distinctiveness and derive from that in any direct way answers to the issues. To do that would misrepresent the nature of Christian faith. Also when Christians engage in the discussion of public issues, they should not be expected to refer often, if at all, to the story by which they live. This story has influenced their sensibilities, opinions, values, and commitments, but it is these, and not the story itself, that relate directly to the issue at hand. And these are not uniquely correlated with the story and its center. The efforts to establish too tight a connection distorts the faith.

Of course, there are other Christians, hardly represented in this collection, who see matters quite differently. For some, prooftexting is a valid approach. Others understand their task as representing the official and authoritative teaching of their church.

Quite distinctive in relation to all this is the Jewish approach. There, without any system for definitively settling issues as to the Jewish position – by conciliar action or papal decree, for example – the community is guided and given an adequate measure of unity by its living tradition. The interaction of personal judgments about the issues with objectivity of historical reporting, in the process of interpreting that tradition today, is studied in the fascinating essay by Green.

What contributions do theologians in general – and these essays in particular – make to bioethics? I have argued: no contribution that in principle cannot be made by others. But that does not mean that they make no contribution at all, or even that they make no distinctive contribution.

Theologians generally will seek to set specific decision-making in a wider context. Note Gilkey's approach! They will want to see the people and communities involved holistically. They will not accept the problematic as given but will look instead for the assumptions beneath the description and for the context in which questions are being asked. They will press for broadening of context and the inclusion of more voices. They will question the wisdom of focusing exclusively on local problems when the conditions and practices generating those problems are not addressed. They will be sensitive to nuances in personal relations and community life. Again, not *all* theologians will do these things – and not *only* theologians will do them. But the involvement of theologians will increase the likelihood that these themes will be involved.

In these brief comments I have underscored the role of academic *disci-*

*plines*. In my view, it is unhealthy. As a Christian I derive from the Christian story – indirectly, of course – an antipathy to this compartmentalization of thought and this reification of what is compartmentalized into *disciplines*. Christian thinkers would be well-advised to redirect their energies from the defining and promotion of *disciplines* to dealing with the urgent issues of the day.

*School of Theology at Claremont,*
*Claremont, Calif., U.S.A.*

# NOTES ON CONTRIBUTORS

James F. Childress, Ph.D., is Commonwealth Professor of Religious Studies, and Professor of Medical Education, University of Virginia, Charlottesville, Virginia.

John B. Cobb, Jr., Ph.D., is Ingraham Professor of Theology, School of Theology; Avery Professor of Religion, Claremont Graduate School; and Director, Center for Process Studies, School of Theology, Claremont, California, U.S.A.

H. Tristram Engelhardt, Jr., Ph.D., M.D., is Professor, Departments of Medicine and Community Medicine, and Member of the Center for Ethics, Medicine, and Public Issues, Baylor College of Medicine, Houston, Texas.

Margaret Farley, Ph.D., is Associate Professor of Christian Ethics, Yale Divinity School, Yale University, New Haven, Connecticut.

William K. Frankena, Ph.D., is Emeritus Professor of Philosophy, University of Michigan, Ann Arbor, Michigan.

Langdon Gilkey, Ph.D., is Shailer Mathews Professor of Theology, Divinity School, University of Chicago, Chicago, Illinois.

Ronald M. Green, Ph.D., is Professor and Chairman, Department of Religion, Dartmouth College, Adjunct Faculty, Department of Community and Family Medicine, Dartmouth College Medical School, Hanover, New Hampshire.

Charles Hartshorne, Ph.D., is Emeritus Professor of Philosophy, University of Texas, Austin, Texas.

Stanley Hauerwas, Ph.D., is Professor of Theological Ethics, Divinity School, Duke University, Durham, North Carolina.

Mark Juergensmeyer, Ph.D., is Associate Professor of Religious Studies, Graduate Theological Union, and University of California, Berkeley, California.

Paul Lehmann, Ph.D., is Emeritus Professor of Systematic Theology, Union Theological Seminary, New York, New York.

Richard A. McCormick, S.J., is Rose F. Kennedy Professor of Christian Ethics, Kennedy Institute of Ethics, Georgetown University, Washington, D.C.



*E. E. Shelp (ed.), Theology and Bioethics,* 309–310.

Basil Mitchell, Ph.D., is Nolloth Professor of the Philosophy of the Christian Religion, University of Oxford, Oxford, England.

James B. Nelson, Ph.D. is Professor of Christian Ethics, United Theological Seminary of the Twin Cities, New Brighton, Minnesota.

George P. Schner, S.J., is Director of Basic Degree Studies, and Assistant Professor of Fundamental Theology, Regis College, Toronto School of Theology, Toronto, Ontario, Canada.

Earl E. Shelp, Ph.D., is Associate Professor, Theology and Ethics, Institute of Religion, and Assistant Professor of Medical Ethics, Department of Community Medicine, and Member of the Center for Ethics, Medicine, and Public Issues, Baylor College of Medicine, Houston, Texas.

David H. Smith, Ph.D., is Professor and Chairman of Religious Studies, and Director of the Poynter Center, Indiana University, Bloomington, Indiana.

Douglas Sturm, Ph.D., is Professor of Religion and Political Science, Department of Religion, Bucknell University, Lewisburg, Pennsylvania.

LeRoy Walters, Ph.D., is Director, Center for Bioethics, Kennedy Institute of Ethics, Georgetown University, Washington, D.C.

# INDEX

*The Philosophy and Medicine Book Series*

*Editors*

H. Tristram Engelhardt, Jr. and Stuart F. Spicker

1. **Evaluation and Explanation in the Biomedical Sciences**
   1975, vi + 240 pp. ISBN 90-277-0553-4
2. **Philosophical Dimensions of the Neuro-Medical Sciences**
   1976, vi + 274 pp. ISBN 90-277-0672-7
3. **Philosophical Medical Ethics: Its Nature and Significance**
   1977, vi + 252 pp. ISBN 90-277-0772-3
4. **Mental Health: Philosophical Perspectives**
   1978, xxii + 302 pp. ISBN 90-277-0828-2
5. **Mental Illness: Law and Public Policy**
   1980, xvii + 254 pp. ISBN 90-277-1057-0
6. **Clinical Judgment. A Critical Appraisal**
   1979, xxvi + 278 pp. ISBN 90-277-0952-1
7. **Organism, Medicine, and Metaphysics**
   *Essays in Honor of Hans Jonas on his 75th Birthday, May 10, 1978*
   1978, xxvii + 330 pp. ISBN 90-277-0823-1
8. **Justice and Health Care**
   1981, xiv + 238 pp. ISBN 90-277-1207-7
9. **The Law-Medicine Relation: A Philosophical Exploration**
   1981, xxx + 292 pp. ISBN 90-277-1217-4
10. **New Knowledge in the Biomedical Sciences**
    1982, xviii + 244 pp. ISBN 90-277-1319-7
11. **Beneficence and Health Care**
    1982, xvi + 264 pp. ISBN 90-277-1377-4
12. **Responsibility in Health Care**
    1982, xxiii + 285 pp. ISBN 90-277-1417-7
13. **Abortion and the Status of the Fetus**
    1983, xxxii + 349 pp. ISBN 90-277-1493-2
14. **The Clinical Encounter**
    1983, xvi + 309 pp. ISBN 90-277-1593-9
15. **Ethics and Mental Retardation**
    1984, xvi + 254 pp. ISBN 90-277-1630-7
16. **Health, Disease, and Causal Explanations in Medicine**
    1984, xxx + 250 pp. ISBN 90-277-1660-9
17. **Virtue and Medicine: Explorations in the Character of Medicine**
    1985, xx + 363 pp. ISBN 90-277-1808-3
18. **Medical Ethics in Antiquity: Philosophical Perspectives on Abortion and Euthanasia**
    1985, xxvi + 242 pp. ISBN 90-277-1825-3
19. **Ethics and Critical Care Medicine**
    1985, xxii + 236 pp. ISBN 90-277-1820-2
20. **Theology and Bioethics: Exploring the Foundations and Frontiers**
    1985, xxiv + 314 pp. ISBN 90-277-1857-1